Engineering Applications of Computational Methods

Volume 18

The book series Engineering Applications of Computational Methods addresses the numerous applications of mathematical theory and latest computational or numerical methods in various fields of engineering. It emphasizes the practical application of these methods, with possible aspects in programming. New and developing computational methods using big data, machine learning and AI are discussed in this book series, and could be applied to engineering fields, such as manufacturing, industrial engineering, control engineering, civil engineering, energy engineering and material engineering.

The book series Engineering Applications of Computational Methods aims to introduce important computational methods adopted in different engineering projects to researchers and engineers. The individual book volumes in the series are thematic. The goal of each volume is to give readers a comprehensive overview of how the computational methods in a certain engineering area can be used. As a collection, the series provides valuable resources to a wide audience in academia, the engineering research community, industry and anyone else who are looking to expand their knowledge of computational methods.

This book series is indexed in both the **Scopus** and **Compendex** databases.

Chao Lu · Liang Gao · Xinyu Li · Lvjiang Yin

Intelligence Optimization for Green Scheduling in Manufacturing Systems

Chao Lu
School of Computer Science
China University of Geosciences
Wuhan, Hubei, China

Xinyu Li
State Key Laboratory of Intelligent
Manufacturing Equipment and Technology
Huazhong University of Science
and Technology
Wuhan, Hubei, China

Liang Gao
State Key Laboratory of Intelligent
Manufacturing Equipment and Technology
Huazhong University of Science
and Technology
Wuhan, Hubei, China

Lvjiang Yin
Hubei University of Automotive
Technology
Shiyan, Hubei, China

ISSN 2662-3366 ISSN 2662-3374 (electronic)
Engineering Applications of Computational Methods
ISBN 978-981-99-6986-9 ISBN 978-981-99-6987-6 (eBook)
https://doi.org/10.1007/978-981-99-6987-6

This Springer imprint is published by the registered company Springer Nature Singapore Pte Ltd.
The registered company address is: 152 Beach Road, #21-01/04 Gateway East, Singapore 189721, Singapore

Paper in this product is recyclable.

Preface

Being the cornerstone of the industrial sector, the manufacturing industry assumes a pivotal role in the national economy. However, with the increasingly serious environmental issues worldwide, green manufacturing has become particularly important. Among them, energy saving has become the key content of green manufacturing, but only relying on technical equipment and design level to achieve energy saving and efficiency is high cost and limited space for improvement. In contrast, scheduling optimization is a key technology of manufacturing system. It can achieve high efficiency and green production of manufacturing system through reasonable resource allocation, so as to achieve the purpose of energy saving, efficiency increase, emission reduction, and consumption reduction. Therefore, production scheduling has great theoretical significance and extensive practical value for realizing energy saving and emission reduction in manufacturing industry.

Based on the above reasons, this book conducts systematic and in-depth research on various types of shop scheduling problems. Specifically, it focuses on single-machine green scheduling, permutation flow shop green scheduling, hybrid flow shop green scheduling, job shop green scheduling, flexible job shop green scheduling, welding shop green scheduling, distributed permutation flow shop green scheduling, and distributed hybrid flow shop green scheduling problems. Different algorithms are designed to solve each problem according to their distinct characteristics. The feasibility and accuracy of the proposed algorithms are verified through experiments. The feature of this book is not only that the problems studied come from the actual production enterprises, but also that it further enriches and improves the theories and methods of shop scheduling. This indicates the practicality of the problems described in this book and the frontiers of the proposed scheduling optimization algorithms.

The research work of this book was supported by the National Natural Science Foundation of China (Nos. 52175490, 51805495), and the authors would like to express heartfelt gratitude for their support. During the writing process of this book, graduate students Yu Fei, Liu Feige, Chen Junkang, Zou Yuan, and others contributed a lot to the formatting work, and the author would like to express gratitude to them. This book references a large number of literature, and the author has tried to mark them as much as possible. If there is any negligence or omission, the author extends their sincere apologies and gratitude to the relevant authors.

The problem of green shop scheduling is still in continuous development. Despite the author's efforts to strive for perfection, there are inevitably omissions in this book due to the limited level and ability of the authors. Welcome readers to criticize and correct.

Wuhan, China
June 2023

Chao Lu
Liang Gao
Xinyu Li
Lvjiang Yin

Contents

Chapter 1
System Overview

Three industrial revolutions have occurred in human history since the mid-eighteenth century, resulting in rapid economic and social development. However, the development of industry has resulted in massive consumption of energy and resources, and the conflict between human and nature has grown dramatically. In the twenty-first century, humanity faces an unprecedented global energy and resource crisis, ecological and environmental crisis, and climate change crisis. To that end, the United Nations (UN) passed the 2030 Sustainable Development Goals (SDGs) at the 70th UN Conference at the dawn of the Fourth Industrial Revolution, aiming to balance economic, social, and environmental aspects and calling on all countries around the world to achieve these goals by 2030 [1]. Industry 4.0 is a national strategy proposed by Germany at the dawn of the Fourth Industrial Revolution [2] that focuses on the use of information technology to promote industrial change and improve manufacturing intelligence. As a branch of industry, manufacturing directly reflects a country's productivity level and plays an important role in economic and social development [3]. However, manufacturing consumes more than 37% of global resources and emits numerous pollutants [4, 5]. Therefore, manufacturing sectors must adopt more efficient methods and technologies to achieve energy savings and emission reductions for realizing sustainable development [6].

There are two main ways to improve energy efficiency and reduce emissions in manufacturing systems. The first is to design products or machines that use less energy and emit fewer pollutants. However, designing environmentally friendly products or machines necessitates a significant amount of human resources, a large capital investment, and a lengthy development time, making it difficult to achieve satisfactory results quickly. Another example is the use of green scheduling technology, which has the potential to significantly improve energy efficiency while costing almost nothing. Green shop scheduling problems (GSSPs) are further extensions of classical shop scheduling problems, which seek to achieve economic and environmental goals by increasing productivity while lowering energy consumption and pollutant

C. Lu et al., *Intelligence Optimization for Green Scheduling in Manufacturing Systems*, Engineering Applications of Computational Methods 18,
https://doi.org/10.1007/978-981-99-6987-6_1

emissions through rational resource allocation, as well as the use of more efficient operating methods and job sequences.

Manufacturing environments have a direct impact on GSSP solutions. The basic manufacturing environments are single machine, parallel machine, open shop, flow shop, job shop, flexible job shop, welding shop, distributed shop, and so on. The most basic shop scheduling problem is when a single machine handles all jobs and is referred to as a single-machine scheduling problem. A parallel machine is one that can process the job on more than one machine. There are three types of parallel machines: identical machines, uniform machines, and unrelated machines. Open shop problems (OSPs) are jobs that require multiple steps to be processed in no particular order. Jobs must be processed on different machines in an arbitrary order. Flow shop scheduling problems (FSPs) and job-shop scheduling problems (JSPs) can be applied to shops based on whether the jobs run in the same order or not [7]. If at least one stage includes more than one machine [8], such multi-stage jobs are referred to as hybrid flow shop problems (HFSPs) [9] or flexible job shop problems (FJSPs) [10]. All of the above types of shops are processed in a single shop and are scheduled in a centralized or semi-distributed manner. However, as Industry 4.0 has evolved [11, 12] the problem of shop scheduling for large-scale integrated multi-shops has gradually emerged and is known as distributed shop scheduling problems (DSPs) [13].

The methods for solving GSSPs are classified as exact and approximate [9, 14, 15]. Exact methods that are commonly used include the branch-and-bound method, integer programming methods, cut-plane methods, and dynamic planning algorithms [16]. Wang et al. [17] used an improved e-constrained approach to calculate the exact Pareto front for a small-scale energy-efficient two-stage scheduling problem. To optimize the parallel machine scheduling problem in the preemption case for up to 24 job instances, Liu et al. [18] proposed a branch-and-bound algorithm. It is impractical to solve the problem precisely due to the complexity of GSSPs [19]. Approximation methods, classified into heuristics and metaheuristics, are more common approaches in solving GSSPs compared with the limitations of the exact method. The approximate method means that a solution very close to the optimal solution is obtained at a reasonable time [20, 21]. Heuristics seek to solve a problem more quickly when classical methods are too slow or to find an approximate solution or when classical methods are unable to find any exact solutions [22]. A variety of heuristics have been developed that have proven to be effective, efficient, simple to implement, and reusable in various shop scheduling conditions [7]. Wang et al. [23] proposed two heuristic algorithms to solve the bi-objective single-machine batch scheduling problem of minimizing makespan and energy consumption while taking machine utilization and economic cost into account, breaking the problem down into three subproblems: how to batch jobs, how to sequence batches, and how to determine the temperature of each batch. Mansouri et al. [24] proposed a heuristic and a quick trade-off analysis of makespan and energy consumption. The metaheuristic is a higher-level heuristic that seeks to offer the best possible answer to an optimization problem, particularly when data is lacking or imperfect or when computational resources are constrained. Metaheuristics come in a variety of forms. Based on the encoding properties of the solution, Li et al. [25] created a six-time local search scheme and integrated simulated annealing into the algorithm. To specifically eliminate antibodies with high crowding values, they also advanced a novel population diversity heuristic. The threshold value of power was optimized using a local search strategy, while makespan and immediately available power were optimized using a hybrid non-dominated sorting genetic algorithm (NSGA-II) [26].

From the above studies, it can be seen that through the study of GSSP, manufacturing production can be guided, so as to achieve the purpose of saving energy consumption and improving production capacity. In this book, we introduce the idea of green scheduling into a variety of different manufacturing environments, while introducing different approximation algorithms to solve such a problem.

References

1. Roblek, V., et al.: The Fourth industrial revolution and the sustainability practices: a comparative automated content analysis approach of theory and practice. Sustainability **12**(20), 8497 (2020)
2. Xu, L.D., Xu, E.L., Li, L.: Industry 4.0: state of the art and future trends. Int. Prod. Res. **56**(8), 2941–2962 (2018)
3. Mansouri, S.A., Aktas, E., Besikci, U.: Green scheduling of a two-machine flowshop: trade-off between makespan and energy consumption. Eur. J. Oper. Res.Oper. Res. **248**(3), 772–788 (2016)
4. Gao, K.Z., et al.: A review of energy-efficient scheduling in intelligent production systems. Complex Intell. Syst. **6**(2), 237–249 (2020)
5. Singh, A., et al.: A simulation based approach to realize green factory from unit green manufacturing processes. J. Clean. Prod. **182**, 67–81 (2018)
6. Lin, J.P., et al.: A two-stage framework for the multi-user multi-data center job scheduling and resource allocation. IEEE Access **8**, 197863–197874 (2020)
7. Branke, J., et al.: Automated design of production scheduling heuristics: a review. IEEE Trans. Evol. Comput.Evol. Comput. **20**(1), 110–124 (2016)
8. Allahverdi, A., et al.: A survey of scheduling problems with setup times or costs. Eur. J. Oper. Res.Oper. Res. **187**(3), 985–1032 (2008)
9. Ruiz, R., Vazquez-Rodriguez, J.A.: The hybrid flow shop scheduling problem. Eur. J. Oper. Res.Oper. Res. **205**(1), 1–18 (2010)
10. Amjad, M.K., et al.: Recent research trends in genetic algorithm based flexible job shop scheduling problems. Math. Probl. Eng. 9270802 (2018)
11. Muhuri, P.K., Shukla, A.K., Abraham, A.: Industry 4.0: A bibliometric analysis and detailed overview. Eng. Appl. Artif. Intell.Artif. Intell. **78**, 218–235 (2019)
12. Zezulka, F., et al.: Industry 4.0-An Introduction in the phenomenon, in IFAC papersonline. 14th IFAC Conference on Programmable Devices and Embedded Systems (PDES), pp. 8–12 (2016)
13. Lu, S.H., Kumar, P.R.: Distributed scheduling based on due dates and buffer priorities. IEEE Trans. Autom. ControlAutom. Control **36**(12), 1406–1416 (1991)
14. Peng, J., et al.: Review on scheduling algorithms for MOFJSP. China Mech. Eng. **25**(23), 3244–3254 (2014)
15. Singh, H., Oberoi, J.S., Singh, D.: The taxonomy of dynamic multi-objective optimization of heuristics algorithms in flow shop scheduling problems: a systematic literature review. Int. J. Ind. Eng. Theor. Appl. Pract. **27**(3), 429–462 (2020)
16. Calis, B., Bulkan, S.: A research survey: review of AI solution strategies of job shop scheduling problem. J. Intell. Manuf.Intell. Manuf. **26**(5), 961–973 (2015)

17. Wang, S.J., et al.: An energy-efficient two-stage hybrid flow shop scheduling problem in a glass production. Int. Prod. Res. **58**(8), 2283–2314 (2020)
18. Liu, Z.H., Lee, W.C., Wang, J.Y.: Resource consumption minimization with a constraint of maximum tardiness on parallel machines. Comput. Ind. Eng.. Ind. Eng. **97**, 191–201 (2016)
19. Ruiz, R., Maroto, C.: A comprehensive review and evaluation of permutation flowshop heuristics. Eur. J. Oper. Res.Oper. Res. **165**(2), 479–494 (2005)
20. Adak, Z., Akan, M., Bulkan, S.: Multiprocessor open shop problem: literature review and future directions. J. Comb. Optim.Optim. **40**(2), 547–569 (2020)
21. Muhamad, A.S., Deris, S.: An artificial immune system for solving production scheduling problems: a review. Artif. Intell. Rev.. Intell. Rev. **39**(2), 97–108 (2013)
22. Ouelhadj, D., Petrovic, S.: A survey of dynamic scheduling in manufacturing systems. J. Sched. **12**(4), 417–431 (2009)
23. Wang, S.J., et al.: Bi-objective optimization of a single machine batch scheduling problem with energy cost consideration. J. Clean. Prod. **137**, 1205–1215 (2016)
24. Allahverdi, A., Gupta, J., Aldowaisan, T.: A review of scheduling research involving setup considerations. Omega-Int. J. Manage. Sci. **27**(2), 219–239 (1999)
25. Li, J.Q., et al.: Improved artificial immune system algorithm for type-2 fuzzy flexible job shop scheduling problem. IEEE Trans. Fuzzy Syst. **29**(11), 3234–3248 (2021)
26. Gondran, M., et al.: Bi-objective optimisation approaches to Job-shop problem with power requirements. Expert Syst. Appl. **162**, 113753 (2020)

Chapter 2
Green Scheduling in Single-Machine Environment

2.1 Brief Introduction

Today's manufacturing companies are facing the challenges of just-in-time (JIT) production and energy conservation. Therefore, the study on JIT production and energy consumption has a pivotal role in the manufacturing sector. In addition, energy savings can be achieved by applying operational methods and turning off/on idle machines, but it also increases the complexity of the problem. As a result, a large part of research still focuses on small-scale problems in single-machine environments where only one goal exists. However, the scheduling problem is a multi-objective optimization problem in practical applications. In this chapter, a single-machine scheduling model with controllable processing and sequential dependent setup times is developed to simultaneously minimize the total earliness/tardiness (E/T), cost, and energy consumption. Here, an efficient multi-objective evolutionary algorithm, LMOEA, is proposed to handle this multi-objective scheduling problem. Based on characteristics of this problem, a new solution representation is proposed, which can represent the discrete combinatorial problem as a continuous problem. In addition, the proposed algorithm introduces a multiple local search strategy with an adaptive mechanism to enhance the exploitation capability. The performance of the proposed algorithm is evaluated by comparing it with other multi-objective metaheuristics, such as NSGA-II, SPEA2, OMOPSO, and MOEA/D. Experimental results show that the proposed LMOEA algorithm outperforms its counterparts for this type of scheduling problem.

C. Lu et al., *Intelligence Optimization for Green Scheduling in Manufacturing Systems*, Engineering Applications of Computational Methods 18,
https://doi.org/10.1007/978-981-99-6987-6_2

2.2 Problem Statement and Modeling

2.2.1 Problem Statement

The energy-efficient single-machine scheduling problem in this chapter can be described as follows: n jobs are processed on one machine. The processing time of the jobs on the machine is controllable. Namely, the processing time of the jobs is advanced/delayed by compressing/expanding the normal processing time of the jobs; consequently, the processing time of the jobs is changed. The preparation time of the job is related to the processing sequence. Under different processing times, the energy consumption of machine processing is different, and the impact of machine on/off on energy consumption is considered. This scheduling problem includes three optimization goals: minimizing E/T, penalty cost, and energy consumption.

Before describing this model, we list some notations as follows.

n	Number of jobs
J_j	Job No. j
$J_{[i]}$	Job that is placed at position i in the processing sequence
π	Job processing sequence
p_j	Normal processing time of job j
p_j^c	Crash (minimum allowable) processing time for job j
p_j^e	Extended (maximum allowed) processing time for job j
m_j^c	The maximum compression of the job j, i.e., $m_j^c = p_j - p_j^c$
m_j^e	The maximum extension of job j, i.e., $m_j^e = p_j^e - p_j$
x_j^c	The amount of compression of job j
x_j^e	The amount of expansion of job j
p_j^a	The actual processing time of job j, $p_j^a = p_j - x_j^c + x_j^e$
a_j	Unit compression costs for job j
b_j	Unit extension cost of job j
d_j	Due date of job j
c_j	Completion time of job j
T_j	The tardiness of job $j, i.e.T_j = \max(0, C_j - d_j)$
E_j	The prematureness of job $j, i.e., E_j = \max(0, d_j - C_j)$
t_j	Unit late penalty factor for job j
e_j	Unit prematureness penalty factor for job j
S_j	Start time of job j
s_{ij}	Setup time between job i and job j
M	An arbitrary large positive number

In this scheduling problem, some assumptions are as follows.

- When there is a setup time s_{ij} between job i and job j, the machine is idle.
- Each machine can only process one job at a time.
- Preemption of job is not allowed.
- The processing time is discrete.
- The normal processing time can be compressed by the number of x_j^c, which requires a unit cost of compression.
- The normal processing time can be expanded by the number of x_j^e, which requires a unit cost of expansion.
- Jobs that have normal processing times do not incur additional costs.

2.2.2 *Mathematical Modeling*

This scheduling problem contains three objectives, i.e., E/T, penalty cost, and energy consumption. The mathematical model consists of a production model and an energy consumption model. The objective of the production model is the minimization of total *E/T* and cost. Energy consumption is the objective of the energy consumption model. They are given in the following parts.

In the production model, we propose a bi-objective mathematical model to simultaneously minimize the total *E/T* and penalty cost. The model is formulated as follows.

$$\min\ f_1 = ET = \sum_{j=1}^{n} \left(t_{[j]} T_{[j]} + e_{[j]} E_{[j]}\right) \tag{2.1}$$

$$\min\ f_2 = cost = \sum_{j=1}^{n} \left(a_{[j]} x_{[j]}^c + b_{[j]} x_{[j]}^e\right) \tag{2.2}$$

$$S_{[j]} + p_{[j]} - x_{[j]}^c + x_{[j]}^e + s_{[j][j+1]} \le S_{[j+1]},\ j = 1, \ldots, n-1 \tag{2.3}$$

$$S_{[j]} + p_{[j]} - x_{[j]}^c + x_{[j]}^e - d_{[j]} \le T_{[j]},\ j = 1, \ldots, n \tag{2.4}$$

$$d_{[j]} - S_{[j]} - p_{[j]} + x_{[j]}^c - x_{[j]}^e \le E_{[j]},\ j = 1, \ldots, n \tag{2.5}$$

$$m_{[j]}^c \ge x_{[j]}^c,\ j = 1, \ldots, n \tag{2.6}$$

$$m_{[j]}^e \ge x_{[j]}^e,\ j = 1, \ldots, n \tag{2.7}$$

$$M\lambda \ge x_{[j]}^c,\ j = 1, \ldots, n \tag{2.8}$$

$$M(1-\lambda) \geq x^{e}_{[j]},\ j=1,\ldots,n \tag{2.9}$$

$$T_{[j]} \geq 0,\ E_{[j]} \geq 0,\ x^{c}_{[j]} \geq 0,\ x^{e}_{[j]} \geq 0,\ S_{[j]} \geq 0,\ \lambda = 0 \text{ or } 1 \tag{2.10}$$

Equation (2.1) represents the sum of tardiness and earliness. Equation (2.2) represents the total penalty cost of compression and expansion. Constraint (2.3) defines that the sum of the completion time of job at the jth position and the setup time between the jobs at jth and $(j+1)$ positions is less than or equal to the starting time of job at the $(j+1)$ position. Constraints (2.4) and (2.5) define the lateness and earliness departure of jobs that must be minimized. Constraints (2.6) and (2.7) limit the amount of compression and expansion of each job, respectively. Constraints (2.8) and (2.9) simultaneously guarantee that the processing time of each job is either compressed or expanded. The constraint (2.10) denotes the non-negativity of the variables.

If the machine does not work for a long time, it should be switched off to save energy. If the total energy required to turn off/on the machine (i.e., E_{sc}) is greater than the idle energy during the interval time (expressed by $E^{[j][j+1]}_{idle} = T_B \times P_{idle}$) between jobs at j and $j+1$ position, it is better to leave the machine idle. The break-even duration T_B is used to determine whether the machine is in the idle state or in stop state. The following is the definition of T_B.

$$T_B = \frac{E_{sc}}{P_{idle}} = \frac{(P_{close} + P_{start}) \cdot T_{sc}}{P_{idle}} \tag{2.11}$$

where P_{idle} is idle power, P_{run} is the processing energy power, E_{sc} is the total energy consumption when machine is turned off and on, P_{close} and P_{start} denote the power required to swith off and on the machine. If T_{idle} exceeds T_B, then it is beneficial to swicth off the machine.

Assuming that the energy level is constant during the working phase, the energy consumption during idle phase can be expressed by the following equation: $E_{idle} = P_{idle} \times T_{makespan}$, where $T_{makespan}$ is the completion time of a schedule, i.e., makespan. The additional energy consumption for processing jobs is $E_{run} = P_{run} \times T_{run}$, where $T_{run} = \sum p^{a}_{j},\ j = 1,\ldots,n$. The total energy savings at a given time can be determined in the following way.

$$E_{save} = \sum_{j=1}^{n-1} \lambda_{[j]}\left[P_{idle} \times \left(S_{[j+1]} - C_{[j]}\right) - E_{sc}\right] \tag{2.12}$$

$$\lambda_{[j]} = \begin{cases} 1, & \text{if } S_{[j+1]} - C_{[j]} \geq T_B \\ 0, & \text{otherwise} \end{cases}$$

Therefore, the total energy consumption can be expressed by the following equation.

$$\min f_3 = \text{energy} = E_{\text{idle}} + E_{\text{run}} - E_{\text{save}} \tag{2.13}$$

To further clarify the energy consumption required for the turn off/on machine, Fig. 2.1 shows an example of calculating the total energy consumption. The processing time and setup time of the jobs are shown in Table 2.1. The energy power for processing and idle is 2 kW and 1 kW, respectively. 2 kW power is required to start-up and shutdown forE_{sc}. To simplify the energy calculation, fixed or actual processing times are used in this problem, even if the processing times of the jobs are controllable. For a given sequence of jobs$\pi = (J_1, J_3, J_2)$, as shown in Fig. 2.1, the total idle energy consumption is$E_{\text{idle}} = P_{\text{idle}} \times T_{\text{makespan}} = 1 \times 11 = 11$. The energy consumption during the processing of the jobs is$E_{\text{run}} = P_{\text{run}} \times T_{\text{run}} = 2 \times 6 = 12$. According to Eq. (2.12), the idle time T_{idle} between J_1 and $J_3 (T_{\text{idle}} = S_{[j+1]} - C_{[j]} = 2 - 1 = 1)$ is less than the scheduled equilibrium time $T_B = 2$ and this machine should continue working normally. Similarly, the idle time ($T_{\text{idle}} = 4$) between J_3 and J_2 is greater than$T_B = 2$, so this machine should be turned off to save energy. The energy savings can be calculated using the following formula.$E_{\text{save}} = 1 \times [1 \times (9 - 5) - 2] = 2$. The total energy consumption for this schedule is$f_3 = E_{\text{idle}} + E_{\text{run}} - E_{\text{save}} = 11 + 12 - 2 = 21$. So, the amount of energy consumption, besides being related to the operation mode, also has a strong relationship with the switching mode. It is possible to turn the machine off or on during a scheduled job without leaving it idle.

Therefore, three optimization objectives for total cost, total cost, and total energy consumption are presented in this chapter.

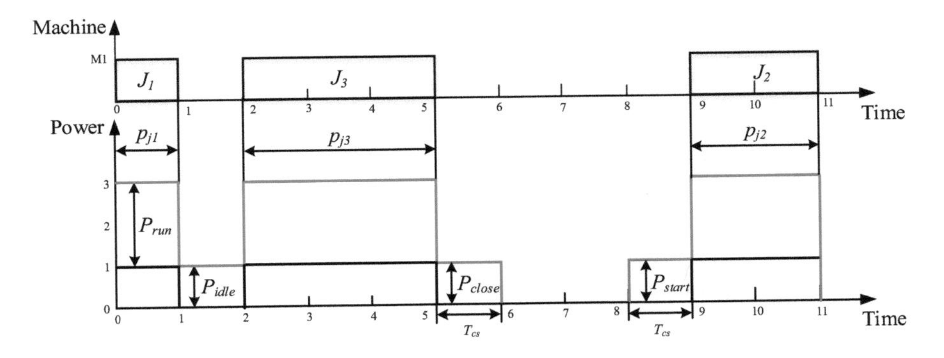

Fig. 2.1 Gantt chart and its corresponding power curve

Table 2.1 3-job example

Jobs	Setup time			Actual processing time
	1	2	3	
1	–	1	1	1
2	1	–	2	2
3	5	4	–	3

$$
\begin{aligned}
&\min f(x) = [f_1(x), f_2(x), f_3(x)]^{\mathrm{T}} \\
&f_1(x) = ET \\
&f_2(x) = \text{cost} \\
&f_3(x) = \text{energy}
\end{aligned}
\tag{2.14}
$$

2.3 Proposed Algorithm

The purpose of this chapter is to develop a hybrid local MOEA (LMOEA) for the above multi-objective scheduling problem. In the LMOEA algorithm, a more rational crossover and local search are used to balance exploration and exploitation.

In LMOEA, we give two main strategies to deal with this scheduling problem. (1) a new encoding scheme based on real numbers to transform the discrete problem into a continuous one; (2) a multiple local search strategy makes each candidate solution closer to the optimal one. The first strategy is used to express the solution representation and decode the solution. The second one is used to find a better solution in the neighborhood of the offspring. The pseudocode of the proposed LMOEA is shown in Algorithm 2.1.

Algorithm 2.1: the proposed LMOEA

1. The external archive: **Create_Front**();// creates an empty Pareto front archive
2. population ← **Initialization**(m, population_Size);// m is the range of decision variables
3. local search probability P ← 1/ns;//ns is the total number of local strategies, here ns = 3
4. **while** (evaluation_number < max_evaluation_number)
5.parents ← **Selection**(population);// binary tournament selection
6.offspring ← **Crossover**(P, parents);
7.offspring ← **Mutation**(Pm, offspring);
8.**if**(evaluation number > max_evaluation_number/2)
9. Offspring **multiple_local_search_based_on_adaptive_scheme**(offspring,[P_1,P_2,P_3]);
10.**end**
11.**Evaluate_Fitness**(offspring);
12.**if** (offspring dominates parent)
13. **Insert_external_archive**(offspring);
14.**end if**
15.**Recalculate_selection_probability** [P_1,P_2,P_3];
16.population ← offspring;
17. evaluation_number++;
18.**end while**
19.**output** PS and the PF
20. **End for**

The line 6 in Algorithm 2.1 is described in Sect. 2.3.2. The line 7 is also described in Sect. 2.3.3. Section 2.3.4 illustrates the specific principles of local search, as in lines 8–10 in Algorithm 2.1. The local search strategy is also given in Sect. 2.3.4. A specific description on replacement strategy is given in Sect. 2.3.5, lines 12 to 14.

External files are used to store non-dominated solutions found so far. To preserve the diversity of non-dominated solutions in the external file, the crowding distance technique is adopted, and solutions with lower crowding distance are discarded when the number of non-dominated solutions exceeds the size of the external file.

In addition, there are three important differences between our proposal and previous studies of hybrid MOEAs.

- As we know, one algorithm does not outperform any other for every problem. Therefore, the proposed LMOEA combines multiple techniques, including harmony search (HS), genetic algorithm (GA), and differential evolution (DE) strategies. Although all three metaheuristics are population-based metaheuristics, they have different search methods and directions in the search space, thus enhancing the diversity of the population and the stability of the algorithm. The reason why we chose these three strategies will be explained in Sect. 2.2.2.
- The proposed algorithm is based on multiple techniques such as HS, GA, and DE, so it is necessary to use a new hybrid selection method for the search. The new hybrid scheme is characterized by the possibility of switching between static and adaptive options. The method aims to avoid repeating search operations that lead the algorithm into local optimization. The method combines static and adaptive selection strategies that are effective in reducing local optima and finding desirable solutions in underutilized regions.

The following section describes in detail our proposed algorithm, which contains the following: a new representation of the scheduling problem, crossover operators, mutation operators, various local search methods based on adaptive mechanisms, and substitution strategies.

2.3.1 Encoding and Decoding

The key to using MOEA for the problem is its solution encoding. A new real number encoding scheme (solution representation) is proposed in this chapter. This solution representation is in line with the literature [2, 3]. In [2], they developed a solution of length $2n$ (n is the number of jobs), while our proposal is n. In [3], they used a maximum position value (LPV) rule to represent the job processing order. The encoding approach described in this chapter has the following three main advantages.

(1) Simpler chromosome structure. Although this proposed encoding scheme contains only a one-dimensional structure, it contains two components, namely the amount of compression or expansion of the job processing time and the job order.
(2) By adopting the recommended coding scheme, the discrete optimization problem can be transformed into a continuous optimization problem.

(3) The constraint handling mechanism is easier and not needed at all. The algorithm using the traditional encoding scheme requires some special reorganization operators to avoid the unimplementable approach. However, the method is able to find practical solutions with the proposed encoding mechanism.

The basic principle of this coding scheme is that the n jobs under the given conditions, n real values, representing the number of compressions or expansions and the order of jobs, not belonging to their own category $[m_c, m_e]$ are generated randomly in a uniform distribution. Rounding of integers indicates the actual compression or expansion of job processing times. For distinguishing between compression and expansion of work processing time, positive integers represent the actual amount of expansion and negative integers represent the actual amount of compression. It is important to know that the possibilities of generating positive and negative are the same. The fractional part of the real number refers to the order of processing working on a computer. A more detailed description of the proposed coding scheme. Figure 2.2 shows the expression of the scheme. For a specified 4-task lady single-computer scheduling problem, we assume that its maximum compressed and expanded amounts are $m_c = (-3, -4, -2, -4)$ and $m_e = (2, 3, 2, 4)$, respectively. A chromosome vector $(-1.325, 2.420, -1.761, 3.067)$ is generated arbitrarily within the range of the compressed or expanded quantities. This chromosome can be converted into two meanings: (1) the actual number of vectors $m = (-1, 2, -2, 3)$ for the compression or expansion of the work processing time; (2) the sequence of the work, i.e., $\pi = (J_4, J_1, J_2, J_3)$. The specific point, the integer part of $(-1, 2, -2, 3)$ is obtained by rounding the real numbers and represents the actual amount of compression or expansion of the job processing time. In this case, the integer value '-1' in the 1st position of the integer part means that the amount of compression of J_1 is one unit of time. Similarly, the value '3' in the 4th digit means that the amount of expansion of J_4 is 3 units of time. The fractional part $(0.325, 0.420, 0.761, 0.067)$ refers to the order of work. The numbers in the fractional part are ordered according to a non-decreasing order, i.e., $0.067 < 0.325 < 0.420 < 0.761$. The corresponding workflow is as follows (J_4, J_1, J_2, J_3). Using this coding scheme, it is easy to identify the actual amount of compression or expansion in the work and the working order of the work. Therefore, based on the above analysis, the decoding mechanism is very simple and efficient.

2.3.2 *Crossover Operator*

Crossover is one of the most critical operators to generate new offspring solutions. Effective crossover operations are required to renew the population and to pass on the good genes of the parents to the next generation. In recent years, many different crossover operators have been proposed to deal with continuous optimization problems, such as uniform crossover [4], simulated binary crossover (SBX) [5], and hybrid crossover (BLX-α) [6]. We know that arithmetic crossover and permutation-based crossover are more suitable for discrete problems, but they are not applicable

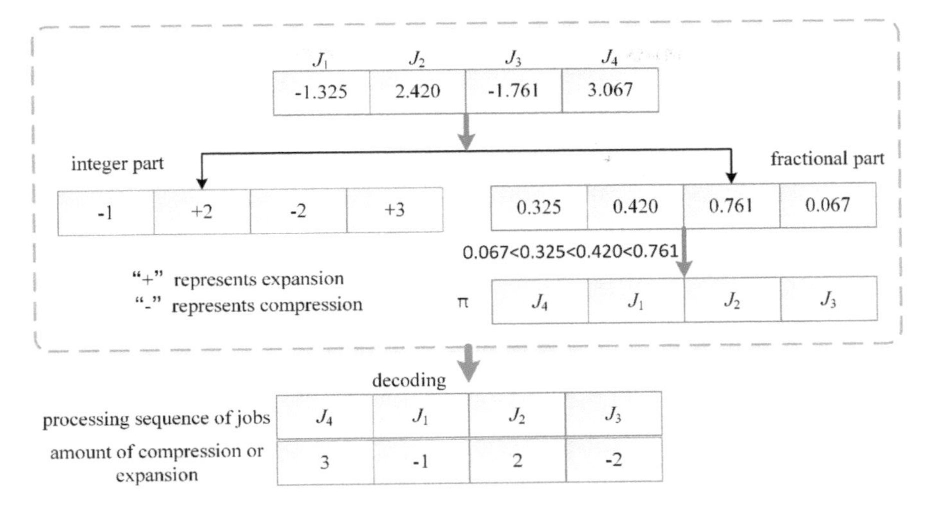

Fig. 2.2 Representation of the solution

to the encoding scheme in this study. Again, the effectiveness of crossover operators depends on the particular problem. Following the experimental study conducted with various experimental examples in Sect. 2.4.3, the final SBX was applied to the LMOEA proposed in this chapter.

2.3.3 *Mutation Operator*

In MOEA, variation is also a very important operation to adjust specific genes at a very low chance. Moreover, this can help us to avoid local optima. On this problem, the mutation operator consists of two mutation techniques. The first mutation technique, its purpose, is to adjust the amount of compression or expansion of the job processing time within a certain range. The second mutation technique, called the swap mutation operator, switches the order of the jobs. The first mutation technique can change the integer values in its scope to achieve its purpose. The other technique can use two genes to exchange the fractional part in order to switch the order of job. When mutating an individual, either the former mutation is chosen or the latter mutation operation is chosen with a probability of 0.5. To explain the process of the mutation operator in more detail, an example of how to implement a mixed mutation is shown in Fig. 2.3. For the first mutation operator, as shown in Fig. 2.3a, the mutation positions are first assigned equally from 1 to 4. We assume that a second gene is selected to generate a new integer at random from its scope. As for the other mutation operator, as in Fig. 2.3b, in the random selection of two genes (e.g., the second and fourth genes), their fractional parts are flooded with "......" More precisely, the original fractional values of the second and fourth genes are 0.420 and 0.067. According to the encoding

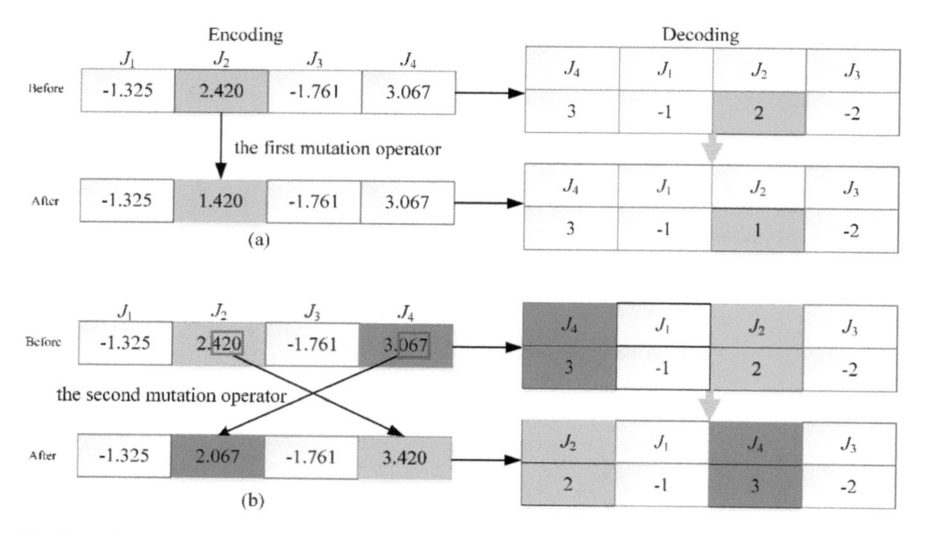

Fig. 2.3 Mutation operators

scheme proposed above, the original sequence of the job is $\pi = (J_4, J_1, J_2, J_3)$. After applying the second mutation operation, the fractional values of the second and fourth genes are 0.067 and 0.420, respectively. Accordingly, the sequence of the job becomes $\pi = (J_2, J_1, J_4, J_3)$, corresponding to a compressed or extended amount of job processing time of $m = (2, -1, 3, -2)$.

2.3.4 *Local Search*

The method is effective in enhancing the performance of the algorithm. In this study, we give three different local search strategies with adaptive selection mechanism in order to obtain better approximate solutions. In a certain time, in strategy 1 (S1), we insert a randomly selected fractional part into another job position with maximum setup time to reduce the energy consumption. Strategy 2 (S2) is another local search technique, which reduces the total tardiness/earliness departure by swapping the maximum late or early departure from the job with a lesser late or early departure. Strategy 3 (S3) can reduce costs by reducing the actual processing time from the corresponding regular processing time. Multiple local search can be seen in the flowchart in Fig. 2.4.

The selection of the appropriate local search operator for the specified time makes use of an adaptive selection mechanism. The selection probability of each local search strategy can be obtained from the record of local operations performed on each solution. When initialized, the selection probability of each local search policy is 1/3. After that, when a local search policy is selected, that policy will assign a

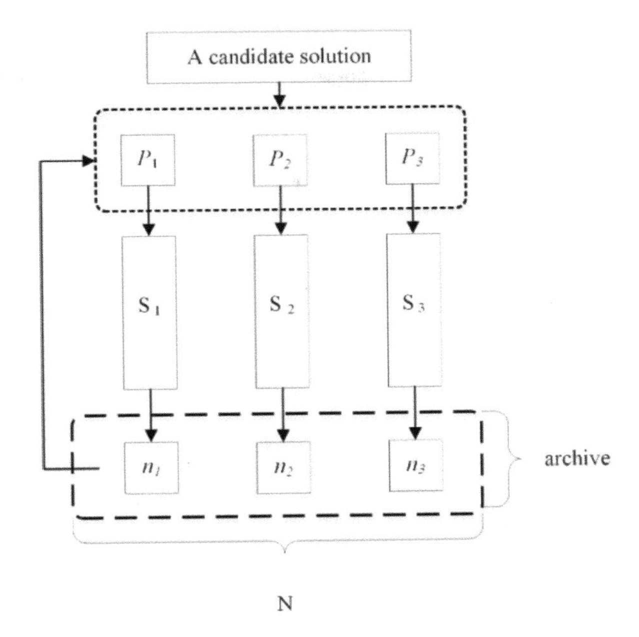

Fig. 2.4 Flowchart of multiple local search

new solution. Let P_i denote the selection probability of the ith policy, P_1 be the probability of S1, P_2 be the probability of S2, and P_3 be the probability of S3. The adaptive selection mechanism can be described as follows. First, if the external profile of the currently found non-dominated solutions is updated, the selection rate $P_i = n_i/|N|$ is calculated for each local strategy, where n_i is the number of non-dominated solutions obtained by the ith local strategy in the external profile. $|N|$ represents the current size of the external profile. Then a strategy is selected using the roulette wheel method. To avoid the case of using the same local search operator during the search, if this is the case, each local strategy will be assigned a probability of 1/3. According to the above analysis, a complex problem can be easily solved by a multi-local operator based on this adaptive mechanism.

2.3.5 Replacement Strategy

Multi-objective optimization differs from single-objective optimization. In multi-objective optimization, each solution is equal to its non-dominant level (e.g., 1 represents the best level, 2 represents the second best level, and so on). Next, within each level or rank, the order is defined between solutions using a crowded distance technique that shows the sum of the distances of the individuals closest to each objective. To achieve a wide spread of the Pareto front for the purpose, experimental solutions with large crowding distances are preferred over those with fewer crowding distances. Our alternative solutions are determined based on rank and crowding distance. In

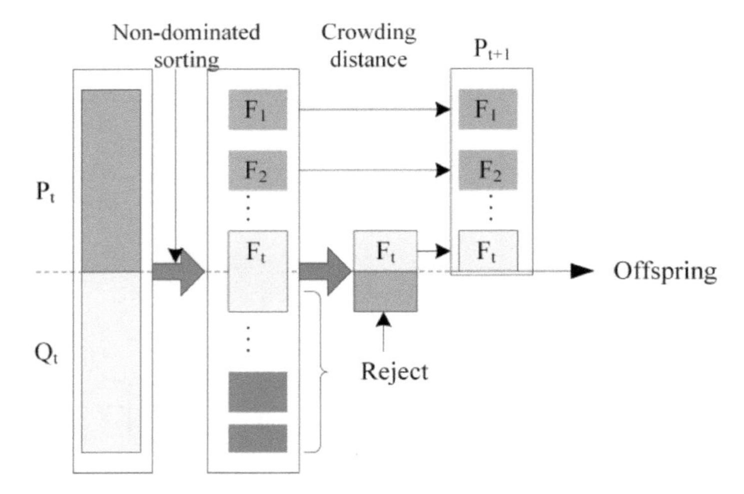

Fig. 2.5 Construction of population $P_t + 1$

other words, if the solution in the external file is dominated by the offspring, or both are non-dominated levels and have the worst crowding distance in a population consisting of the solution in the external file and the offspring, then this will be replaced. Otherwise, the children will be discarded. However, when the update starts, the external profile is empty and the PS in the initial population is inserted directly into the profile (Fig. 2.5).

2.4 Experiments

In this section, we did a set of computational experiments. To evaluate the performance of LMOEA, this experiment used java code and ran on an Intel Core i5 1.6 GHz PC with 1 GB of RAM. This section aims to measure the performance of the proposed LMOEA with respect to the addressed problem. In the next section, we will present the test problems, performance metrics, parameter settings, and experimental studies.

2.4.1 Test Instances

This experiment involves three problem size (i.e., small, medium, and large), according to the number of jobs. The due date for each job is $d_j = k \times p_j, j = 1, 2, \ldots, n$ where k is the control factor. The value of k will be set to 1, 2, ..., 5, which is in line with the trend toward looser due date. The units are in minutes. The other parameters of the data are shown in Table 2.2. This example with n jobs

Table 2.2 Distribution of datasets

Input variables	Value
Number of jobs (n)	10, 50, 100
Normal processing time (p_j)	~Slum (20, 80)
Collision processing time (p_j^c)	~DU($-p_j/3$, 0)
Expansion processing time (p_j^e)	~DU(0, $p_j/5$)
Due time (d_j)	$d_j = k \times p_j, k = 1, 2, 3, 4, 5$
Unit penalty compression cost (a_j)	~U(0.5, 2.5)
Unit punitive expansion cost (b_j)	~U(0.5, 2.5)
Penalty factor for early arrival and late arrival (e_j or t_j)	0.1
Power at idle (p_{idle})	2.2 kW
Additional power during processing operation (p_{run})	7.5 kW
Total energy when turning the machine off and on (E_{sc})	6.72 kWh
Setup time between adjacent jobs (S_{ij})	~DU(0, 10)

and k parameters is denoted by the symbol 'Case_n_k'. Let's say that 'Case_10_1' represents that the problem has 10 jobs and $k = 1$.

2.4.2 Performance Metrics

Several indices of the Pareto front (PF), such as Spread [1, 7], GD [8], and IGD [9] should be taken as follows.

(1) Spread (Δ). This metric serves as a distribution indicator. It has the ability to measure the distribution condition along the front found. The definition for this metric is:

$$\Delta = \frac{\sum_{j=1}^{n_o} d_j^e + \sum_{i=1}^{|\text{PF}|} \left| d_i - \overline{d} \right|}{\sum_{j=1}^{n_o} d_j^e + |\text{PF}| \cdot \overline{d}} \tag{2.15}$$

where d_i denotes the Euclidean distance between the i-th member in the obtained PF and its nearest member, the symbol $\overline{d}$ denotes the average Euclidean distance of all the d_i, the Euclidean distance between the extreme solution in the j-th objective direction and the corresponding extreme solution in the PF* is recorded by the symbol d_j^e, the total number of members in PF is |PF|, and n_o represents the total number of objectives. The distribution along PF will be more uniform if the spread value is lower.

(B) Generational Distance (GD). The convergence performance is represented by the GD metric. It is the average distance from obtaining PF to the PF*. This metric is written as follows:

$$\mathrm{GD} = \frac{\sqrt{\sum_{i=1}^{|\mathrm{PF}|} D_i^2}}{|\mathrm{PF}|} \tag{2.16}$$

where |PF| represents the total number of members in the PF found, D_i is the Euclidean distance between the i-th member in PF and the nearest member in PF*. A low GD value means a good convergence performance. The acquired metric results could have several dimensions, which would impact the outcome of the data analysis. A normalizing method is used in this metric to reduce the impact of various dimensions.

(C) Inverted Generational Distance (IGD). Even if it is a GD variant, it represents a comprehensive indicator. The average distance between the PF* and PF produced from the algorithm is calculated. IGD is described as follows:

$$\mathrm{IGD} = \frac{\sum_{\mathbf{x} \in \mathrm{PF}^*} \mathrm{dist}(\mathbf{x}, \mathrm{PF})}{|\mathrm{PF}^*|} \tag{2.17}$$

where dist($\mathbf{x}$, PF) is the minimum Euclidean distance between the $\mathbf{x}$ ($\mathbf{x}$ is the member in PF*) and the member in PF, the total number of all members in PF* is denoted as $|\mathrm{PF}^*|$. If $|\mathrm{PF}^*|$ is large enough, IGD may be employed as an indicator for evaluating the overall performance of convergence and diversity. It is preferable to have fronts with lower IGD values. A normalization approach is also used in this metric.

The best results are highlighted in bold in the table. Because all candidate MOEAs are randomized algorithms, the following statistical analyses are required for confidential comparisons. First, a Kolmogorov–Smirnov test is performed to check whether the results conform to a normal distribution. If the results are not normally distributed, a t-test is used to check the results for each algorithm; otherwise, a Wilcoxon rank sum test is performed on the mean. Two statistical tests on the algorithms are used to measure the significant differences in the results obtained by the different algorithms. The confidence level was set to 95% (corresponding to a p-value below 0.05) for all tests. The symbol "+" indicates a significant improvement in the performance of the LMOEA algorithm proposed in this paper over the second best method. And the "−" LMOEA algorithm shows a significant difference compared with the best algorithm. The symbol "=" LMOEA algorithm is not significantly different from the best and second best MOEA. In the bottom row of these tables, there is a corresponding metric dominance performance ratio. For example, 3/15 represents the corresponding algorithm for 15 problems that is better than their competitor on 3 problems.

2.4.3 LMOEA Against Other MOEAs

For evaluating the performance of the algorithm in scheduling problems, this section will be compared with other MOEAs such as NSGA-II [1], SPEA2 [8], OMPSO [9], and MOEA/D [10]. In this experiment, to have a fair comparison, the number of function evaluations for all MOEAs was set to 25,000. The same encoding scheme was used for all MOEAs. Further, the initial populations of the above MOEAs were generated randomly. Other parameters are shown in Table 2.3. Thirty separate runs of each algorithm were conducted on each test instance.

Tables 2.4, 2.5 and 2.6 show the statistics for the GD, Spread, and IGD indicators. From these tables, we can see clearly that this LMOEA solves these scheduling instances better than the other algorithms for the IGD and GD. To be more precise. As shown in Table 2.4, in 8 out of 15 examples, the proposed LMOEA is significantly better than the other four MOEAs in terms of GD metrics. Note that OMOPSO cannot get better results in all instances. Regarding the Spread, LMOEA has a greater advantage compared to NSGA-II and OMOPSO. Further, LMOEA has a stronger competition for SPEA2 and MOEA/D concerning Spread metric. Table 2.6 shows that LMOEA has a significantly better IGD values compared to other MOEAs. It is important to note that no MOEA can maintain a consistent and good behavior in all cases. It is concluded that LMOEA is a good method to handle such problems.

The average and standard deviation of CPU time in seconds for each algorithm are shown in Table 2.7. To better understand the performance of the five algorithms, we select the CPU time for all test instances and express it graphically. Figure 2.6 validates the conclusions from the numerical analysis. It can be seen that when all cases are solved, SPEA2 is the fastest and OMOPSO is the slowest. With more machines at each stage, each algorithm requires an increase in CPU time, however, the CPU time of LMOEA grows slower compared to the other algorithms. In this chapter, we give three different local search strategies (S1, S2, and S3) based on adaptive selection mechanisms so that better approximate solutions can be obtained. Therefore, LMOEA would benefit from a local search strategy, and this strategy is designed to

Table 2.3 Parameter settings in MOEAs

Algorithm	Parameters			
LMOEA	Population size: 100	Crossover rate. 0.9	Mutation rate. 0.1	File size: 100
NSGA-II	Population size: 100	Crossover rate. 0.9	Mutation rate. 0.1	
SPEA2	Population size: 100	Crossover rate. 0.9	Mutation rate. 0.1	File size: 100
OMOPSO	Colony size: 100		Mutation rate. 0.1	Number of leaders: 100
MOEA/D	Population size: 500	CR = 1.0, $F = 0.5$, NR = 2	Mutation rate. 0.1	Residential area size T: 20

Table 2.4 Mean and standard deviation values of GD indicators on all algorithms

Case	NSGA-II	SPEA2	OMOPSO	MOEA/D	LMOEA
	Mean/standard value	Mean/standard value	Mean/standard value	Mean/standard value	Mean/ standard value
Case_10_1	$7.0 \times e^{-3}$ /8.7 $\times e^{-3}$	$1.1 \times e^{-2}$ /2.0 $\times e^{-2}$	$2.4 \times e^{-2}$ /1.4 $\times e^{-2}$	$6.7 \times e^{-2}$ /7.7 $\times e^{-3}$	**$6.6 \times e^{-3}$/5.8 $\times e^{-3}$ +**
Case_10_2	$1.0 \times e^{-2}$ /1.0 $\times e^{-2}$	$8.9 \times e^{-3}$ /1.1 $\times e^{-2}$	$4.3 \times e^{-2}$ /2.6 $\times e^{-2}$	$3.2 \times e^{-2}$ /1.2 $\times e^{-2}$	**$5.0 \times e^{-3}$/6.3 $\times e^{-3}$ +**
Case_10_3	$4.2 \times e^{-3}$ /3.3 $\times e^{-3}$	$3.7 \times e^{-3}$ /2.5 $\times e^{-3}$	$1.2 \times e^{-2}$ /8.4 $\times e^{-3}$	$4.2 \times e^{-3}$ /3.3 $\times e^{-3}$	$4.2 \times e^{-3}$/1.7 $\times e^{-3}$ –
Case_10_4	$7.0 \times e^{-3}$ /3.8 $\times e^{-3}$	$6.5 \times e^{-3}$ /4.3 $\times e^{-3}$	$1.3 \times e^{-2}$ /8.5 $\times e^{-3}$	$6.3 \times e^{-3}$ /3.7 $\times e^{-3}$	**$5.2 \times e^{-3}$/1.9 $\times e^{-3}$ +**
Case_10_5	$5.4 \times e^{-3}$ /1.3 $\times e^{-3}$	$4.0 \times e^{-3}$ /1.9 $\times e^{-3}$	$8.9 \times e^{-3}$ /4.7 $\times e^{-3}$	$5.6 \times e^{-3}$ /2.7 $\times e^{-3}$	$5.4 \times e^{-3}$/2.2 $\times e^{-3}$ –
Case_50_1	$5.1 \times e^{-2}$ /8.1 $\times e^{-2}$	$3.2 \times e^{-2}$ /6.6 $\times e^{-2}$	$3.0 \times e^{-1}$ /3.3 $\times e^{-1}$	$3.1 \times e^{-2}$ /6.0 $\times e^{-2}$	$8.6 \times e^{-2}$/1.1 $\times e^{-1}$ –
Case_50_2	$3.9 \times e^{-2}$ /3.2 $\times e^{-2}$	$8.3 \times e^{-2}$ /2.9 $\times e^{-1}$	$2.3 \times e^{-1}$ /2.1 $\times e^{-1}$	$4.8 \times e^{-2}$ /3.6 $\times e^{-2}$	**$3.2 \times e^{-2}$/3.5 $\times e^{-2}$ +**
Case_50_3	$1.3 \times e^{-2}$ /1.9 $\times e^{-2}$	$3.5 \times e^{-2}$ /7.8 $\times e^{-2}$	$1.1 \times e^{-1}$ /1.0 $\times e^{-1}$	$2.3 \times e^{-2}$ /2.1 $\times e^{-2}$	$2.0 \times e^{-2}$/3.2 $\times e^{-2}$ –
Case_50_4	$1.0 \times e^{-1}$ /1.3 $\times e^{-1}$	$1.0 \times e^{-1}$ /2.2 $\times e^{-1}$	$6.1 \times e^{-1}$ /5.8 $\times e^{-1}$	$1.1 \times e^{-1}$ /1.2 $\times e^{-1}$	**$2.0 \times e^{-2}$/3.9 $\times e^{-2}$ +**
Case_50_5	$3.7 \times e^{-2}$ /4.5 $\times e^{-2}$	$1.9 \times e^{-1}$ /4.1 $\times e^{-1}$	$1.5 \times e^{-1}$ /1.4 $\times e^{-1}$	$1.7 \times e^{-1}$ /3.5 $\times e^{-1}$	$1.4 \times e^{-1}$/1.8 $\times e^{-1}$ –
Case_100_1	$6.2 \times e^{-3}$ /5.6 $\times e^{-3}$	$6.1 \times e^{-3}$ /1.7 $\times e^{-3}$	$2.9 \times e^{-2}$ /1.5 $\times e^{-2}$	$6.0 \times e^{-3}$ /5.2 $\times e^{-3}$	**$5.6 \times e^{-3}$/2.6 $\times e^{-3}$ +**
Case_100_2	$5.2 \times e^{-3}$ /4.4 $\times e^{-3}$	$4.4 \times e^{-3}$ /1.8 $\times e^{-3}$	$2.0 \times e^{-2}$ /7.2 $\times e^{-3}$	$5.0 \times e^{-3}$ /4.1 $\times e^{-3}$	**$4.1 \times e^{-3}$/1.2 $\times e^{-3}$ +**
Case_100_3	$9.8 \times e^{-3}$ /8.4 $\times e^{-3}$	$2.4 \times e^{-3}$ /1.8 $\times e^{-3}$	$2.9 \times e^{-2}$ /1.0 $\times e^{-2}$	$3.1 \times e^{-3}$ /1.4 $\times e^{-3}$	$5.7 \times e^{-3}$/1.8 $\times e^{-3}$ –
Case_100_4	$5.8 \times e^{-3}$ /4.2 $\times e^{-3}$	$4.5 \times e^{-3}$ /1.8 $\times e^{-3}$	$1.9 \times e^{-2}$ /6.8 $\times e^{-3}$	$5.8 \times e^{-3}$ /4.2 $\times e^{-3}$	**$4.0 \times e^{-3}$/9.6 $\times e^{-4}$ +**
Case_100_5	$7.6 \times e^{-3}$ /8.5 $\times e^{-3}$	$4.9 \times e^{-3}$ /2.5 $\times e^{-3}$	$9.0 \times e^{-2}$ /4.5 $\times e^{-2}$	$3.6 \times e^{-3}$ /4.5 $\times e^{-2}$	$2.2 \times e^{-1}$/1.2 $\times e^{-2}$ –
Hit rate	2/15	3/15	0/15	2/15	8/15

improve the exploitation capability of the algorithm. In contrast, OMOPSO does not contain any heuristic information about the reduction of the running time.

The performance of these algorithms is illustrated graphically. Figure 2.7 shows the PF approximations obtained by different MOEAs in solving Case_100_3. Figure 2.7a shows the PF in three dimensions, and it is difficult to say which MOEA is the best. LMOEA obtains a better PF approximation in two dimensions compared to other MOEAs as shown in Fig. 2.7b. This indicates a good convergence performance

Table 2.5 Mean and standard deviation of propagation metrics for NSGA-II, SPEA2, OMOPSO, MOEA/D, and LMOEA

Case	NSGA-II	SPEA2	OMOPSO	MOEA/D	LMOEA
	Mean/standard value	Mean/standard value	Mean/standard value	Mean/standard value	Mean/ standard value
Case_10_1	$8.3 \times e^{-1}$/1.2 $\times e^{-1}$	$8.1 \times e^{-1}$ /1.6 $\times e^{-1}$	$9.5 \times e^{-1}$ /1.3 $\times e^{-1}$	$7.4 \times e^{-1}$ /1.2 $\times e^{-1}$	$8.9 \times e^{-1}$/1.5 $\times e^{-1}$ –
Case_10_2	$1.0 \times e^{-1}$/2.2 $\times e^{-1}$	$8.6 \times e^{-1}$ /2.3 $\times e^{-1}$	1.12/1.5 $\times e^{-1}$	$8.4 \times e^{-1}$ /1.8 $\times e^{-1}$	1.03/2.0 $\times e^{-1}$ –
Case_10_3	$9.8 \times e^{-1}$ /7.1 $\times e^{-2}$	$9.4 \times e^{-1}$ /9.7 $\times e^{-1}$	$9.7 \times e^{-1}$ /1.2 $\times e^{-1}$	$9.2 \times e^{-1}$ /8.6 $\times e^{-2}$	$9.8 \times e^{-1}$/6.5 $\times e^{-2}$ –
Case_10_4	$9.7 \times e^{-1}$ /8.2 $\times e^{-2}$	$8.5 \times e^{-1}$ /7.9 $\times e^{-1}$	$9.8 \times e^{-1}$ /1.4 $\times e^{-1}$	$9.1 \times e^{-1}$ /8.9 $\times e^{-2}$	$8.6 \times e^{-1}$/8.5 $\times e^{-2}$ –
Case_10_5	$8.8 \times e^{-1}$ /7.2 $\times e^{-2}$	$9.1 \times e^{-1}$ /9.1 $\times e^{-1}$	$9.2 \times e^{-1}$ /9.1 $\times e^{-2}$	$8.8 \times e^{-1}$ /9.2 $\times e^{-2}$	**$8.7 \times e^{-1}$/9.3 $\times e^{-2}$ +**
Case_50_1	$9.7 \times e^{-1}$ /2.1 $\times e^{-1}$	1.0/3.2 $\times e^{-1}$	1.0/2.0 $\times e^{-1}$	1.0/2.2 $\times e^{-1}$	1.0/3.5 $\times e^{-1}$ –
Case_50_2	$9.2 \times e^{-1}$ /1.4 $\times e^{-1}$	$7.7 \times e^{-1}$ /2.2 $\times e^{-1}$	1.1/2.5 $\times e^{-1}$	$9.6 \times e^{-1}$ /2.7 $\times e^{-1}$	$9.6 \times e^{-1}$/2.9 $\times e^{-2}$ –
Case_50_3	$8.4 \times e^{-1}$ /1.3 $\times e^{-1}$	$9.0 \times e^{-1}$ /3.2 $\times e^{-1}$	1.0/2.2 $\times e^{-1}$	$9.2 \times e^{-1}$ /3.1 $\times e^{-1}$	**$8.2 \times e^{-1}$/2.6 $\times e^{-1}$ +**
Case_50_4	$9.8 \times e^{-1}$ /2.4 $\times e^{-1}$	$8.9 \times e^{-1}$ /2.7 $\times e^{-1}$	1.1/2.1 $\times e^{-1}$	$9.9 \times e^{-1}$ /3.4 $\times e^{-1}$	**$8.5 \times e^{-1}$/2.8 $\times e^{-1}$ +**
Case_50_5	$9.3 \times e^{-1}$ /1.8 $\times e^{-1}$	$9.4 \times e^{-1}$ /3.1 $\times e^{-1}$	1.0/1.7 $\times e^{-1}$	$9.4 \times e^{-1}$ /3.1 $\times e^{-1}$	**$9.1 \times e^{-1}$/3.2 $\times e^{-1}$ +**
Case_100_1	$9.0 \times e^{-1}$ / 1.6e-1	$7.5 \times e^{-1}$ /1.8 $\times e^{-1}$	$9.6 \times e^{-1}$ /1.8 $\times e^{-1}$	$8.6 \times e^{-1}$ /1.8 $\times e^{-1}$	$8.1 \times e^{-1}$/1.8 $\times e^{-1}$ –
Case_100_2	$8.6 \times e^{-1}$ /9.0 $\times e^{-2}$	$7.3 \times e^{-1}$ /9.7 $\times e^{-1}$	$8.6 \times e^{-1}$ /9.3 $\times e^{-2}$	$7.1 \times e^{-1}$ /8.8 $\times e^{-2}$	$8.0 \times e^{-1}$/6.1 $\times e^{-2}$ –
Case_100_3	$8.5 \times e^{-1}$ /6.1 $\times e^{-2}$	$7.3 \times e^{-1}$ /7.7 $\times e^{-2}$	$8.2 \times e^{-1}$ /9.1 $\times e^{-2}$	$7.8 \times e^{-1}$ /7.9 $\times e^{-1}$	$7.6 \times e^{-1}$/7.3 $\times e^{-2}$ –
Case_100_4	$1.0 \times e^{-1}$ /2.8 $\times e^{-1}$	$8.3 \times e^{-1}$ /2.9 $\times e^{-1}$	$8.4 \times e^{-1}$ /8.2 $\times e^{-2}$	$8.6 \times e^{-1}$ /3.1 $\times e^{-1}$	$8.4 \times e^{-1}$/2.9 $\times e^{-1}$ –
Case_100_5	1.04/2.8 $\times e^{-1}$	$9.2 \times e^{-1}$ /2.9 $\times e^{-1}$	$8.5 \times e^{-1}$ /9.9 $\times e^{-2}$	$9.2 \times e^{-1}$ /2.9 $\times e^{-1}$	1.1/1.5 $\times e^{-1}$ –
Hit rate	1/15	5/15	1/15	4/15	4/15

of the LMOEA. This also implies that the two objectives of total *E*/*T* and cost are contradictory to each other. As shown in Fig. 2.7c, d, although SPEA2, OMOPSO, and LMOEA are more close toward PF* than NSGA-II and MOEA/D, the coverage of the proposed LMOEA is larger than that of SPEA2 and OMOPSO. It means that LMOEA obtains a rich variety of PF. SPEA2 is also applicable to a wide area, but tends to converge to local areas. The performance of LMOEA benefits from a variety

Table 2.6 Mean and standard deviation values of IGD indicators on NSGA-II, SPEA2, OMOPSO, MOEA/D, and LMOEA

Case	NSGA-II	SPEA2	OMOPSO	MOEA/D	LMOEA
	Mean/standard value	Mean/standard value	Mean/standard value	Mean/standard value	Mean/ standard value
Case_10_1	$4.9 \times e^{-3}$/2.0 $\times e^{-3}$	$5.0 \times e^{-3}$ /1.8 $\times e^{-3}$	$5.8 \times e^{-3}$ /1.7 $\times e^{-3}$	$5.1 \times e^{-3}$ /1.8 $\times e^{-3}$	**$4.3 \times e^{-3}$/1.8 $\times e^{-3}$ +**
Case_10_2	$6.1 \times e^{-3}$/2.3 $\times e^{-3}$	$7.0 \times e^{-3}$ /2.6 $\times e^{-3}$	$6.5 \times e^{-3}$ /2.6 $\times e^{-3}$	$6.8 \times e^{-3}$ /2.5 $\times e^{-3}$	**$5.4 \times e^{-3}$/2.2 $\times e^{-3}$ +**
Case_10_3	$4.6 \times e^{-3}$/2.1 $\times e^{-3}$	$5.4 \times e^{-3}$ /1.7 $\times e^{-3}$	$5.8 \times e^{-3}$ /2.1 $\times e^{-3}$	$5.2 \times e^{-3}$ /1.6 $\times e^{-3}$	**$3.6 \times e^{-3}$/1.2 $\times e^{-3}$ +**
Case_10_4	$4.0 \times e^{-3}$/1.2 $\times e^{-3}$	$5.1 \times e^{-3}$ /1.8 $\times e^{-3}$	$4.8 \times e^{-3}$ /1.3 $\times e^{-3}$	$4.5 \times e^{-3}$ /1.9 $\times e^{-3}$	**$3.5 \times e^{-3}$/1.3 $\times e^{-3}$ +**
Case_10_5	$4.5 \times e^{-3}$/1.2 $\times e^{-3}$	$5.5 \times e^{-3}$ /1.5 $\times e^{-3}$	$5.3 \times e^{-3}$ /1.4 $\times e^{-3}$	$5.1 \times e^{-3}$ /1.2 $\times e^{-3}$	**$4.5 \times e^{-3}$/1.9 $\times e^{-3}$ +**
Case_50_1	$6.0 \times e^{-3}$/7.1 $\times e^{-4}$	$7.3 \times e^{-3}$ /7.5 $\times e^{-4}$	$9.1 \times e^{-3}$ /1.1 $\times e^{-3}$	$6.3 \times e^{-3}$ /7.2 $\times e^{-4}$	**$5.7 \times e^{-3}$/9.8 $\times e^{-4}$ +**
Case_50_2	$5.0 \times e^{-3}$/1.1 $\times e^{-3}$	$6.0 \times e^{-3}$ /7.0 $\times e^{-4}$	$7.6 \times e^{-3}$ /1.3 $\times e^{-3}$	$5.1 \times e^{-3}$ /1.5 $\times e^{-3}$	**$4.6 \times e^{-3}$/6.4 $\times e^{-4}$ +**
Case_50_3	$4.8 \times e^{-3}$/7.8 $\times e^{-4}$	$6.1 \times e^{-3}$ /6.1 $\times e^{-4}$	$8.4 \times e^{-3}$ /1.8 $\times e^{-3}$	$6.1 \times e^{-3}$ /5.7 $\times e^{-4}$	**$4.5 \times e^{-3}$/5.1 $\times e^{-4}$ +**
Case_50_4	$5.8 \times e^{-3}$/9.1 $\times e^{-4}$	$6.6 \times e^{-3}$ /7.4 $\times e^{-4}$	$1.0 \times e^{-3}$ /1.7 $\times e^{-3}$	$6.7 \times e^{-3}$ /7.2 $\times e^{-4}$	**$5.4 \times e^{-3}$/8.3 $\times e^{-4}$ +**
Case_50_5	$5.2 \times e^{-3}$/8.4 $\times e^{-4}$	$6.6 \times e^{-3}$ /5.6 $\times e^{-4}$	$7.6 \times e^{-3}$ /1.0 $\times e^{-3}$	$6.5 \times e^{-3}$ /5.1 $\times e^{-4}$	**$5.0 \times e^{-3}$/1.1 $\times e^{-4}$ +**
Case_100_1	$7.4 \times e^{-3}$/1.4 $\times e^{-3}$	$9.7 \times e^{-3}$ /9.4 $\times e^{-4}$	$8.4 \times e^{-3}$ /1.1 $\times e^{-3}$	$8.0 \times e^{-3}$ /7.3 $\times e^{-4}$	**$6.7 \times e^{-3}$/1.0 $\times e^{-3}$ +**
Case_100_2	$1.1 \times e^{-2}$/2.7 $\times e^{-3}$	$1.2 \times e^{-2}$ /2.3 $\times e^{-3}$	**$9.2 \times e^{-3}$ /1.0 $\times e^{-3}$**	$3.0 \times e^{-3}$ /2.7 $\times e^{-3}$	$9.5 \times e^{-3}$/2.3 $\times e^{-3}$ –
Case_100_3	$8.0 \times e^{-3}$/1.3 $\times e^{-3}$	$9.7 \times e^{-3}$ /8.9 $\times e^{-4}$	$9.2 \times e^{-3}$ /1.2 $\times e^{-3}$	$7.8 \times e^{-4}$ /5.9 $\times e^{-4}$	**$7.1 \times e^{-3}$/1.1 $\times e^{-3}$ +**
Case_100_4	$8.3 \times e^{-3}$ /1.3 $\times e^{-3}$	$1.0 \times e^{-2}$ /9.9 $\times e^{-4}$	$8.7 \times e^{-3}$ /9.6 $\times e^{-4}$	$2.1 \times e^{-2}$ /1.9 $\times e^{-3}$	**$7.6 \times e^{-3}$/1.1 $\times e^{-3}$ +**
Case_100_5	$6.8 \times e^{-3}$ /8.7 $\times e^{-4}$	$8.3 \times e^{-3}$ /1.2 $\times e^{-3}$	$8.5 \times e^{-3}$ /1.1 $\times e^{-3}$	$7.3 \times e^{-3}$ /1.7 $\times e^{-3}$	**$6.4 \times e^{-3}$/1.1 $\times e^{-3}$ +**
Hit rate	0/15	0/15	1/15	0/15	14/15

of local search strategies based on adaptive mechanisms, with which LMOEA is able to find better choices in its neighborhood.

Based on the results of the above experiments, LMOEA is a promising method. Further, we have some ideas.

(1) In the proposed LMOEA, different crossover operators for operation will produce different effects. In LMOEA, the SBX operator is the best operator to handle the scheduling instances.

Table 2.7 Mean and standard deviation CPU time (in seconds) for each algorithm

Case	NSGA-II	SPEA2	OMOPSO	MOEA/D	LMOEA
	Mean/standard value	Mean/standard value	Mean/standard value	Mean/standard value	Mean/standard value
Case_10_1	0.66/0.37	**0.64/0.**32	17.07/1.71	1.29/**0.20**	7.34/0.52
Case_10_2	0.71/0.31	**0.61/0.**27	15.42/1.28	1.24/0.21	7.31/0.65
Case_10_3	0.69/0.32	**0.65/0.**39	14.80/1.30	1.23/**0.21**	7.15/0.67
Case_10_4	0.70/0.32	**0.63/0.**29	15.84/1.49	1.25/0.24	7.63/0.91
Case_10_5	0.69/0.32	**0.63/0.**27	15.61/1.58	1.25/**0.20**	7.21/0.63
Case_50_1	2.16/0.51	**1.29/0.33**	119.24/5.67	2.98/0.37	9.06/0.89
Case_50_2	2.21/0.48	**1.29/0.34**	129.34/8.98	3.23/0.44	9.18/0.72
Case_50_3	2.13/0.44	**1.31/0.27**	117.90/6.09	3.50/0.44	9.02/0.70
Case_50_4	2.18/0.38	**1.28/0.**37	119.64/6.53	3.44/0.37	8.63/0.62
Case_50_5	2.18/0.36	**1.28/0.**34	115.43/7.43	3.43/**0.33**	8.59/0.88
Case_100_1	5.60/0.48	**2.78/0.**37	389.00/15.97	6.95/0.79	11.61/0.83
Case_100_2	5.49/**0.34**	**2.86/0.**37	387.85/15.51	6.19/0.48	11.60/0.92
Case_100_3	5.54/0.46	**2.74/0.36**	387.41/16.46	6.19/0.43	11.68/0.92
Case_100_4	5.72/0.53	**2.79/0.44**	418.13/24.10	6.23/0.46	11.53/0.92
Case_100_5	5.54/0.40	**2.74/0.39**	395.40/16.74	6.47/0.53	11.75/1.02

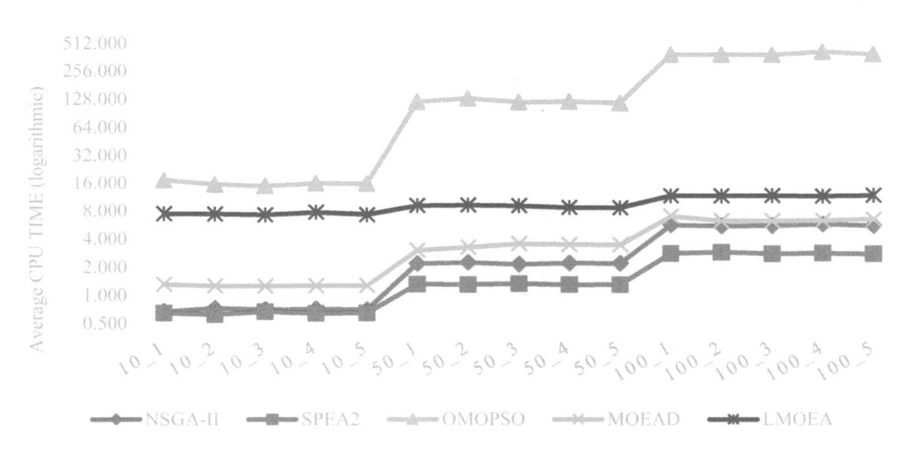

Fig. 2.6 CPU time cost for all test instances

(2) Compared with using only one local search strategy, the combination of multiple local search techniques and adaptive mechanisms can largely improve the behavior of LMOEA.

(3) According to Fig. 2.7, the targets for total *E*/*T*, penalty cost, and energy consumption are contradictory.

Fig. 2.7 Final solutions obtained for different MOEAs with the best IGD values from different perspectives

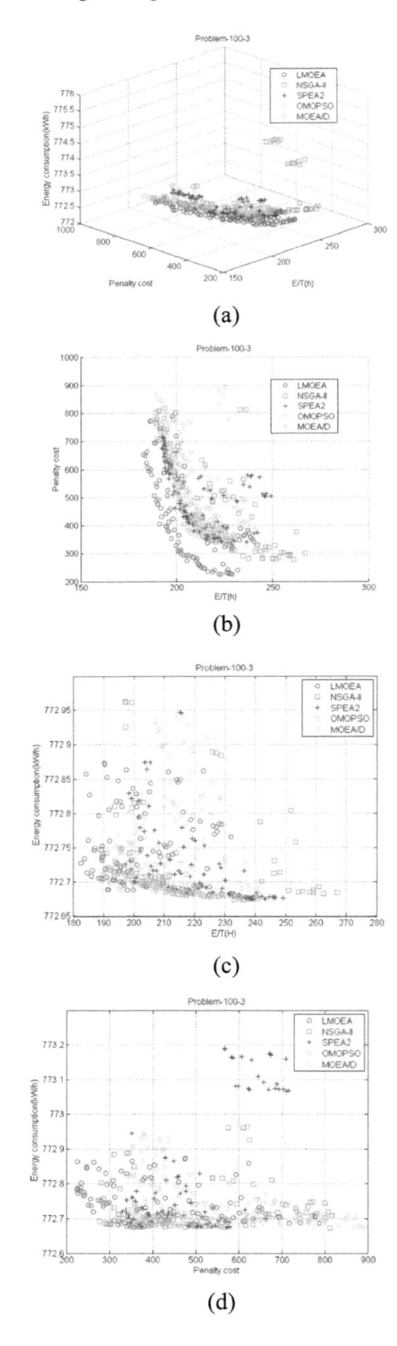

(4) In most scheduling instances, the proposed LMOEA can outperform its competitors such as NSGA-II, OMOPSO, SPEA2, and MOEA/D.

2.5 Conclusion

This chapter presents a multi-objective mathematical model that considers productivity (i.e., *E*/*T* and cost) and energy consumption under single-machine scheduling problem. In this new model, controllable processing times are applied in order to approximate JIT production and also to reduce energy consumption by turning off/on strategy. The scheduling problem is solved using the LMOEA algorithm, which differs from other MOEA algorithms in that three adaptive local search-based strategies are used to improve the search performance of the algorithm. However, there are still some limitations. First, LMOEA has not yet introduced heuristic information based on problem attributes. Second, the algorithm includes several parameters that have an impact on the performance of the algorithm. To measure the performance of the LMOEA algorithm for solving these problems, we have done some experiments. First, the different crossovers in all problems are compared to select the SBX operator as a suitable crossover operator for LMOEA. Secondly, the effect of crossover and mutation probabilities is also profiled. Finally, the computational experimental results show that for most of the scheduling problems, the LMOEA recommended in this chapter outperforms its competitors, such as NSGA-II, SPEA2, OMOPSO, and MOEA/D.

Regarding future work, the proposed algorithm needs to be evaluated in different scheduling environments, such as open shop scheduling environments, to test its wide range of applicability. Another future research topic is to combine existing new techniques with deep learning to improve the search efficiency of the algorithm. Further, the speed of operations related to processing time will be considered in future research.

References

1. Deb, K., Pratap, A., Agarwal, S., Meyarivan, T.: A fast and elite multi-objective genetic algorithm. NSGA-II. IEEE Trans. Evol. Comput. **6**(2), 182–197 (2002)
2. Nearchou, A.C.: Scheduling with controlled processing time and compression cost using population-based heuristics. Int. J. Prod. Res. **48**(23), 7043–7062 (2010)
3. Wang, L., Pan, Q.-K., Tasgetiren, M.F.: A hybrid harmonic search algorithm for scheduling problems in blocking envelope flow shops. Comput. Ind. Eng. **61**(1), 76–83 (2011)
4. Spears, W.M., De Jong, K.A.: . On the merits of parametric unified crossover. In: Proceedings of the Fourth International Conference on Genetic Algorithms, 13–16 July 1991, pp. 230–230). Published by Morgan-Kaufmann Publishing Company, San Diego, CA, USA (1991)
5. Deb, K., Agrawal, R.B.: Simulated binary hybridization in continuous search spaces. Complex Syst. **9**(2), 115–148 (1995)
6. Eshelman, L.J.: Chapter real coded genetic algorithms and interval chemistry. Found. Genetic Algorithms **2**, 187–202 (1993)

7. Shen, X.N., Yao, X.: Mathematical modeling and multi-objective evolutionary algorithms applied to dynamic flexible job shop scheduling problem. Inf. Sci. **298**, 198–224 (2015)
8. Zitzler, E., Laumanns, M., Thiele, L.: SPEA2: An improved intensity Pareto evolutionary algorithm for multiobjective optimization. In: Proceedings Evolutionary Methods Design, Optimization, Control Application in Industrial Problem, pp. 95–100. CIMNE, Barcelona, Spain (2001)
9. Sierra, M., Coello Coello, C.: Improving PSO-based multi-objective optimization using crowding, mutation and ∈ dominance. In: Coello Coello, C., Hernández Aguirre, A., Zitzler, E. (eds.) Evolutionary Multi-Criterion Optimization, vol. 34e, pp. 505–519. Springer Berlin Heidelberg (2005)
10. Qingfu, Z., Hui, L.: MOEA/D: a decomposition-based multi-objective evolutionary algorithm. IEEE Trans. Evol. Comput. **11**(6), 712–731 (2007)

Chapter 3
Green Scheduling in Permutation Flow Shop Environment

3.1 Brief introduction

Due to the wide range of industrial applications, permutation flow shop scheduling problems (PFSPs) have attracted wide attention. The majority of research, however, typically ignores setup and transportation time, which consequently results in a significant gap between theoretical study and practical application. Meanwhile, the emergence of sustainable manufacturing has given prominence to energy conservation. Thus, we study an energy-efficient PFSP with sequence-dependent setup and controllable transportation time from a real-world manufacturing company. First and foremost, based on a thorough analysis, a unique multi-objective mathematical model that takes into account both makespan and energy usage is developed. In order to solve this problem, a hybrid multi-objective backtracking search algorithm (HMOBSA) is developed. In addition, a novel energy-saving scenario is proposed to guarantee both energy conservation and equipment longevity.

3.2 Problem Statement and Modeling

3.2.1 Problem Statement

An energy-efficient permutation flow shop scheduling problem with controllable transportation time (PFSPCT) and sequence-dependent setup time (SDST) from a real-world manufacturing system can be described as follows. There is a set of n jobs, $\mathcal{J} = \{1, 2, \cdots, n\}$ and a set of m machines, $\mathcal{M} = \{1, 2, \cdots, m\}$. The collection of jobs is processed in the same order by m machines. Job preemption is not permitted, and each machine can process at most one job at a time. Each machine's position is

C. Lu et al., *Intelligence Optimization for Green Scheduling in Manufacturing Systems*, Engineering Applications of Computational Methods 18,
https://doi.org/10.1007/978-981-99-6987-6_3

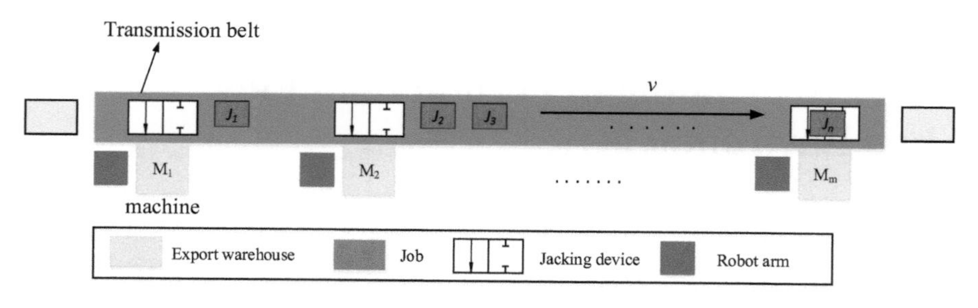

Fig. 3.1 A permutation flow shop problem with controllable transportation time layout

fixed. Furthermore, transportation time and SDST are taken into account at the same time.

There is a finite and discrete set of speeds $\mathcal{S} = \{v_1, v_2, \cdots, v_L\}$ for the transmission belt, and job transportation time can be reduced by raising the transmission belt's speed. The amount of time it takes to convey a job lowers as speed increases. In the meanwhile, the usage of energy rises. The SDST can be categorized into two groups: non-anticipatory and anticipatory [1]. This study takes into account setup and transportation times with non-anticipatory, where setup can only begin once the transportation stage is over [2]. The simplified layout graph of the PFSPCT is illustrated in Fig. 3.1.

3.2.2 *Mathematical Modeling*

The related notations used throughout the study are given as follows:

- j: job j, $j \in \mathcal{J}$.
- n: the number of jobs.
- m: the number of machines.
- $\pi(j)$: the job in the j-th position in a sequence.
- $C_{\max}$: the makespan of the schedule, i.e., the completion time of the last job transported to the warehouse in the schedule.
- C_k: the completion time of all jobs on machine $k \in \mathcal{M}$.
- $C_{k,j}$: the completion time of $\pi(j)$ on machine $k \in \mathcal{M}$.
- $S_{k,j}$: the start time of $\pi(j)$ on machine k.
- $WC_{k,j}$: the completion time of $\pi(j)$ transported from machine k to $k+1$, $k \in \{1, \cdots, m-1\}$.
- $WS_{k,j}$: the starting time of $\pi(j)$ before the preparation operation on the machine k.
- $wt_{k,j}$: the idle time of $\pi(j)$ on machine k, i.e., $wt_{k,j} = (S_{k,j+1} - C_{k,j})$, $j \in \{1, \cdots, n-1\}$.
- Wt_k: the set of idle time of jobs on machine k. That is, $wt_{k,j} \in Wt_k$, $j \in \{1, \cdots, n-1\}$.

T: the maximum number of times for turning off each machine, where $T \leq n-1$.
$WT_{k,T}$: the set of the maximal idle time with T size on machine k.
t_{on}: the time required to turn on a machine.
t_{off}: the time required to turn off a machine.
$v_{\max}$: the maximal transportation speed.
$v_{\min}$: the minimal transportation speed.
$p_{k,j}$: the processing time of job j on machine k.
$r_{k,j}$: the setup time of the robot arm to process consecutive jobs $\pi(j-1)$ and $\pi(j)$ on machine k, $j \in \{2, 3, \cdots, n\}$.
$r_{k,1}$: the setup time of the robot arm to process the first job on machine k.
$T_{k,v}$: the transportation time of the job from machine k to machine $k+1$ at a transportation speed v, $k \in \{1, \cdots, m-1\}$.
$T_{k,\overline{v}}$: the normal transportation time of the job from machine k to machine $k+1$ at a normal speed $\overline{v} \in \mathcal{S}$, $\overline{v}$ is the minimum available speed.
$T_{0,v}$: the transportation time required to convey jobs from the warehouse to the first machine at a speed v.
$T_{m,v}$: the transportation time required to convey jobs from machine m to the warehouse at a transportation speed v.
$x_{k,v}$: the compressed amount of the transportation time of the job from machine k at a transportation speed $v \in \mathcal{S}$. Specifically, $x_{k,v} = T_{k,\overline{v}} - T_{k,v}$.
t_k: the number of turning off the machine k.
P_k^{idle}: the idle power of machine k.
P_v^T: the power of the transmission belt at a speed v.
P_k^s: the setup power of robot arm k.
$E_{\text{off-on}}$: the energy consumption requirement for turning off, then turning on the machine.
T_B: the breakeven duration, $T_B = \max\left(E_{\text{off-on}}/P_k^{\text{idle}}, t_{\text{on}} + t_{\text{off}}\right)$.
π: a feasible sequence.
M_j: the mass of job j.
X_{ij}: it is equal to 1 if job j is the i-th position in the sequence; otherwise, it is equal to 0.
Y_v: it is set to 1 if a given speed v of the transmission belt is selected; otherwise, it is equal to 0.
Z_{kj}: it is equal to 1 if the machine k is shut down after finishing $\pi(j)$; otherwise, it is equal to 0.

The PFSPCT requires the makespan and energy efficiency to be optimized concurrently. To make computation easier, we give a fixed value to the input power throughout the manufacturing process. The total energy consumption consists of energy consumption EC_1 for the setup stage, energy consumption EC_2 for the transportation stage, energy consumption EC_3 for the machine idle stage, energy consumption EC_4 for the processing stage, and EC_5 during periods of public energy consumption (i.e., light and air-conditioner use). There are solely three of them that are connected to the order of jobs and the pace of conveyance. The other categories of energy consumption are irrelevant to various schedule scenarios and transportation speeds so that they are fixed values. By reducing energy consumption including EC_1, EC_2, and EC_2, the goal of lowering overall energy consumption can be fulfilled.

Therefore, the goal of minimizing total energy consumption can be simplified into three parts: EC_1, EC_2, and EC_3. For EC_1, the setup duration is determined by the differences in the properties (i.e., size, shape, texture, and weight) between the two subsequent tasks [3]. This distinction between successive tasks has an immediate effect on the EC_1. EC_1 can be described by using the formula below:

$$\mathrm{EC}_1 = \sum_{k=1}^{m}\sum_{j=1}^{n} P_k^s \cdot r_{k,j} \tag{3.1}$$

Besides, EC_2 is susceptible to the transportation time. Obviously, the transportation speed between two successive machines together with transportation velocity decides the amount of transportation time [4, 5]. As the distance between machines is usually fixed in real-world issue, only the transportation speed is taken into account. EC_2 can be computed as:

$$\mathrm{EC}_2 = \sum_{v \in S} P_v^T \cdot C_{\max} \cdot Y_v + \Delta E \tag{3.2}$$

While ΔE stands for additional energy consumption, which is taken into account in EC_2 to maintain a constant speed for the transmission belt, ΔE is transformed to kinetic energy E_k and heat energy Q in accordance with the law of conservation of energy.

The bulk of earlier research did not take transportation energy consumption into account. The unit "Joule" of the following equation is supposed to be multiplied by $2.8 \times e^{-7}$ in order to preserve consistency of the unit of energy consumption (namely, kWh). Therefore, ΔE can be calculated as:

$$\begin{aligned}\Delta E = E_k + Q &= \left(\frac{1}{2}\sum_{i=1}^{m} M_i v^2 + \frac{1}{2}\sum_{i=1}^{m} M_i v^2\right) \cdot 2.8 \times 10^{-7} \\ &= 2.8 \times 10^{-7} \sum_{i=1}^{m} M_i v^2 \end{aligned} \tag{3.3}$$

Finally, EC_3 is connected to the schedule scenario. The breakeven duration T_B is used to decide whether the machine should be closed or not when the machine is idle [6]. The turn off/on technique could cut energy use dramatically. However, the maximum allowable number of times (T) to switch off a machine is introduced since repeatedly employing the turn off/on method could reduce the service life of a machine [7]. We suggest a better energy-saving technique to further reduce energy consumption and extend the lifespan of the machine. To demonstrate the proposed energy-saving scenario, an example with four jobs processed on machine k

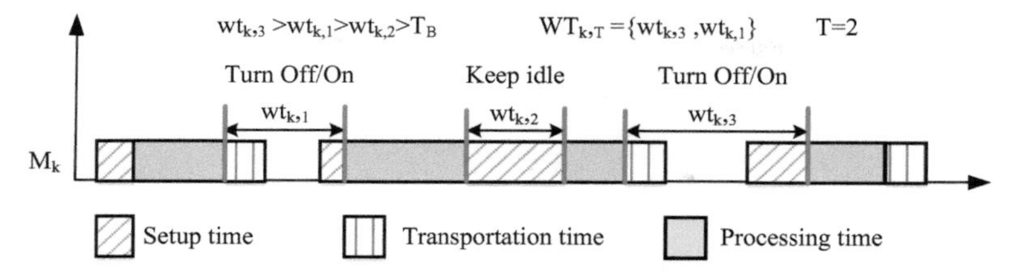

Fig. 3.2 An energy-saving strategy

is shown in Fig. 3.2. Three idle time elements $(\text{wt}_{k,1}, \text{wt}_{k,2}, \text{wt}_{k,3})$ satisfy the possible requirement of turning off the machine (i.e., $\text{wt}_{k,j} > T_B$) and $T = 2$. A non-ascending order is used to sort all idle time components for machine k, especially, (wt_3, wt_1, wt_2). The first two idle time components are chosen as a set $\text{WT}_{k,2} = \{\text{wt}_3, \text{wt}_1\}$. If $wt_{k,j}$ is greater than T_B and the set $\text{WT}_{k,2}$ contains $\text{wt}_{k,j}$, the machine is shut down; otherwise, it is kept idle. To clarify, turn off/on method on machine k can be used to conducted he idle time elements $\text{wt}_{k,3}$ and $\text{wt}_{k,1}$, while the machine is left idle at $\text{wt}_{k,2}$.

As a result, the energy model of EC_3 is given as follows:

$$\begin{aligned}\text{EC}_3 &= \sum_{k=1}^{m} P_k^{\text{idle}} \cdot C_{k,n} - E_{\text{save}} \\ &= \sum_{k=1}^{m} P_k^{\text{idle}} \cdot (C_{k,n} - S_{k,1}) \\ &\quad - \sum_{k=1}^{m} \sum_{j=1}^{n-1} (P_k^{\text{idle}} \cdot \text{wt}_{k,j} - E_{\text{off-on}}) \cdot Z_{k,j}\end{aligned} \tag{3.4}$$

$$Z_{k,j} = \begin{cases} 1, \text{ if } \text{wt}_{k,j} \geq T_B \wedge \text{wt}_{k,j} \in \text{WT}_{k,T} \\ 0, \text{ otherwise} \end{cases}, k \in \mathcal{M};\ j \in \{1, 2, \cdots, n-1\} \tag{3.5}$$

With the objectives of minimizing makespan and energy consumption, the mathematical model for the PFSPCT is stated as follows:

$$\begin{cases} \min f_1 = C_{\max} = \max\{C_k | k = 1, 2, \cdots, m\} \\ \min f_2 = \sum_{h=1}^{3} EC_h \end{cases} \tag{3.6}$$

$$\sum_{j=1}^{n} X_{\text{ij}} = 1, i = 1, 2, \cdots, n \tag{3.7}$$

$$\sum_{i=1}^{n} X_{\text{ij}} = 1, j = 1, 2, \cdots, n \tag{3.8}$$

$$\sum_{v \in S} Y_v = 1 \tag{3.9}$$

$$C_{\max} \geq WC_{m,n} \tag{3.10}$$

$$C_{1,1} \geq \sum_{v \in S} T_{0,v} Y_v + r_{1,1} + \sum_{j=1}^{n} p_{1,j} X_{1,j} \tag{3.11}$$

$$C_{k,i} \geq C_{k-1,i} + \sum_{v \in S} T_{k-1,v} Y_v + r_{k,i} + \sum_{j=1}^{n} p_{k,j} X_{i,j}, k \in \{2, \cdots, m\}; i \in \{1, \cdots, n\} \tag{3.12}$$

$$C_{k,i} \geq C_{k,i-1} + r_{k,i} + \sum_{j=1}^{n} p_{k,j} X_{i,j}, k \in \{1, \cdots, m\}; i \in \{2, \cdots, n\} \tag{3.13}$$

$$WC_{1,1} \geq C_{1,1} + \sum_{v \in S} T_{1,v} Y_v \tag{3.14}$$

$$WC_{k,i} \geq WC_{k-1,i} + r_{k,i} + \sum_{j=1}^{n} p_{k,j} X_{i,j} + \sum_{v \in S} T_{k,v} Y_v, k \in \{2, \cdots, m\}; i \in \{1, \cdots, n\} \tag{3.15}$$

$$WC_{k,i} \geq C_{k,i-1} + r_{k,i} + \sum_{j=1}^{n} p_{k,j} X_{i,j} + \sum_{v \in S} T_{k,v} Y_v, k \in \{1, \cdots, m\}; i \in \{2, \cdots, n\} \tag{3.16}$$

$$WC_{k,i} \geq C_{k,i} \geq 0, k \in \{1, 2, \cdots, m\}; i \in \{1, 2, \cdots, n\} \tag{3.17}$$

$$T_{k,\overline{v}} - T_{k,v} \geq x_{k,v}, k \in \{1, 2, \cdots, m\}; j \in \{1, 2, \cdots, n\}; v \in \mathcal{S} \tag{3.18}$$

$$v_{\min} \leq v_l \leq v_{\max}, \sum_{j=1}^{n-1} Z_{k,j} \geq T, l \in \{1, 2, \cdots, L\}; k \in \{1, 2, \cdots, m\} \tag{3.19}$$

$$X_{i,j} = \{0, 1\}, Y_v = \{0, 1\}, i, j \in \{1, 2, \cdots, n\}; v \in \mathcal{S} \tag{3.20}$$

The assignment of jobs to positions in the sequence on machines is given in **constraints (3.7)** and **(3.8)**. According to **constraint (3.9)**, the transmission belt can only run at one speed at once. **Constraint (3.10)** means that the makespan is equal to or larger than the completion time of a schedule that contains both the transportation time of the final job to the warehouse and the completion time of all tasks. The first job's setup time, job processing time, and transportation time from the warehouse to the first machine are all factored into the first task's completion time thanks to **constraint (3.11)**. **Constraints (3.12)** and **(3.13)** ensure that operation of the i-th job cannot be started on machine k before the preceding operations, which include the transportation of the job from machine k-1 to machine k and the setup. **Constraint (3.14)** makes sure that the completion time of the first job consists of the setup time, the job processing time, and the transportation time from one machine to another. **Constraints (3.15)** and **(3.16)** satisfy that the i-th job cannot be put to process on machine k until its previous operations on machine k and the transportation of the job from machine k-1 to machine k is finished. The completion time of each operation carried from one machine to another must be more than the completion time of the corresponding operation, which must be greater than zero, according to **constraint (3.17)**. The amount of compression time for a particular transportation speed is constrained by **constraint (3.18)**. **Constraint (3.19)** decides that the transportation speed should be within its allowed range and determines the allowed number of shutting down times for each machine. Decision variables X and Y are guaranteed to be binary by **constraint (3.20)**.

3.3 Proposed Algorithm

For the multi-objective PFSPCT, we design a hybrid multi-objective backtracking search with the genetic algorithm (HMOBSA). The fact that GA is a successful strategy for tackling NP-hard problems is the main justification for using this hybrid metaheuristic. In addition, dual populations in BSA [8, 9] can be utilized to store the information for different populations and ensure the diversity of population. According to these characteristics, we specially proposed a hybrid multi-objective algorithm based on BSA and GA for this kind of scheduling problem. The flowchart of the proposed HMOBSA for the energy-efficient PFSPCT is given in Fig. 3.3.

3.3.1 Encoding Representation

The job sequence and transportation speed are both indicated in the PFSPCT's encoding representation. A feasible solution representation is composed of a permutation vector and a transportation speed as follows:

$$\text{individual} = [\pi(1), \pi(2), \cdots, \pi(n), v] \tag{3.21}$$

where $\pi(n)$ denotes the n-th job to be processed on the machine, and v represents the chosen speed of the transmission belt.

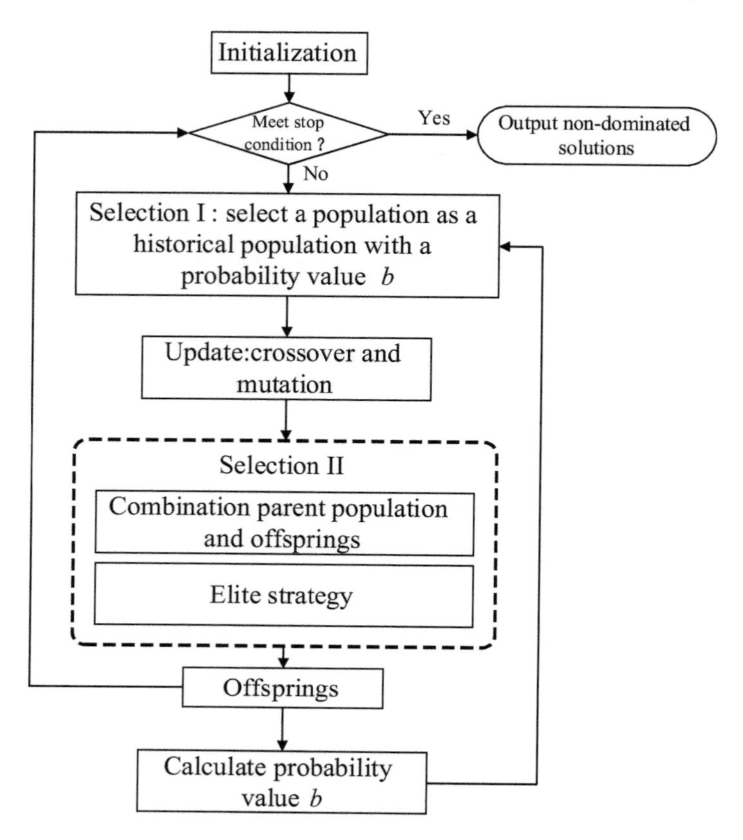

Fig. 3.3 Flowchart of the proposed HMOBSA

3.3.2 *Initialization*

According to the previously mentioned encoding mechanism, the initial population can be generated in a random way and is expressed below:

$$P = \begin{bmatrix} \pi(1,1), \pi(1,2), \cdots, \pi(1,n), v(1) \\ \vdots \quad \vdots \quad \vdots \quad \vdots \quad \vdots \\ \pi(i,1), \pi(i,2), \cdots, \pi(i,n), v(i) \\ \vdots \quad \vdots \quad \vdots \quad \vdots \quad \vdots \\ \pi(N,1), \pi(N,2), \cdots, \pi(N,n), v(N) \end{bmatrix} \tag{3.22}$$

where $\pi(i, n)$ represents the n-th job of the i-th individual and N indicts the size of population.

3.3.3 Selection-I

The historical population old P is determined by HMOBSA's selection-I stage in order to calculate the search direction. The definition of the initial historical population (old P) is below.

$$\text{old}\,P = \begin{bmatrix} \pi(1,1), \pi(1,2), \cdots, \pi(1,n), v(1) \\ \vdots \quad \vdots \quad \vdots \quad \vdots \quad \vdots \\ \pi(i,1), \pi(i,2), \cdots, \pi(i,n), v(i) \\ \vdots \quad \vdots \quad \vdots \quad \vdots \quad \vdots \\ \pi(N,1), \pi(N,2), \cdots, \pi(N,n), v(N) \end{bmatrix} \tag{3.23}$$

HMOBSA performs equation at the beginning of each iteration to generate old P, and the equation is given below:

$$\text{if}\, a \langle b \,\text{then old}\, P := P | a, b \in [0, 1] \tag{3.24}$$

where: = is the update operation. This formula ensures that HMOBSA has a population. This population is either the parent population or the historical population (old P). When old P is chosen, a permutation function changes the individuals' order as follows.

$$\text{old}\, P := \text{permutation}(\text{old}\, P) \tag{3.25}$$

Notably, the permutation function is executed on the job sequence segment of the population rather than the speed segment of the population.

From the Eq. (3.24), we can clearly know that the dual population based on a random selection solution may help the algorithm maintain the population's diversity. However, this scheme can't guarantee a good convergence toward the optimal answers because of its strong emphasis on variety which would lead to a pure random search. As a result, we come up with a self-adaptive selection scheme as follows:

Step (1) After the population is updated, calculate the selection probability of P as old P, namely $b = n/|A|$, where n is the number of non-dominated solutions from the current historical population, and $|A|$ represents the total number of the non-dominated solutions from the dual population at its current iteration.

Step (2) Use the roulette-wheel approach to select a population.

To prevent selecting all the solutions from the same population, the population P has a minimum selection probability $b_{\min}$. That is, if $b < b_{\min}$, then set $b = b_{\min}$, where $b_{\min} = 0.2$.

3.3.4 Crossover and Mutation

Due to its efficiency for combinatorial optimization issues, the update operation mainly uses the crossover and mutation operators of GA. In this research, to ensure the feasibility of the solutions, the partially matched crossover (PMX) [10] is applied, which could guarantee the practicality of the solutions. The PMX can produce two good solutions from two parents, one from the population and the other from the historical population. The PMX is exclusively applied to the job sequence segment, and it doesn't have any impact on the speed of the transmission belt.

In this study, the mutation operator includes two kinds of mutation approaches that are chosen with a probability of 0.5. The first mutation approach can change the pace of transportation within its range, and the second technique can exchange the order of jobs. It should be noted that the first mutation strategy only applies to the transportation speed segment, whereas the second mutation technique is only allowed in the job sequence segment.

3.3.5 Selection-II

This phase of the HMOBSA is distinct from that of the basic BSA when solving MOPs, the newly created solutions are assessed in terms of both fitness values (i.e., makespan and energy consumption). Especially, all solutions are sorted according to the non-dominated sorting technique [11]. As a result, each solution has a rank that is equal to its non-dominance level. Then, within each front, a crowding distance approach is employed to establish an order among individuals. To achieve wide-spread Pareto fronts, individuals with the large crowding distances are better than individuals with the smaller crowding distances for solutions that belong to the same non-dominated level. The newly generated population Q_t is connected with the existing parent population P_t to form $Q_t \cup P_t$ solution vectors. Then, the non-dominated sorting strategy and crowding distance technique are performed on the combined population $Q_t \cup P_t$. Eventually, choose the best population with size N (N is population size) from the combined population for the following update.

3.4 Case Study and Discussion

A case from a real-world workshop for manufacturing the connecting rods of a motor engine at a Chinese Motor Company is used to test the proposed mathematical model and HMOBSA. The HMOBSA is contrasted with the NSGA-II [11] and MOEA/D in order to show the effectiveness of the HMOBSA [12]. All algorithms are coded in Java. Experimental tests are performed on a computer with Intel Core i5, 2.39 GHz, 4 GB RAM, and a Windows 8 operating system.

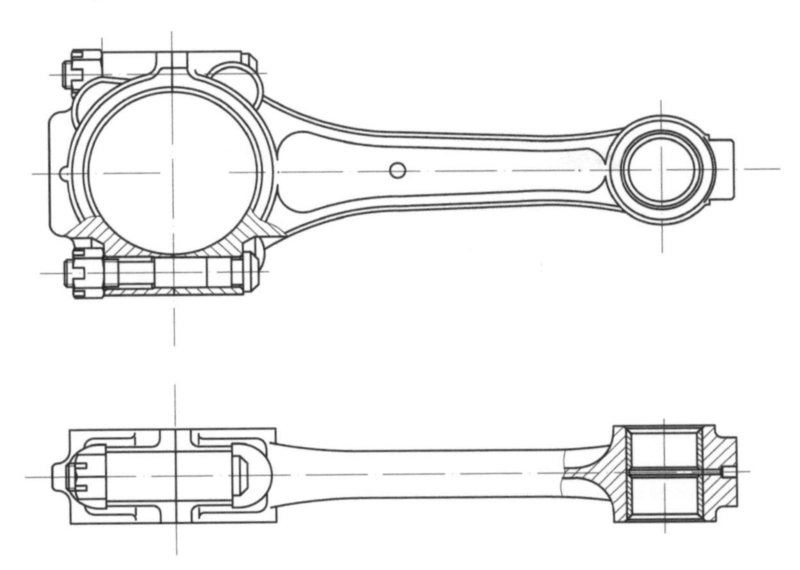

Fig. 3.4 A connecting rod

3.4.1 Case Introduction

The following case exists in a real-world workshop of a Chinese Motor Company that processes connecting rods as presented in Fig. 3.4. The connecting rod demands 9 operations on 9 machines, as shown in Fig. 3.5. There are 12 different types of connecting rods in various sizes. Four different speeds can be set for the transmission belt. The minimal transportation speed is treated as the normal speed. The relevant information that contains job number, processing time, power, and setup time are provided in Tables 3.1, 3.2, 3.3, 3.4, 3.5 and 3.6. To simplify the calculation, the energy consumption requirement for turning off/on the machine is $E_{\text{off_on}} = \{2, 3, 4, 5, 6, 7\}$Wh, $t_{on} + t_{off} = 2$ seconds, and T is set to 3.

3.4.2 Parameter Settings

In order to ensure a fair comparison and assess the proposed HMOBSA's effectiveness, the parameter settings are adjusted to be as effective as possible for each MOEA. Due to space restrictions, we undertake a preliminary experiment on one example with several parameter settings. The parameter settings of HMOBSA, NSGA-II, and MOEA/D are summarized in Table 3.7.

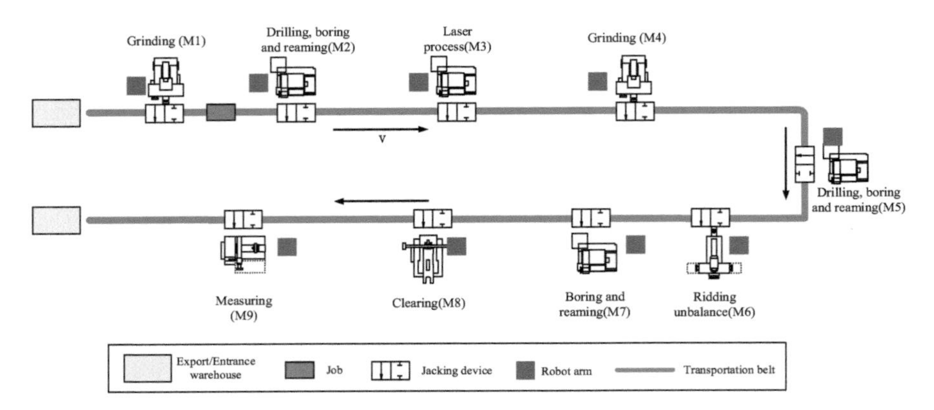

Fig. 3.5 PFSPCT layout and process flow

Table 3.1 Idle power of each machine and setup power of the corresponding robot arm

Power (kW)	M1/ robot 1	M2/ robot 2	M3/ robot 3	M4/ robot 4	M5/ robot 5	M6/ robot 6	M7/ robot 7	M8/ robot 8	M9/ robot 9
P_i^{idle}	2.6	3.5	3.0	2.0	2.6	3.8	2.3	2.5	2.2
P_i^s	3.5	2.5	2.5	3.0	2.0	2.5	2.0	3.0	2.0

Table 3.2 Mass of jobs

Mass (kg)	Job1	Job2	Job3	Job4	Job5	Job6	Job7	Job8	Job9	Job10	Job11	Job12
M_i	7.5	8.5	10.5	9.0	8.0	5.0	12.0	6.0	15.0	9.0	11.0	6.5

Table 3.3 The setup time of the first job to processed on each machine

Setup time (second)	Job1	Job2	Job3	Job4	Job5	Job6	Job7	Job8	Job9	Job10	Job11	Job12
M1	4	8	5	3	8	9	4	6	5	4	3	8
M2	5	3	6	6	8	5	9	8	8	4	3	2
M3	4	9	5	9	8	6	5	6	8	3	9	5
M4	2	8	8	6	3	9	4	9	7	9	2	4
M5	9	7	5	8	2	8	3	2	5	2	9	7
M6	5	8	9	8	7	8	2	6	5	4	7	3
M7	6	9	2	9	8	9	5	9	6	4	3	7
M8	9	7	3	4	9	4	3	6	5	8	2	8
M9	3	5	8	2	3	9	8	9	4	6	7	9

Table 3.4 Setup time between jobs

Setup time (second)	Job 1	Job 2	Job 3	Job 4	Job 5	Job 6	Job 7	Job 8	Job 9	Job 10	Job 11	Job 12
Job 1	–	10	5	13	10	15	14	12	18	17	10	6
Job 2	20	–	12	15	19	15	14	10	7	15	19	13
Job 3	10	11	–	10	13	14	15	15	16	12	8	12
Job 4	13	18	16	–	15	9	7	16	14	19	18	15
Job 5	9	13	14	13	–	10	12	17	14	12	16	8
Job 6	13	14	13	7	9	–	14	15	10	8	12	11
Job 7	18	14	12	11	12	8	–	5	9	8	13	10
Job 8	16	17	9	12	15	13	7	–	13	5	10	14
Job 9	11	5	5	13	10	12	16	15	–	8	6	9
Job 10	4	11	9	5	14	7	6	9	14	–	12	8
Job 11	8	5	7	13	16	8	15	14	13	10	–	6
Job 12	6	8	6	e	7	6	13	8	4	5	7	–

Table 3.5 Processing time of jobs on machine

Job processing time (second)	Process 1	Process 2	Process 3	Process 4	Process 5	Process 6	Process 7	Process 8	Process 9
Job1	35	40	43	35	20	108	26	41	38
Job2	39	45	23	32	32	110	35	30	35
Job3	42	39	48	40	35	100	27	24	38
Job4	20	30	33	35	26	96	39	36	30
Job5	30	28	40	22	30	80	35	38	37
Job6	35	35	36	40	37	90	30	20	38
Job7	29	24	29	50	36	88	22	46	47
Job8	41	30	34	23	23	100	29	36	29
Job9	23	24	39	30	33	1e	18	34	37
Job10	20	38	24	35	19	95	30	39	41
Job11	18	30	35	31	37	99	28	35	34
Job12	25	28	29	26	29	102	29	33	35

3.4.3 Comparison of HMOBSA and the Other Two Algorithms

For this study, the HMOBSA is compared with NSGA-II and MOEA/D in 50 independent runs. Standard metrics such as Spread [11], GD, and IGD [13, 14] should

Table 3.6 Transportation time of jobs on machine

Speed (m/s)	Transportation time (second)									
	Warehouse to M1	M1 to M2	M2 to M3	M3 to M4	M4 to M5	M5 to M6	M6 to M7	M7 to M8	M8 to M9	M9 to warehouse
0.5	16	18	20	19	21	22	18	16	19	16
1	8	9	10	9.5	10.5	11	9	8	9.5	8
1.5	5.33	6	6.67	6.33	7	7.33	6	5.33	6.33	5.33
2	4	4.5	5	4.75	5.25	5.5	4.5	4	4.75	4

Table 3.7 Parameter settings of HMOBSA, NSGA-II, and MOEA/D

HMOBSA	NSGA-II	MOEA/D
Population size: 50	Population size: 50	Population size: 50
Historical population size: 50	NFE: 20,000	NFE: 20,000
Number of function evaluation (NFE): 20,000	Crossover (PMX) rate: 0.9	Archive size: 100
Crossover (PMX) rate: 0.9	Swap mutation rate: 0.1	Crossover (PMX) rate: 0.9
Swap mutation rate: 0.1		Swap mutation rate: 0.1

be used to evaluate the effectiveness of the proposed algorithm. Because the true Pareto front (PF*) of the considered problem are unclear, the non-dominated solutions for each MOEA in all independent runs are viewed as the PF* for that instance. Table 3.8 shows the statistical metric results gained by three MOEAs under various $E_{\text{off-on}}$ conditions. The results of the best metric are underlined in bold text. Because these MOEAs are stochastic, the statistical test ought to offer significant difference comparisons. The obvious difference between the outcomes produced by various algorithms is measured by a Wilcoxon signed-rank test. All tests have a 95% confidence level, which is equal to a $\alpha = 0.05$). If the proposed HMOBSA performs noticeably better or worse than its competitors, the signs " + " or "−" indicate that fact. There are no obvious differences between the proposed algorithm and its compared algorithm, as shown by the sign " = ".

For this case, Table 3.8 shows that the HMOBSA outperforms NSAG-II and MOEA/D for three metrics, particularly for GD and IGD. The following are the primary reasons for the HMOBSA's superior performance: Firstly, the dual population technique can improve population diversity because of its different search directions. Secondly, the purpose of good convergence performance can be achieved by self-adaptive selection mechanism which can assist in selecting a suitable population as the parent population and thus raise the search efficiency. According to the results, we can know that HMOBSA is mildly superior to the compared algorithms for the Spread metric, but there is no significant difference between them. Because the HMOBSA clearly outperforms its competitors in terms of the GD metric, the self-adaptive selection strategy enhances convergence performance (i.e., GD metric).

Table 3.8 Mean and standard deviation value of three metrics for NSGA-II, MOEA/D, and HMOBSA

$E_{\text{off-on}}$(Wh)	MOEAs	GD	Spread	IGD
		Mean (std)	Mean (std)	Mean (std)
2	NSGA-II	6.9e − 03(4.1e − 03)	1.6e + 00(1.1e − 01)	2.4e–03(1.3e − 03)
	MOEA/D	6.6e − 03(8.7e − 03)	1.6e + 00(1.6e − 01)	2.4e–03(2.7e − 03)
	HMOBSA	**6.0e − 03(3.7e − 03)+**	**1.6e + 00(1.4e − 01)=**	**2.0e − 03(1.1e − 03)+**
3	NSGA-II	6.7e − 03(6.3e − 03)	1.6e + 00(1.1e − 01)	2.6e–03(9.7e − 04)
	MOEA/D	7.7e − 03(3.7e − 03)	1.6e + 00(1.7e − 01)	2.7e–03(4.2e − 03)
	HMOBSA	**6.2e − 03(5.1e − 03)+**	**1.6e + 00(1.7e − 01)=**	**2.2e − 03(1.1e − 03)+**
4	NSGA-II	5.4e − 03(2.9e − 03)	1.6e + 00(2.5e − 01)	2.1e–03(7.3e − 04)
	MOEA/D	5.9e − 03(6.8e − 03)	1.6e + 00(1.7e − 01)	2.6e − 03(1.4e − 03)
	HMOBSA	**5.0e − 03(2.9e − 03)+**	**1.6e + 00(2.5e − 01)=**	**2.0e − 03(9.4e − 04)=**
5	NSGA-II	8.8e − 03(3.7e − 03)	**1.6e + 00(1.2e − 01)**	2.7e–03(1.5e − 03)
	MOEA/D	7.5e − 03(1.1e − 03)	1.6e + 00(1.5e − 01)	2.6e − 03(3.1e − 03)
	HMOBSA	**6.8e − 03(3.2e − 03)+**	1.6e + 00(1.2e − 01)=	**2.1e − 03(1.6e − 03)+**
6	NSGA-II	7.0e − 03(3.8e − 03)	**1.6e + 00(1.2e − 01)**	2.9e–03(1.3e − 03)
	MOEA/D	7.1e − 03(1.1e − 03)	1.6e + 00(1.5e − 01)	2.8e − 03(3.1e − 03)
	HMOBSA	**6.6e − 03(3.4e − 03)+**	1.6e + 00(1.3e − 01)=	**2.4e − 03(9.8e − 04)+**
7	NSGA-II	8.7e-03(5.0e-03)	1.6e + 00(1.4e − 01)	2.5e–03(1.0e − 03)
	MOEA/D	9.5e-03(9.9e-03)	1.6e + 00(1.7e − 01)	2.6e − 03(4.1e − 03)
	HMOBSA	**7.2e − 03(3.7e − 03)+**	**1.6e + 00(1.3e − 01)=**	**2.3e − 03(1.0e − 03)+**

Additionally, in terms of overall performance (i.e., IGD), the HMOBSA outperforms the other algorithms considerably. Generally speaking, the self-adaptive dual population strategy has a favorable impact on the behavior of the proposed algorithm.

To compare the performance of these algorithms clearly and intuitively, we choose the Pareto fronts acquired from one run of each algorithm to give a graphical representation for the situation ($E_{\text{off-on}} = 2\text{Wh}$) as shown in Fig. 3.6. In terms of solution quality and coverage, the proposed HMOBSA is able to outperform its competitors. However, the other MOEAs tend to converge to a local optimum because of their weak convergence to the Pareto optimal solutions. In addition, it is also noteworthy that the decrease in energy usage loses performance in terms of makespan. For instance, the maximum energy consumption is identified at the point A where the makespan is at its minimum in Fig. 3.6. The maximal makespan at point B is where the least amount of energy is used. Table 3.9 provides a summary of the specifications of two sample points A and B in Pareto fronts for $E_{\text{off-on}} = 2\text{Wh}$. Notably, the maximum speed of transportation is accompanied by the maximum energy consumption, and vice versa.

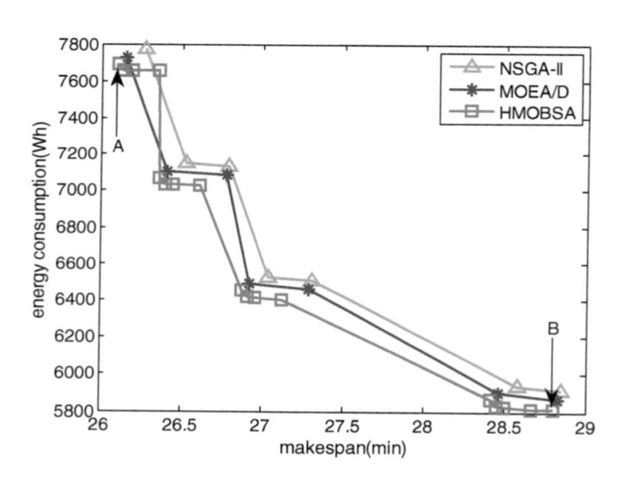

Fig. 3.6 Pareto front obtained by NSGA-II, MOEA/D, and HMOBSA algorithms when $E_{\text{off-on}} = 2\text{Wh}$

Table 3.9 Corresponding results by HMOBSA on the scheduling problem in Fig. 3.6

No.	Solution	Decoding	f_1	f_2
A	[11,10,1,8,4,5,3,6,7,9,0,2,2]	$\pi = [1\text{–}12]$	26.1 min	7690Wh
		$v = 2$ m/s		
B	[10,11,8,1,4,5,3,6,7,9,0,2,0.5]	$\pi = [1\text{–}12]$	28.8 min	5813Wh
		$v = 0.5$ m/s		

3.4.4 Analysis of Energy-Saving Scenario

In this section, we analyzed energy-saving scenarios. Scenario 1 presents our proposed energy-saving strategy, while Scenario 2 depicts the energy-saving approach discussed in the literature [7]. The two previously mentioned energy-saving strategies increased the machine's lifespan while simultaneously reducing energy consumption. However, scenarios 1 and 2 have different contributions in terms of the maximum energy saving. Two energy-saving scenarios have the same job sequence and transportation speed which can ensure a fair comparison. Figure 3.7 shows the results produced by the HMOBSA for two energy-saving scenarios under $E_{\text{off-on}} = \{2, 3, 4, 5\}$Wh. The graph clearly shows that scenario 1 outperforms scenario 2 in terms of energy consumption under the same situation. Table 3.10 displays the corresponding energy consumption and makespan. This outcome demonstrates the superiority of the proposed energy-saving scenario.

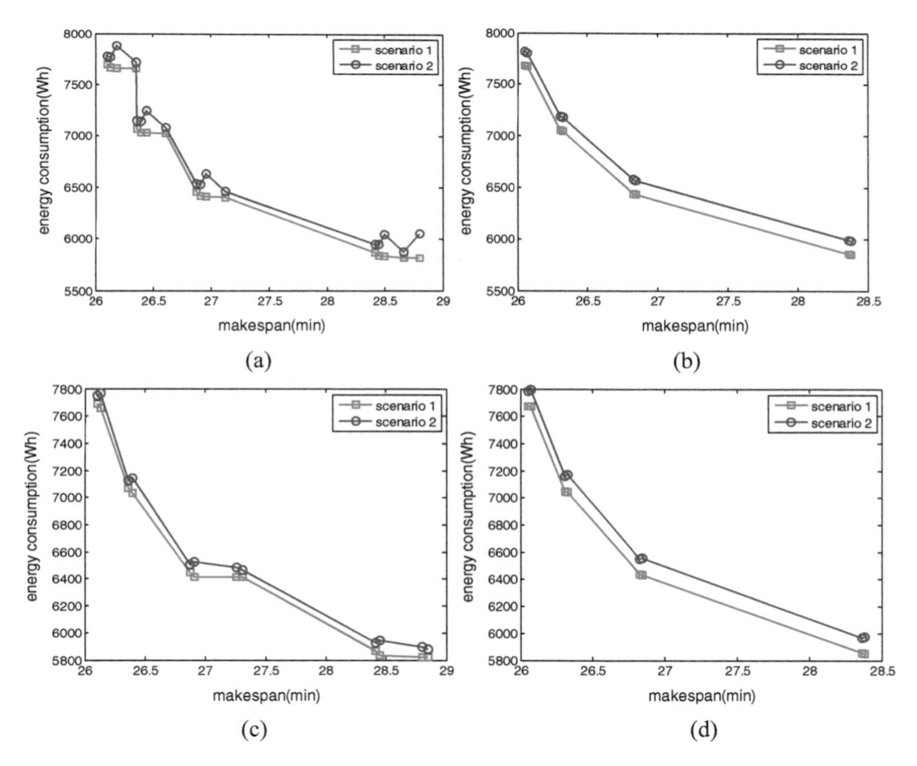

Fig. 3.7 Pareto front obtained by HMOBSA with scenario 1 and scenario 2, **a** $E_{\text{off-on}} = 2$ Wh, **b** $E_{\text{off-on}} = 3$ Wh, **c** $E_{\text{off-on}} = 4$ Wh, **d** $E_{\text{off-on}} = 5$ Wh

Table 3.10 Pareto front including the same makespan while different energy consumption by HMOBSA with different scenarios and E_{off-on}

$E_{\text{off-on}} = 2$ Wh				$E_{\text{off-on}} = 3$ Wh				$E_{\text{off-on}} = 4$ Wh				$E_{\text{off-on}} = 5$ Wh			
Scenario 1		Scenario 2		Scenario 1		Scenario 2		Scenario 1		Scenario 2		Scenario 1		Scenario 2	
f_1	f_2	f_1	f_2	f_1	f_2	f_1	f_2	f_1	f_2	f_1	f_2	f_1	f_2	f_1	f_2
26.10	**7690**	26.e	7773	**26.05**	**7675**	26.05	7815	**26.10**	**7690**	26.e	7752	**26.05**	**7675**	26.05	7787
26.35	**7655**	26.35	7716	**26.07**	**7670**	26.07	7801	**26.14**	**7659**	26.14	7769	**26.07**	**7670**	26.07	7797
26.39	**7033**	26.39	7144	**26.31**	**7051**	26.31	7191	**26.39**	**7033**	27.39	7144	**26.31**	**7051**	26.31	7163
26.61	**7024**	26.61	7085	**26.33**	**7046**	26.33	7178	**26.91**	**6417**	26.91	6527	**26.33**	**7046**	26.33	7173
26.88	**6450**	26.88	6532	**26.82**	**6437**	26.82	6576	**27.26**	**6414**	27.26	6485	**26.82**	**6437**	26.82	6549
27.13	**6402**	27.13	6463	**26.84**	**6431**	26.84	6563	**28.45**	**5835**	28.45	5945	**26.84**	**6431**	26.84	6558
28.42	**5869**	28.42	5951	**28.37**	**5857**	28.37	5997	**28.80**	**5823**	28.80	5896	**28.37**	**5857**	28.37	5969
28.80	**5813**	28.80	6051	**28.38**	**5851**	28.38	5982	**28.85**	**5820**	28.85	5878	**28.38**	**5851**	28.38	5978

3.5 Chapter Conclusion

In this chapter, we explore a real-world flow shop scheduling problem with the controllable transportation time and SDST from the perspective of energy efficiency. This difficulty can be considered as a multi-objective optimization issue that takes both makespan and energy consumption into consideration. To deal with this problem, we formulate the mathematical model and propose a new hybrid multi-objective backtracking search optimization algorithm with an energy-saving scenario. The proposed HMOBSA is superior to the other two well-known multi-objective metaheuristics, according to experimental results, and it is quite successful in solving this real-world case. In addition, the findings have the following profound effects on manufacturing in the real world: (1) The makespan and energy consumption have a contradictory relationship. The preference solutions for the makespan are picked when a decision maker favors that criterion, and vice versa. (2) With regard to energy usage, the proposed energy-saving scenario can significantly reduce energy consumption during the same period.

References

1. Allahverdi, A., Ng, C.T., Cheng, T.C.E., et al.: A survey of scheduling problems with setup times or costs[J]. Eur. J. Oper. Res. **187**(3), 985–1032 (2008)
2. Ahmadizar, F., Shahmaleki, P.: Group-shop scheduling with sequence-dependent set-up and transportation times[J]. Appl. Math. Model. **38**(21–22), 5080–5091 (2014)
3. Vallada, E., Ruiz, R.: A genetic algorithm for the unrelated parallel machine scheduling problem with sequence dependent setup times[J]. Eur. J. Oper. Res. **211**(3), 612–622 (2011)
4. Chao, L., Xiao, S., Li, X., et al.: An effective multi-objective discrete grey wolf optimizer for a real-world scheduling problem in welding production[J]. Adv. Eng. Softw. **99**, 161–176 (2016)
5. Lu, S.M.: A low-carbon transport infrastructure in Taiwan based on the implementation of energy-saving measures[J]. Renew. Sustain. Energy Rev. **58**, 499–509 (2016)
6. Yildirim, M.B., Mouzon, G.: Single-machine sustainable production planning to minimize total energy consumption and total completion time using a multiple objective genetic algorithm[J]. IEEE Trans. Eng. Manage. **59**(4), 585–597 (2011)
7. Dai, M., Tang, D., Giret, A., et al.: Energy-efficient scheduling for a flexible flow shop using an improved genetic-simulated annealing algorithm[J]. Robot. Comput. Integr. Manuf. **29**(5), 418–429 (2013)
8. Civicioglu, P.: Backtracking search optimization algorithm for numerical optimization problems[J]. Appl. Math. Comput. **219**(15), 8121–8144 (2013)
9. Lu, C., Gao, L., Li, X., et al.: Energy-efficient multi-pass turning operation using multi-objective backtracking search algorithm[J]. J. Clean. Prod. **137**, 1516–1531 (2016)
10. Goldberg D.E., Alleles, R.L.: Loci and the traveling salesman problem[J]. Inventiones Mathematicae (1985)
11. Deb, K., Pratap, A., Agarwal, S., et al.: A fast and elitist multiobjective genetic algorithm: NSGA-II[J]. IEEE Trans. Evol. Comput. **6**(2), 182–197 (2002)
12. Hui, L., Zhang, Q.: Multiobjective optimization problems with complicated Pareto sets, MOEA/D and NSGA-II[J]. IEEE Trans. Evol. Comput. **13**(2), 284–302 (2009)
13. Zitzler, E., Thiele, L.: Multiobjective evolutionary algorithms: a comparative case study and the strength Pareto approach[J]. IEEE Trans. Evol. Comput. **3**(4), 257–271 (1999)

14. Zhou, W., Chow, T., Chow, T.S.: A local multiobjective optimization algorithm using neighborhood field[J]. Struct. Multidisc. Optim. **46**(6), 853–870 (2012)

Chapter 4
Green Scheduling in Hybrid Flow Shop Environment

4.1 Brief Introduction

The hybrid flow shop scheduling problem (HFSP) has received a lot of attention over the past few decades. Production efficiency is the criterion that is most frequently employed. With people's increased awareness of environmental protection, green criteria, such as energy consumption and carbon emission, have gained more and more attention. However, people pay little attention to noise pollution which can cause health and emotional problems. So, we investigate a multi-objective HFSP that takes into account noise pollution as well as production efficiency and energy consumption. First and foremost, we develop a novel mixed-integer programming model for this multi-objective HFSP. One energy conservation/noise reduction strategy is embedded in this model to achieve green scheduling. To solve this issue, a new multi-objective cellular grey wolf optimizer (MOCGWO) is then put forth. The proposed MOCGWO combines the benefits of cellular automata (CA) for diversification and variable neighborhood search (VNS) for intensification, balancing exploration, and exploitation. Finally, to confirm the efficiency and effectiveness of the proposed MOCGWO, we conduct comparison experiments with other well-known multi-objective evolutionary algorithms. The experimental results clearly show that the proposed MOCGWO is significantly superior to its competitors on this problem.

C. Lu et al., *Intelligence Optimization for Green Scheduling in Manufacturing Systems*, Engineering Applications of Computational Methods 18,
https://doi.org/10.1007/978-981-99-6987-6_4

4.2 Problem Statement and Modeling

4.2.1 Problem Statement

Before describing the multi-objective HFSP, it is necessary to provide a simple explanation of the MOP. A MOP can be expressed as follows without losing its generality:

$$\min f(\mathbf{x}) = \min[f_1(\mathbf{x}), f_2(\mathbf{x}), \ldots, f_m(\mathbf{x})] \tag{4.1}$$

where $\mathbf{x}$ is a solution vector, R^n represents the feasible solution space. We denote two feasible solutions as $\boldsymbol{a}$ and $\boldsymbol{b}$. One solution $\boldsymbol{a}$ dominates another solution $\boldsymbol{b}$ if and only if $f_i(\boldsymbol{a}) \leq f_i(\boldsymbol{b})$ for each index $i \in \{1, \ldots, m\}$ and $f_l(\boldsymbol{a}) < f_l(\boldsymbol{b})$ for at least one index $l \in \{1, \ldots, m\}$. If a solution $\mathbf{x}^* \in R^n$ is not dominated by any other solution, then $\mathbf{x}^*$ is a Pareto optimal solution. We call a set of Pareto optimal solutions Pareto optimal set (PS*). The optimal Pareto front (PF*) is the corresponding objective point when the Pareto optimal set is mapped in objective space. The primary objective of multi-objective optimization is to find PF*.

The HFSP covered by this study can be defined as follows. There are n jobs ($J = \{1, 2, \ldots, n\}$), each job should be processed through s stages in the same route. Each stage i ($i = 1, \ldots, s$) contains a collection of uniform parallel machines $\left(M_{i,1}, M_{i,2}, \ldots, M_{i,k}, \ldots, M_{i,m_i}\right)$, where $M_{i,k}$. is the k-th machine at stage i. Each job $j \in J$. has a set of operations $\left(O_{1,j}, O_{2,j}, \ldots, O_{i,j}, \ldots, O_{s,j}\right)$, where $O_{i,j}$. represents the operation of the job j at stage i. One operation $O_{i,j}$ can be processed by one of m_i ($i = 1, \ldots, s$) uniform parallel machines, where the machines have the same function and different processing speed. Each operation $O_{i,j}$ has a processing requirement $p_{i,j}$. If operation $O_{i,j}$ is assigned on machine $M_{i,k}$ with a speed $v_{i,k}$, then it needs $p_{i,j,k} = p_{i,j}/v_{i,k}$ units of time to finish. Our goal is to devise a schedule that minimizes the makespan, noise pollution, and total energy consumption at the same time. Figure 4.1 depicts a typical hybrid flow shop layout.

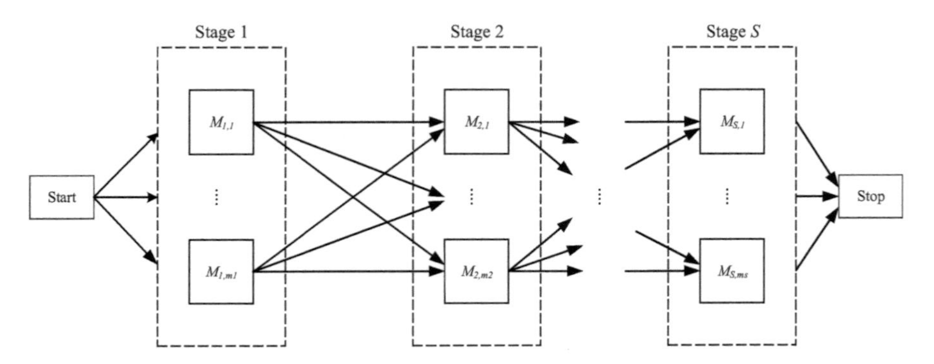

Fig. 4.1 A hybrid flow shop layout

The HFSP's relevant assumption is given as below:

(1) One job can only be processed by exactly one machine at each stage.
(2) One machine can process at most one operation at a time.
(3) There is unlimited storage or buffer capacity between two consecutive stages.
(4) All jobs are independent and available for processing at time 0.
(5) The transportation times and setup times between two successive stages can be ignored.
(6) Preemption and machine breakdown are prohibited.

4.2.2 Mathematical Modeling

The relevant notations and decision variables are provided below to create a mixed-integer programming model for the multi-objective HFSP.

(1) **Notations**

n: the number of all the jobs.
s: the number of all the stages.
j, j': index of job, $j, j' = 1, 2, \ldots, n$.
i, i': index of stage, $i, i' = 1, 2, \ldots, s$.
m_i: the total number of parallel machines at stage i.
k: index of the machine at each stage, $k = 1, 2, \ldots, m_i$.
$M_{i,k}$: the k-th machine at stage i.
C_j: the completion time of job j.
$C_{i,j}$: the completion time of the operation O_{ij}.
$C_{\max}$: the makespan of the schedule, i.e., the completion time of the last operation.
$O_{i,j}$: the operation of job j at stage i.
$v_{i,k}$: the speed of the machine k at stage i, $k = 1, 2, \ldots, m_i$.
$p_{i,j}$: the normal processing time of the operation O_{ij}.
$p_{i,j,k}$: the actual processing time of the operation O_{ij} on machine k, namely $p_{i,j,k} = p_{i,j}/v_{i,k}$.
$Id_{i,j,k}$: the idle time of the operation O_{ij} before it is assigned on machine k.
$P_{i,k}^{\text{idle}}$: the idle power of a machine k at stage i (kW).
$P_{i,k}^{\text{work}}$: the work power of a machine k at stage i (kW).
$N_{i,j}^{w}$: the equivalent continuous sound pressure level of $O_{i,j}$ (dB(A)) during working mode.
$N_{i,j}^{\text{idle}}$: the equivalent continuous sound pressure level of $O_{i,j}$ (dB(A)) during idle mode.
T: the maximal allowable number to shut down a machine at each stage.
T_b: the breakeven duration.
L: a very large positive number.

(2) Decision variables

$S_{i,j}$: the starting time of the operation O_{ij}.
π: a job permutation.

$$x_{i,j,k} = \begin{cases} 1, \text{ if } O_{i,j} \text{ is processed by machine } k \\ 0, \text{ otherwise} \end{cases}$$

$$y_{i,j,j'} = \begin{cases} 1, \text{ if } O_{i,j} \text{ precedes } O_{i,j'} \text{ on the same stage} \\ 0, \text{ if } O_{i,j'} \text{ precedes } O_{i,j} \text{ on the same stage} \end{cases}$$

$$z_{i,i',j} = \begin{cases} 1, \text{ if } O_{i,j} \text{ precedes } O_{i,j'} \text{ on the same stage} \\ 0, \text{ if } O_{i,j'} \text{ precedes } O_{i,j} \text{ on the same stage} \end{cases}$$

$$h_{i,j,k} = \begin{cases} 0, \text{ if a machine } k \text{ preparing to process } O_{i,j} \text{ is shut down} \\ 1, \text{ otherwise} \end{cases}$$

The primary goal of this issue is to reduce the makespan $C_{\max}$. It can be described by using the formula below:

$$\min f_1 = C_{\max} = \max\{C_j | j = 1, \ldots, n\} \tag{4.2}$$

Lessening noise pollution is the second goal. It might be a display of a logarithmic non-linear summation of the noise levels produced by at least one source of noise. Below is the noise pollution function [1]:

$$L_{\text{eq}} = 10 \cdot \lg \frac{1}{T_e} \int_0^{T_e} 10^{0.1 \cdot L_t} \mathrm{d}t \tag{4.3}$$

where L_{eq} refers to the equivalent continuous sound pressure level during the measured time interval [the unit is dB(A)], the exposure time of noise is recorded as T_e, L_t is the equivalent continuous sound pressure level in a measured time t.

In this case, noise pollution is caused by two types of noise sources: working stage and idle stage. To reduce noise and overall energy consumption without interfering with task sequencing, in the idle state machines can be shut off. The span of machines may be impacted by using the turn off/on method frequently [2]. Consequently, a fresh approach to energy conservation and noise reduction is put forward. It can be described as follows:

Step 1: For each stage i ($\forall i = 1, 2, \ldots, s$), perform the following procedures on the machine $M_{i,k}$ ($k = 1, 2, \ldots, m_i$) at the stage i.

Step 2: Compute the current idle time $Id_{i,j,k}$ for preparing to process $O_{i,j}$ on machine $M_{i,k}$ from left to right. Then put these $Id_{i,j,k}$ values into a set $Wt_{i,k}$. Finally, sort these values in a non-ascending order.

Step 3: Select the former T elements from the $Wt_{i,k}$ as a new set $WT_{i,k,T}$. This set $WT_{i,k,T}$ includes all idle times during which a machine is likely to be shut down.

Step 4: For a machine $M_{i,k}$ at stage i, shut down this machine at the current stage (namely, $h_{i,j,k} = 0$) if the $Id_{i,j,k}$ simultaneously satisfies the following condition:

$$h_{i,j,k} = \begin{cases} 0, \ if\, Id_{i,j,k} \geq T_B \wedge Id_{i,j,k} \in WT_{i,k,T} \\ 1, \quad \text{otherwise} \end{cases},$$

$$i = 1, \ldots, s;\ j = 1, \ldots, n;\ k = 1, \ldots, m_i \tag{4.4}$$

Thus, Eq. (4.3) can be converted into the following formula.

$$\begin{aligned} \min\ f_2 &= L_{\mathrm{eq}} \\ &= 10 \cdot \lg \frac{\sum_{i=1}^{s} \sum_{j=1}^{n} \sum_{k=1}^{m_i} \left(p_{i,j,k} \cdot 10^{0.1 \cdot N_{i,j}^{w}} + Id_{i,j,k} \cdot 10^{0.1 \cdot N_{i,j}\mathrm{idle} \cdot h_{i,j,k}}\right) \cdot x_{i,j,k}}{\sum_{i=1}^{s} \sum_{j=1}^{n} \sum_{k=1}^{m_i} \left(p_{i,j,k} + Id_{i,j,k}\right) \cdot x_{i,j,k}} \end{aligned} \tag{4.5}$$

Reduce the overall energy usage of all the machines is the third goal. It is made up of all machines' energy consumption E_w in the working mode, and energy consumption E_{id} in the idle mode. Furthermore, we can also apply the previously mentioned energy conservation/noise reduction strategy to reduce the energy consumption in the idle mode. As a result, the third goal can be stated as follows:

$$\min f_3 = E_w + E_{\mathrm{id}} = \sum_{i=1}^{s} \sum_{j=1}^{n} \sum_{k=1}^{m_i} \left(P_{i,k}^{w} \cdot p_{i,j,k} + h_{i,j,k} \cdot P_{i,k}^{\mathrm{idle}} \cdot Id_{i,j,k}\right) \cdot x_{i,j,k} \tag{4.6}$$

In conclusion, the following is the formulation of a mathematical model for the HFSP using mixed-integer programming that aims to reduce makespan, noise pollution, and energy consumption:

$$\begin{cases} \min\ f_1 = C_{\max} \\ \min\ f_2 = 10 \cdot \lg \frac{\sum_{i=1}^{s} \sum_{j=1}^{n} \sum_{k=1}^{m_i} \left(p_{i,j,k} \cdot 10^{0.1 \cdot N_{i,j}^{w}} + Id_{i,j,k} \cdot 10^{0.1 \cdot N_{i,j}\mathrm{idle} \cdot h_{i,j,k}}\right) \cdot x_{i,j,k}}{\sum_{i=1}^{s} \sum_{j=1}^{n} \sum_{k=1}^{m_i} \left(p_{i,j,k} + Id_{i,j,k}\right) \cdot x_{i,j,k}} \\ \min\ f_3 = \sum_{i=1}^{s} \sum_{j=1}^{n} \sum_{k=1}^{m_i} \left(P_{i,k}^{w} \cdot p_{i,j,k} + h_{i,j,k} \cdot P_{i,k}^{\mathrm{idle}} \cdot Id_{i,j,k}\right) \cdot x_{i,j,k} \end{cases} \tag{4.7}$$

Subject to

$$C_{\max} \geq C_{i,j}, \forall i = 1, 2, \ldots, s;\ j = 1, 2, \ldots, n \tag{4.8}$$

$$\sum_{k=1}^{m_i} x_{i,j,k} = 1, \forall i = 1, 2, \ldots, s;\ j = 1, 2, \ldots, n \tag{4.9}$$

$$\begin{aligned} S_{i,j'} \geq\ & S_{i,j} \\ & + \sum_{k=1}^{m_i} \frac{p_{i,j}}{v_{i,k}} x_{i,j,k} \\ & - L \cdot \left(1 - y_{i,j,j'}\right), \forall i = 1, 2, \ldots, s;\ j, j' = 1, 2, \ldots, n;\ j \neq j' \end{aligned} \tag{4.10}$$

$$S_{i,j} \geq S_{i,j'} + \sum_{k=1}^{m_i} \frac{p_{i,j'}}{v_{i,k}} x_{i,j',k} - L \cdot y_{i,j,j'}, \forall i = 1, 2, \ldots, s; \; j, j' = 1, 2, \ldots, n; \; j \neq j' \tag{4.11}$$

$$S_{i',j} \geq S_{i,j} + \sum_{k=1}^{m_i} \frac{p_{i,j}}{v_{i,k}} x_{i,j,k} - L \cdot \left(1 - z_{i,i',j}\right), \forall i, i' = 1, 2, \ldots, s; \; j = 1, 2, \ldots, n; \; i \neq i' \tag{4.12}$$

$$S_{i,j} \geq S_{i',j} + \sum_{k=1}^{m_i} \frac{p_{i',j}}{v_{i',k}} x_{i',j,k} - L \cdot z_{i,i',j}, \forall i, i' = 1, 2, \ldots, s; \; j = 1, 2, \ldots, n; \; i \neq i' \tag{4.13}$$

$$x_{i,j,k} \in \{0, 1\}, \forall i = 1, 2, \ldots, s; \; j = 1, 2, \ldots, n; \; k = 1, 2, \ldots, m_i \tag{4.14}$$

$$y_{ij'j} \in \{0, 1\}, \forall i = 1, 2, \ldots, s; \; j, j' = 1, 2, \ldots, n; \; j \neq j' \tag{4.15}$$

$$z_{i,i',j} \in \{0, 1\}, \forall i, i' = 1, 2, \ldots, s; \; j = 1, 2, \ldots, n; \; i \neq i' \tag{4.16}$$

$$h_{i,j,k} \in \{0, 1\}, \forall i = 1, 2, \ldots, s; \; j = 1, 2, \ldots, n; \; k = 1, 2, \ldots, m_i \tag{4.17}$$

According to Eq. (4.7), the goal of this problem is to minimize makespan, noise pollution, and overall energy usage. Constraint (4.8), in conjunction with the minimization makespan objective, requires to be equal to the completion time of the last operation. Each job can only be processed by one machine at a time thanks to constraint (4.9). The precedence relationship between various operations at the same stage is ensured by constraints (4.10) and (4.11). The priority relationship between various operations for the same job is guaranteed by constraints (4.12) and (4.13). Some decision variables are all binary values, according to constraints (4.14) to (4.17). Large-scale multi-objective HFSP cannot be solved by accurate algorithms within a reasonable computing cost due to its NP-hard property. So, to deal with this multi-objective HFSP, a metaheuristic is developed.

4.3 Proposed Algorithm

In order to address this multi-objective HFSP, a multi-objective cellular grey wolf optimizer (MOCGWO) is put forth in this article. The GWO and cellular automatic (CA) are the foundations of the proposed approach. GWO is distinguished by its rapid convergence and high exploitation capability. However, GWO, like other metaheuristics, suffers from a rapid loss in search diversity. This shortcoming results in the search's stagnation and early convergence [3–5]. The CA idea is integrated into this MOGWO to solve this weakness. The following are the primary reasons for creating this MOGWO: The MOGWO can obtain a cellular search structure from CA. Each solution, in a cellular topological structure, has its own neighborhood, and the solutions within a neighborhood form a separate subpopulation. Every subpopulation changes its solutions locally on an independent basis. As a result, the entire population has a greater chance of discovering previously unknown areas of the search space through parallel searching. Therefore, a multi-subpopulation mechanism can enrich search diversity. At the same time, the overlap region between cellular neighbors can improve information sharing among subpopulations. Additionally, this proposed approach incorporates a variable neighborhood search (VNS) method to raise the quality of solutions. In the end, a good balance between exploration and exploitation can be maintained by the proposal.

The following is the CA model for MOGWO:

(1) Cell: The selected candidate solution (the current cell).
(2) Cell space: The set of all the solutions (cells).
(3) Cell state: The solution's information at the iteration t, including current position $\mathbf{x}_i$, three best solutions $\alpha_i, \beta_i, \delta_i$, and so on. For the i-th solution at the iteration t, its state can be denoted by $S_i^t = \{\mathbf{x}_i^t, \alpha_i^t, \beta_i^t, \delta_i^t, \ldots\}$.
(4) Neighborhood: A set of solutions that closed to the current solution $\mathbf{x}_i^t$. Its neighborhood is defined as $G(\mathbf{x}_i^t) = \{\mathbf{x}_{i,1}^t, \ldots, \mathbf{x}_{i,k}^t, \ldots, \mathbf{x}_{i,T}^t\}$, where $\mathbf{x}_{i,k}^t$ is the k-th neighbor solution that closed to the current solution $\mathbf{x}_i^t$ at the iteration t, T is the size of neighborhood.
(5) Transition rule: This rule is regarded analogous to the fitness function in the intelligence optimization algorithms. Namely, $S_i^{t+1} = f\left(S_i^t \cup S_{G(i)}^t\right) = f\left(S_i^t, S_{i,1}^t, S_{i,2}^t, \ldots, S_{i,T}^t\right)$, where $S_{i,k}^t$ is the state of the k-th neighbor solution $\mathbf{x}_k^t$ that is closed to the current solution $\mathbf{x}_i^t$.

As previously stated, HFSP is a MOP and NP-hard problem in this research. MOEA is extremely effective in dealing with such an issue. As a result, a MOCGWO for solving the multi-objective HFSP is offered in this work. Figure 4.2 displays the flowchart for the suggested MOCGWO.

The following is a description of the main steps of the developed MOCGWO.

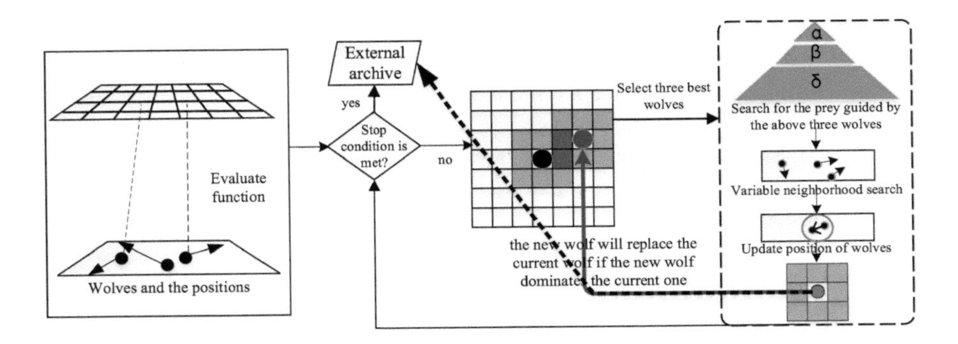

Fig. 4.2 Flowchart of the proposed MOCGWO

Step (1) Initialization: Produce an initial population at random in accordance with the encoding scheme (see Sects. 4.3.1 and 4.3.2), then create an external archive to save the non-dominated solutions discovered so far.

Step (2) Fitness assignment: Calculate the fitness values of all the solutions in the population (see Sect. 4.3.3).

Step (3) Cellular structure assignment: Every solution in the population is set in a lattice structure's cell (i.e., usually a two-dimensional grid structure). The number of cells of a lattice structure is the same as the population size. There is one index for each solution. Instead of corresponding to a decision variable space, this index represents a lattice structure cell position. There are neighbors for each solution. As a result, the population is split up into numerous subpopulations. The candidate solution will be updated only inside its subpopulations. A cellular neighborhood structure is the key characteristic of the hybridizing intelligence optimization algorithm with CA.

Step (4) Social hierarchy: The fitness values can be used to divide each subpopulation into four categories. More details on this social hierarchy mechanism are given in Sect. 4.3.4.

Step (5) Update the population: In each subpopulation, choose the top three solutions based on the fitness values of all the solutions in this population. Only in cases where the update operator is used can solutions communicate with their neighbors.

Step (5.1) Search operation: The alpha, beta, and delta typically direct the hunt's process. However, in a hypothetical search space, the best location of the prey is uncertain. Assume that the alpha, beta, and delta have a superior understanding of the potential location of the prey in order to simulate the hunting behavior of grey wolves. The three greatest answers direct the other candidate wolves to search for the hidden location of the prey. For more details, see Sect. 4.3.5.

Step (5.2) VNS: To find probable answers in the search space, VNS makes use of various neighborhood architectures. Procedure of VNS is given at length in Sect. 4.3.6.

Step (6) Update the external archive: The Pareto fronts in the external archive are updated based on the Pareto dominance concept. Specifically, insert current solution $\mathbf{x}$ into external archive and remove all solutions dominated by $\mathbf{x}$ when no solutions in external archive outweigh $\mathbf{x}$. Solutions with higher fitness values are eliminated by employing the truncation operator if the size of the produced non-dominated solutions is bigger than the archive size [6].

Step (7) Stop criterion: In the case that the stop criterion is met, output non-dominated solutions in the external archive. Otherwise, carry on with the aforementioned loop.

The following steps of the suggested algorithm involve several crucial strategies: encoding and decoding, initialization, social hierarchy, search operator, and VNS. These crucial steps are thoroughly described in the sections that follow.

4.3.1 *Encoding and Decoding*

The metaheuristic's performance is significantly impacted by the encoding strategy. The HFSP [7] uses a permutation-based method which is efficient and easy to use. Therefore, we apply a permutation-based encoding scheme to research. There are two steps involved in HFSP decoding stage: machine assignment and job permutation. The job permutation is the job processing sequence in which all jobs are assigned to machines at the beginning. The times at which jobs in the previous stage were completed can be used to decide the order in which jobs are processed in the following stage. In other words, if the job's previous operation is finished first, the job will be processed first. The first available machine rule is used to assign machines. The completion times of jobs at the previous stage are deemed as the release times. According to the First In First Out (FIFO) rule, the jobs are given to the first machine that is available. Generally speaking, the first available machine will result in the earliest completion time for a job.

4.3.2 *Initialization and Dividing the Population into Subpopulations*

According to the permutation-based approach, the starting population is generated at random, and a fixed-size external archive is formed. The non-dominated solutions that have been discovered so far are kept in this external archive. Solutions with higher fitness values are deleted using the truncation operator [7] if the resultant non-dominated solutions' size is more than the archive size. A large number of subpopulations divided from the initial population are placed in a single cellular lattice structure. Within its own neighborhood structure, each subpopulation improves its search strategy individually.

4.3.3 *Fitness Evaluation*

Calculate the population's fitness values. The following is a description of the precise steps involved in calculating fitness value. We compute the strength value and the raw fitness value for each solution. Each answer has a matching strength value that indicates how many other solutions it dominates. The strength values of solutions that dominate the current solution can be added to get the raw fitness value of each solution. The non-dominated solutions have raw fitness values that are all equal to zero. To retain the diversity of solutions, an indication of density is defined below [8]:

$$\text{Density}(i) = \frac{1}{\sigma_i^k + 2} \tag{4.18}$$

where σ_i^k is the Euclidean distance between solution I and its k-th closest neighbor.

A denominator greater than zero and a density value less than one are both guaranteed by the value 2 in Eq. (4.18). In this work, k has been set to 1. In the end, the equation shown below can be used to define the fitness of any solution i:

$$\text{Fitness}(i) = \text{raw\_ fitness}(i) + \text{Density}(i) \tag{4.19}$$

Be aware that non-dominated solutions are those where the raw fitness values are zero. It is advisable to choose a solution with a low fitness value.

4.3.4 Social Hierarchy

The original GWO's central concept is that the search process is directed by the three best solutions (wolves). Instead of focusing on non-dominance level, the subpopulation can be divided into four social hierarchies on the basis of fitness values to create a social hierarchy of solutions. The individual in each subpopulation who has the highest fitness value is regarded as α, and the following second and third best ones are recorded as β and δ. ω represents the remaining solutions.

4.3.5 Search Operator

The classical GWO search operator is made to address issues with continuous optimization. The issue in hand, however, is a common discrete optimization issue. Therefore, it cannot be used directly to this HFSP with multiple objectives. The modified search operator is formulated as follows.

$$\pi_i^{t+1} = \begin{cases} \text{shift}\left(\pi_i^t(j), \left(\pi_\alpha^t(j) - \pi_i^t(j)\right)\right) \text{ if rand} < \frac{1}{3} \\ \text{shift}\left(\pi_i^t(j), \left(\pi_\beta^t(j) - \pi_i^t(j)\right)\right) \text{ else if rand} < \frac{2}{3} \\ \text{shift}\left(\pi_i^t(j), \left(\pi_\delta^t(j) - \pi_i^t(j)\right)\right) \text{ otherwise} \end{cases} ; j = 1, 2, \ldots, n \tag{4.20}$$

where $\pi_i^t(j)$ is the j-th job of the permutation (or solution) i at the iteration t, $shift(x, d)$ means that the element x in the solution can be moved toward the right or left with $|d|$ units, rand is a uniform random number in a range [0, 1]. Figure 4.3 gives an example of the proposed search operator. Its search process is as follows.

Step 1: Sort all solutions according to the fitness value and select the three best solutions as the guiding solutions. Assume that the three best solutions are marked

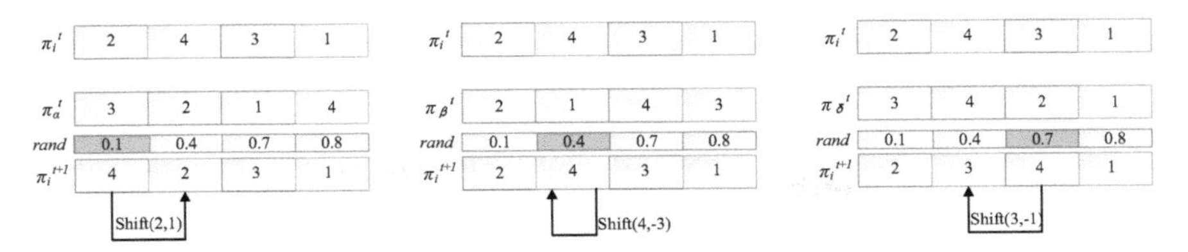

Fig. 4.3 An example of search operator

as $\pi_\alpha^t = [3, 2, 1, 4]$, $\pi_\beta^t = [2, 1, 4, 3]$ and $\pi_\delta^t = [3, 4, 2, 1]$, and a candidate solution i denoted by $\pi_i^t = [2, 4, 3, 1]$ in the current iteration t.

Step 2: Generate a random number in a uniform range [0, 1], and scan each element in π_i^t from left to right. Then, perform the update operation on each element of this candidate solution based on Eq. (4.20). For example, if this random number is 0.1 which is less than 1/3, then the solution will be updated to $\pi_i^t = \text{shift}\left(\pi_i^t(1), \left(\pi_\alpha^t(1) - \pi_i^t(1)\right)\right) = \text{shift}(2, (3-2))$. That is to say, the first element in the solution i is moved to the right with one unit. A temporary new solution is generated, namely, $\pi_i^t = [4, 2, 3, 1]$.

Step 3: Perform step 2 until all the elements in this solution are all scanned.

4.3.6 VNS Strategy

In the area of scheduling, variable neighborhood search (VNS) is a crucial local search heuristic [9]. VNS explores the search space for potential solutions using various neighborhood search architectures. The inner loop of VNS often contains shake and local search function. The shake function allows users to go from one type of local search neighborhood to another, increasing the variety of results. To further raise the quality of solutions, local search might look for the most promising solution in the surrounding area. The pseudocode of this VNS is given in Algorithm 4.1. First, one trial solution x is picked from the population. Then a neighborhood solution $\mathbf{x}'$ is randomly generated using the k-th neighborhood structure N_k of the solution$\mathbf{x}$, where$k \in \{1, 2, \ldots, k_{\max}\}$. To obtain the local optimal solution$\mathbf{x}''$, we apply a local search strategy to$\mathbf{x}'$. If the solutions $\mathbf{x}''$ and $\mathbf{x}$ satisfy any of the following two conditions, the acquired solution $\mathbf{x}''$ will replace the solution $\mathbf{x}$: : (1) the solution $\mathbf{x}''$ dominates the present solution $\mathbf{x}$, , and (2) solutions $\mathbf{x}''$ and x are non-dominated with each other, with the fitness value of solution $\mathbf{x}''$ being greater than that of solution $\mathbf{x}$. The obtained solution $\mathbf{x}''$ will replace the solution $\mathbf{x}$ if the solutions $\mathbf{x}''$ and $\mathbf{x}$ meet any one of the following two conditions: (1) The solution $\mathbf{x}''$. dominates the current solution $\mathbf{x}$, and (2) solutions $\mathbf{x}''$ and $\mathbf{x}$ are both non-dominated with each other and the fitness value of the solution $\mathbf{x}''$ is better than that of solution $\mathbf{x}$. Otherwise, the neighborhood search structure shifts to a different neighborhood search structure and starts local search. The available literature describes a wide variety of neighborhood search structures, including insert, swap, interchange, 2-opt, and inverse. Of all the

VNS strategies, the insert, swap, and inverse operations are the most productive [10]. To save computation time, we fix the number of variable neighborhood search structures. As a result, in this study we employ three efficient neighborhood search structures: insert, swap, and inverse operations. The following is an explanation of the three neighborhood search algorithms:

(1) Insert operation (N_1): randomly generate two different positions on the job permutation, and then select two jobs on the corresponding positions. Finally, insert the latter job before the former one.
(2) Swap operation (N_2): randomly generate two different positions on the job permutation and swap the two corresponding jobs.
(3) Reverse operation (N_3): randomly generate two different positions on the job permutation and reverse a subsequence between the area of two positions.

Algorithm 4.1: Local search based on VNS

Input: a given schedule
Output: a new schedule
1. Select a solution denoted by $\mathbf{x}$ from the current population. And design a set of neighborhood structures of the current solution $\mathbf{x}$
2. **For** $i = 1$ to t_{max} **do**
3. **Set** $k = 1$;
4. **While** $k < k_{max}$ do
5. Perform shake procedure: generate a random solution $\mathbf{x}'$ from the k-th neighborhood N_k of $\mathbf{x}$
6. Perform a local search on N_k to obtain a new solution $\mathbf{x}''$
7. **If** $\mathbf{x}''$ dominates $\mathbf{x}$
8. $\mathbf{x} = \mathbf{x}''$;
9. set $k = 1$;
10. **Else if** $\mathbf{x}''$ and $\mathbf{x}$ are both non-dominated with each other
11. **if** fitness ($\mathbf{x}''$) < fitness ($\mathbf{x}$)
12. $\mathbf{x} = \mathbf{x}''$;
13. set $k = 1$;
14. **End**
15. **Else**
16. $k = k + 1$;
17. **End If**
18. **End while**
19. $i = i + 1$;
20. **End for**

Table 4.1 Parameter settings for instances

Input variables	Distribution
Number of jobs (n)	10, 20, 30, 40, 50, 60, 70, 80, 90, 100
Number of stage (s)	4, 6
Number of available machines at each stage (m_i)	Discrete uniform [1, 5]
Normal processing time (p)	Discrete uniform [30, 50] min
The speed of the machine k at stage i ($v_{i,k}$)	Discrete uniform [1, 5]
Noise during working mode ($N_{i,j}^{w}$)	Discrete uniform [60, 90] db
Noise during idle mode ($N_{i,j}^{\text{idle}}$)	Discrete uniform [30, 50] db
The idle power of a machine at stage i ($P_{i,k}^{\text{idle}}$)	Continuous uniform (0, 2] kw
The loading power of a machine at stage i ($P_{i,k}^{\text{work}}$)	Continuous uniform [2, 5] kw
The breakeven duration (T_b)	e min

4.4 Experiments

This section is devoted to rating MOCGWO's effectiveness. The empirical research in this section covers the following three parts: The instances, parameter settings, and performance metrics are discussed first in the following subsections, after which the experimental investigations are thoroughly examined step by step.

4.4.1 Test Instances

We need examples with various configurations in terms of the problem scale in order to systematically evaluate the behavior of various MOEAs. However, obtaining real-world data from practical applications is quite challenging. Therefore, in this study, these cases are generated at random. These examples are representatives of actual manufacturing industry statistics. The notation "HFSP_*n*_*s*" can be used to represent a instance. For example, "HFSP_20_5" indicates that the HFSP instance has 20 jobs and 5 phases. Table 4.1 provides a summary of the relevant parameter values for these situations. In addition, there are some HFSP benchmarks [11]. These changed benchmarks have new names. For instance, "mj10c5a2" stands for the updated version of "j10c5a2".

4.4.2 Parameter Settings

As we all know, the parameter settings can make substantial influence to the algorithms' performance. As a result, pilot tests were carried out to examine various parameter configurations. The majority of parameter settings in experiments are constants. Some factors, however, such as the maximum number of function evaluations (NFEs), are linked to the scale of problem. Meanwhile, due to their unique features, different algorithms have different parameters. As a result, all of the parameter settings of all of the MOEAs investigated in this study are mentioned below, based on the features of the algorithm and problem:

1. The PMX crossover probability: PMX crossover is a very successful and widely used method for solving such a scheduling problem. The PMX crossover operator with the highest probability will be more likely to explore more area of search space. As a result, the PMX crossover is used in SPEA2, NSGA-II, and MOEA/D, with a crossover probability of 0.9.
2. The swap mutation probability: One of the most widely used approaches in scheduling issues is the swap operation, which is considered a mutation operator. Generally speaking, a mutation operator with a lower probability can aid an optimization algorithm in escaping a local optimum at a later stage of the search procedure. Therefore, 0.2 is chosen as the swap mutation probability for SPEA2, NSGA-II, and MOEA/D.
3. The external archive size and population size: For MOCGWO, SPEA2, and MOE/D, the external archive size is set to 100. The HFSP has three objectives, thus 500 is often the ideal population size for MOEA/D. According to the results of the prior experiment, the population size for the remaining MOEAs is set at 100, which is likewise appropriate for various problem scales.
4. The neighborhood size: According to the preliminary experiment, the neighborhood size for MOEA/D is set to 20. The experiment specifies a neighborhood size of 12 for MOCGWO. Keep in mind that MOEA/D and MOCGWO have clear distinctions between them when it comes to neighborhood definition. In MOEA/D, a MOP is divided into N scalar subproblems, and each of them is connected to weighed vector $\boldsymbol{\lambda} = \left\{\boldsymbol{\lambda}^1, \ldots, \boldsymbol{\lambda}^N\right\}$. $B(k) = \left\{\boldsymbol{\lambda}^1, \ldots, \boldsymbol{\lambda}^T\right\}$ is the definition of the neighborhood for each subproblem k, where T is the neighborhood size. Calculating Euclidean distance between any two weight vectors can be used to determine the neighborhood. The population of MOCGWO is distributed in a cellular structure and is broken up into numerous subpopulations. Each subpopulation is an independent neighborhood. Each solution $\mathbf{x}$ has its neighbors. $G(\mathbf{x}) = \{\mathbf{x}_1, \ldots, \mathbf{x}_k, \ldots, \mathbf{x}_T\}$ is a definition of the neighborhood of $\mathbf{x}$ where T represents the neighborhood size. The neighborhood is determined by calculating Euclidean distance between index of any two solutions.
5. The stop condition: All MOEAs employ the same stop criterion (i.e., the maximum NFEs) for a particular problem in order to allow for a fair comparison. The optimization process will be stopped by the algorithm if the maximum NFEs are met. The problem scale is connected to the maximum NFEs. The maximum NFEs are therefore set to $n \times 1000$ (n is the total number of jobs).

In order to ensure a fair comparison, the same encoding strategy and random seed are used to initialize all of the MOEAs in this paper. All experiments are carried out on a PC with a 2.90 GHz Intel Core i7 processor and 16 GB RAM under a Windows 10 Operating System.

4.4.3 Performance Metrics

A good representative of the obtained PF should approach closely toward the PF* and distribute uniformly along the PF. In this chapter, three popular performance metrics are applied to assess the quality of outcomes produced by these MOEAs. These performance metrics consist of Spread, GD [12], and IGD [13]. The non-dominated solutions found by all metaheuristics for each problem in all the independent runs are treated as PF* (reference points) for that problem because the true PF* of the test problems may be unknown in advance.

4.4.4 Effectiveness of the Cellular Automata

In this section, we run an experiment to evaluate the performance of the CA mechanism in the MOCGWO. MOGWO serves as a representation in this experiment that the MOCGWO does not feature a CA mechanism. Both MOCGWO and MOGWO use the identical parameter settings as described in Sect. 4.4.2. Table 4.2 lists the average and standard deviation for 20 run times of various algorithms. The Wilcoxon signed-rank test findings, including "R+", "R-", and p-value, are displayed in Table 4.3. For problems where the MOCGWO surpasses its rival, "R+" stands for the sum of ranks, while "R−" stands for the inverse [14]. The proposed MOCGWO significantly performs better or worse than its competitor can be denoted as the symbols "+" or "−". There is no discernible difference between them, as indicated by the " = " marker. The counts of significant differences between two methods are listed in the final row in Table 4.3, labeled by " $+/=/-$". For instance, "15/6/4" indicates that MOCGWO performs significantly better or worse than its rival on 15 or 4 issues, respectively. There is no discernible difference on 6 problems. All the tables include bolded text that indicates the best outcomes.

With regard to the Spread metric, Table 4.2 shows that the MOCGWO performs better than the MOGWO on the majority of problems. The related p-values in Table 4.3 imply that there is no significant difference between them on the aforementioned problems, despite the MOCGWO being worse than the MOGWO on "HFSP_20_6", "HFSP_60_4", "HFSP_90_4", and "mj10c5a4" in terms of solutions' distribution. In addition, the evidence that the MOCGWO greatly beats the MOGWO for the Spread metric can be found in Table 4.3, where the MOCGWO has greater "+" counts than the MOGWO on 19 out of the 31 problems for that metric. In terms of the GD metric, the MOCGWO performs significantly better than the MOGWO on 22 out of the 31 issues. Regarding IGD metrics, it is evident that the MOCGWO greatly outperforms the MOGWO on practically all of the issues. The MOCGWO

Table 4.2 Mean and standard deviation values of all the metrics between MOCGWO and MOGWO

	Spread (mean/std)		GD (mean/std)		IGD (mean/std)	
Problems	MOGWO	MOCGWO	MOGWO	MOCGWO	MOGWO	MOCGWO
HFSP_ 10_4	**6.15e − 01/** 3.4e − 02	6.49e − 01/ **3.1e − 02**	**2.45e − 03/** 7.0e − 04	2.56e − 03/ **6.8e − 04**	9.98e − 04/ 2.0e − 04	**7.94e − 04/ 9.7e − 05**
HFSP_ 10_6	6.94e − 01/ **2.5e − 02**	**6.75e − 01/** 3.3e − 02	4.03e − 03/ 9.4e − 04	**3.43e − 03/ 8.0e − 04**	1.04e − 03/ **1.9e − 04**	**6.81e − 04/** 7.2e − 04
HFSP_ 20_4	6.53e − 01/ **4.4e − 02**	**6.52e − 01/** 5.9e − 02	**5.69e − 03/ 1.2e − 03**	6.53e − 03/ 1.4e − 03	5.30e − 03/ 7.8e − 04	**4.77e − 03/ 5.5e − 04**
HFSP_ 20_6	**6.44e − 01/** 7.7e − 02	6.60e − 01/ **3.9e − 02**	7.19e − 03/ 1.7e − 03	**6.88e − 03/ 1.7e − 03**	7.56e − 03/ **8.0e − 04**	**7.14e − 03/** 1.1e − 03
HFSP_ 30_4	6.42e − 01/ 4.7e − 02	**6.28e − 01/ 3.6e − 02**	7.80e − 03/ 2.2e − 03	**6.65e − 03/ 1.4e − 03**	6.07e − 03/ 6.1e − 04	**4.89e − 03/ 6.1e − 04**
HFSP_ 30_6	1.11e + 00/ 5.4e − 02	**1.09e + 00/ 4.5e − 02**	9.39e − 03/ 2.6e − 03	**7.60e − 03/ 2.4e − 03**	1.16e − 02/ 3.7e − 03	**1.03e − 02/ 3.1e − 03**
HFSP_ 40_4	9.09e − 01/ 8.5e − 02	**7.15e − 01/ 5.1e − 02**	8.26e − 03/ 1.8e − 03	**7.64e − 03/ 1.6e − 03**	8.28e − 03/ 1.7e − 03	**5.46e − 03/ 6.2e − 04**
HFSP_ 40_6	7.59e − 01/ 5.4e − 02	**6.73e − 01/ 5.2e − 02**	9.63e − 03/ **1.7e − 03**	**7.77e − 03/** 1.8e − 03	9.24e − 03/ 1.4e − 03	**6.40e − 03/ 9.3e − 04**
HFSP_ 50_4	9.03e − 01/ 6.4e − 02	**7.64e − 01/ 6.3e − 02**	1.15e − 02/ **1.6e − 03**	**7.43e − 03/** 2.0e − 03	8.07e − 03/ 1.1e − 03	**5.48e − 03/ 6.8e − 04**
HFSP_ 50_6	6.85e − 01/ 5.3e − 02	**6.68e − 01/ 4.7e − 02**	9.34e − 03/ 1.7e − 03	**6.15e − 03/ 1.6e − 03**	6.86e − 03/ 8.3e − 04	**5.20e − 03/ 5.8e − 04**
HFSP_ 60_4	**6.05e − 01/** 5.2e − 02	6.29e − 01/ **5.0e − 02**	**6.68e − 03/** 1.4e − 03	7.03e − 03/ **1.2e − 03**	7.10e − 03/ 9.2e − 04	**5.92e − 03/ 8.3e − 04**
HFSP_ 60_6	7.27e − 01/ 7.9e − 02	**7.06e − 01/ 6.4e − 02**	1.30e − 02/ 3.3e − 03	**9.27e − 03/ 3.2e − 03**	1.06e − 02/ 2.7e − 03	**8.63e − 03/ 2.5e − 03**
HFSP_ 70_4	8.32e − 01/ 7.6e − 02	**6.79e − 01/ 5.0e − 02**	1.00e − 02/ 1.5e − 03	**7.28e − 03/ 2.2e − 03**	1.19e − 02/ 1.9e − 03	**8.12e − 03/ 1.3e − 03**
HFSP_ 70_6	6.79e − 01/ 6.7e − 02	**6.68e − 01/ 4.3e − 02**	1.07e − 02/ 2.9e − 03	**6.89e − 03/ 1.8e − 03**	9.74e − 03/ **1.1e − 03**	**6.43e − 03/** 1.3e − 03
HFSP_ 80_4	1.02e + 00/ **8.2e − 02**	**8.68e − 01/** 1.1e − 01	1.31e − 02/ 4.3e − 03	**7.99e − 03/ 2.3e − 03**	1.17e − 02/ 2.2e − 03	**8.13e − 03/ 2.2e − 03**
HFSP_ 80_6	9.58e − 01/ 7.3e − 02	**7.55e − 01/ 5.6e − 02**	1.57e − 02/ 3.5e − 03	**8.41e − 03/ 2.3e − 03**	1.66e − 02/ 1.8e − 03	**9.11e − 03/ 1.4e − 03**
HFSP_ 90_4	**6.06e − 01/ 3.1e − 02**	6.29e − 01/ 4.8e − 02	9.22e − 03/ 2.0e − 03	**7.53e − 03/ 1.8e − 03**	7.17e − 03/ **9.0e − 04**	**5.79e − 03/** 9.3e − 04
HFSP_ 90_6	1.16e + 00/ **9.9e − 02**	**1.04e + 00/** 1.1e − 01	2.35e − 02/ **1.1e − 03**	**1.43e − 02/** 5.0e − 03	2.06e − 02/ 3.1e − 03	**1.97e − 02/ 2.7e − 03**
HFSP_ 100_4	7.18e − 01/ **4.0e − 02**	**6.37e − 01/** 6.7e − 02	1.19e − 02/ 3.1e − 03	**8.30e − 03/ 3.0e − 03**	9.53e − 03/ 1.5e − 03	**7.32e − 03/ 1.5e − 03**
HFSP_ 100_6	6.43e − 01/ **5.2e − 02**	**6.40e − 01/** 5.5e − 02	1.48e − 02/ 3.7e − 03	**1.06e − 02/ 3.1e − 03**	1.24e − 02/ **2.0e − 03**	**9.30e − 03/** 2.8e − 03

(continued)

Table 4.2 (continued)

	Spread (mean/std)		GD (mean/std)		IGD (mean/std)	
Problems	MOGWO	MOCGWO	MOGWO	MOCGWO	MOGWO	MOCGWO
Mj10c5a2	8.56e − 01/ 6.8e − 02	**6.73e − 01/ 5.1e − 02**	4.38e − 03/ 7.2e − 04	**3.45e − 03/ 4.2e − 04**	8.83e − 03/ 7.4e − 04	**7.43e − 04/ 7.5e − 04**
Mj10c5a3	8.98e − 01/ 7.5e − 02	**7.06e − 01/ 4.6e − 02**	9.64e − 03/ 8.8e − 04	**7.59e − 03/ 7.8e − 04**	7.58e − 03/ 6.5e − 04	**6.64e − 03/ 6.7e − 04**
Mj10c5a4	**7.46e − 01/ 6.5e − 02**	7.78e − 01/ 6.8e − 02	**8.78e − 03/ 8.5e − 04**	9.24e − 03/ 4.1e − 03	**6.60e − 03/** 5.1e − 04	6.86e − 03/ **3.2e − 04**
Mj10c5a5	6.90e − 01/ 7.7e − 02	**6.06e − 01/ 5.3e − 02**	1.25e − 02/ 4.8e − 03	**9.91e − 03/ 3.2e − 03**	5.17e − 03/ **2.2e − 03**	**4.25e − 03/** 4.6e − 03
Mj10c5a6	8.74e − 01/ 7.5e − 02	**7.58e − 01/ 6.5e − 02**	1.32e − 02/ 2.4e − 03	**9.63e − 03/ 1.8e − 03**	2.26e − 02/ 2.3e − 03	**1.13e − 02/ 2.1e − 03**
mj15c5a1	9.17e − 01/ 8.8e − 02	**8.53e − 01/ 7.4e − 02**	2.04e − 02/ **2.5e − 03**	**1.24e − 02/** 3.6e − 03	8.91e − 03/ 8.4e − 04	**7.28e − 03/ 2.7e − 04**
mj15c5a2	9.06e − 01/ 9.7e − 02	**7.79e − 01/ 6.8e − 02**	1.21e − 02/ 4.3e − 03	**8.62e − 03/ 3.3e − 03**	1.04e − 02/ 2.1e − 03	**8.48e − 03/ 1.5e − 03**
mj15c5a3	7.79e − 01/ **5.6e − 02**	**6.75e − 01/** 7.4e − 02	2.64e − 02/ **3.2e − 03**	**1.14e − 02/** 3.8e − 03	2.34e − 02/ **2.6e − 03**	**1.10e − 02/** 3.1e − 03
mj15c5a4	9.47e − 01/ 9.3e − 02	**8.69e − 01/ 6.7e − 02**	1.52e − 02/ 4.3e − 03	**9.76e − 03/ 3.0e − 03**	3.26e − 02/ 3.4e − 03	**2.11e − 02/ 1.1e − 03**
mj15c5a5	6.39e − 01/ 4.7e − 02	**5.44e − 01/ 3.4e − 02**	2.25e − 02/ 3.7e − 03	**8.69e − 03/ 2.9e − 03**	8.57e − 03/ 6.4e − 03	**7.38e − 03/ 4.5e − 03**
mj15c5a6	8.85e − 01/ 8.4e − 02	**7.57e − 01/ 5.5e − 02**	1.05e − 02/ 4.5e − 03	**1.24e − 02/ 2.3e − 03**	2.12e − 02/ 4.1e − 03	**1.07e − 02/ 2.3e − 03**

is the method in our comparative experiment that performs the best, as displayed in Tables 4.2 and 4.3. This effectively indicates that the CA mechanism can enhance the algorithm's overall performance. The reason for this superiority of MOCGWO lies in CA as follows: Each solution has its own neighbors in CA, where the population is broken up into numerous subpopulations. In its own neighborhood, each subpopulation individually develops its search progress. As a result, the optimizer can do parallel searches to look for unexplored portions of the search space. Thus, search variety can be increased. The overlap area between nearby communities can also serve as a means of information exchange (an implicit migration mechanism). Good solutions can spread easily among the entire population, which can aid in the search space's exploration. As a result, the capacity to explore can be improved greatly. In the meantime, the search operator is exploiting the area around each solution. Finally, it is possible to keep the balance between exploration and exploitation.

The results of the aforementioned comparative experiment indicate that the MOCGWO is significantly more effective than the MOGWO for the majority of problems, which amply supports the CA strategy's efficacy. Furthermore, it implies that GWO with cellular structure is superior to GWO without any structures.

Table 4.3 Wilcoxon signed-rank test results for each instance (a level of significant $\alpha = 0.05$)

Problems	MOCGWO versus $MOGWO_{no}$								
	Spread			GD			IGD		
	R^+	R^-	*p*-value/win	R^+	R^-	*p*-value/win	R^+	R^-	*p*-value/win
HFSP_10_4	33	177	7.18e − 03/−	100	110	8.52e − 01/−	188	22	**1.94e − 03/ +**
HFSP_10_6	158	52	**4.79e − 02/ +**	156	54	**5.69e − 02/ =**	210	0	**8.86e − 05/ +**
HFSP_20_4	109	101	**8.81e − 01/ =**	48	162	3.33e − 02/−	181	29	**4.55e − 03/ +**
HFSP_20_6	95	115	7.08e − 01/ =	114	96	**7.37e − 01/ =**	151	59	**8.59e − 02/ =**
HFSP_30_4	146	64	**1.26e − 01/ =**	142	68	**1.67e − 01/ =**	198	12	**5.17e − 04/ +**
HFSP_30_6	103	107	9.40e − 01/ =	168	42	**1.87e − 02/ +**	141	69	**1.79e − 01/ =**
HFSP_40_4	208	2	**1.20e − 04/ +**	120	90	**5.75e − 01/ =**	209	1	**1.03e − 04/ +**
HFSP_40_6	195	15	**7.79e − 04/ +**	174	36	**1.00e − 02/ +**	207	3	**1.40e − 04/ +**
HFSP_50_4	205	5	**1.89e − 04/ +**	209	1	**1.03e − 04/ +**	210	0	**8.86e − 05/ +**
HFSP_50_6	126	84	**4.33e − 01/ =**	199	11	**4.49e − 04/ +**	205	5	**1.89e − 04/ +**
HFSP_60_4	64	146	1.25e − 01/ =	84	126	4.33e − 01/ =	185	25	**2.82e − 03/ +**
HFSP_60_6	132	78	**1.31e − 01/ =**	186	24	**2.49e − 03/ +**	164	46	**2.76e − 02/ +**
HFSP_70_4	209	1	**1.03e − 04/ +**	189	21	**1.71e − 03/ +**	208	2	**1.20e − 04/ +**
HFSP_70_6	114	96	**7.36e − 01/ =**	199	11	**4.49e − 04/ +**	210	0	**8.86e − 05/ +**
HFSP_80_4	202	8	**2.93e − 04/ +**	196	14	**6.81e − 04/ +**	188	22	**1.94e − 03/ +**
HFSP_80_6	210	0	**8.86e − 05/ +**	210	0	**8.86e − 05/ +**	210	0	**8.86e − 05/ +**
HFSP_90_4	63	147	1.16e − 01/ =	161	49	**3.66e − 02/ +**	193	17	**1.02e − 03/ +**
HFSP_90_6	184	26	**3.18e − 03/ +**	179	31	**5.73e − 03/ +**	127	83	**4.12e − 01/ =**
HFSP_100_4	191	19	**1.32e − 03/ +**	190	20	**1.51e − 03/ +**	196	14	**6.81e − 04/ +**

(continued)

Table 4.3 (continued)

Problems	MOCGWO versus MOGWO$_{no}$								
	Spread			GD			IGD		
	R$^+$	R$^-$	p-value/win	R$^+$	R$^-$	p-value/win	R$^+$	R$^-$	p-value/win
HFSP_100_6	73	137	2.32e − 01/ =	190	20	**1.51e − 03/ +**	193	17	**1.02e − 03/ +**
Mj10c5a2	200	10	**3.90e − 04/ +**	210	0	**8.86e − 05/ +**	202	8	**2.93e − 04/ +**
Mj10c5a3	210	0	**8.86e − 05/ +**	210	0	**8.86e − 05/ +**	195	15	**7.79e − 04/ +**
Mj10c5a4	100	110	8.52e − 01/ =	80	130	3.50e − 01/ =	83	127	4.12e − 01/ =
Mj10c5a5	210	0	**8.86e − 05/ +**	210	0	**8.86e − 05/ +**	186	24	**2.49e − 03/ +**
Mj10c5a6	210	0	**8.86e − 05/ +**	185	25	**2.82e − 03/ +**	210	0	**8.86e − 05/ +**
mj15c5a1	202	8	**2.93e − 04/ +**	120	90	**5.75e − 01/ =**	208	2	**1.20e − 04/ +**
mj15c5a2	210	0	**8.86e − 05/ +**	208	2	**1.20e − 04/ +**	210	0	**8.86e − 05/ +**
mj15c5a3	210	0	**8.86e − 05/ +**	209	1	**1.03e − 04/ +**	210	0	**8.86e − 05/ +**
mj15c5a4	210	0	**8.86e − 05/ +**	205	5	**1.89e − 04/ +**	210	0	**8.86e − 05/ +**
mj15c5a5	199	11	**4.49e − 04/ +**	202	8	**2.93e − 04/ +**	210	0	**8.86e − 05/ +**
mj15c5a6	210	0	**8.86e − 05/ +**	205	5	**1.89e − 04/ +**	210	0	**8.86e − 05/ +**
+ / = /−	19/11/1			22/7/2			27/4/0		

4.4.5 *Effectiveness of VNS on MOCGWO*

We make a comparison between the MOCGWO and the MOCGWO without VNS for confirming the effectiveness of the developed VNS strategy in the MOCGWO. In this experiment, MOCGWO$_{no}$ represents the MOCGWO without VNS. All parameter values are identical to those in Sect. 4.4.2 in order to ensure a fair comparison. Table 4.4 records statistical results of all the metrics over 20 independent run times. Table 4.5 reports the Wilcoxon signed-rank test results.

Tables 4.4 and 4.5 demonstrate that, in terms of GD, the MOCGWO totally outperforms the MOCGWO$_{no}$ on all the problems. This indicates that the VNS can significantly enhance the algorithm's convergence performance. With regard to the Spread measure, it is evident that the MOCGWO$_{no}$ outperforms the MOCGWO on all but

Table 4.4 Mean and standard deviation value of all metrics between $MOCGWO_{no}$ and MOCGWO

Problems	Spread (Mean/std)		GD (Mean/std)		IGD (Mean/std)	
	$MOCGWO_{no}$	MOCGWO	$MOCGWO_{no}$	MOCGWO	$MOCGWO_{no}$	MOCGWO
HFSP_10_6	**5.66e − 01/** 8.1e − 02	7.09e − 01/ **4.0e − 02**	3.68e − 02/ 8.8e − 03	**3.34e − 03/6.9e − 04**	4.83e − 03/ 4.4e − 04	**1.07e − 03/1.6e − 04**
HFSP_20_4	**5.30e − 01/ 5.1e − 02**	6.41e − 01/ 6.1e − 02	5.33e − 02/ 8.6e − 03	**6.61e − 03/1.6e − 03**	2.46e − 02/ 2.1e − 03	**5.56e − 03/8.2e − 04**
HFSP_20_6	**5.54e − 01/** 7.4e − 02	6.55e − 01/ **4.8e − 02**	8.63e − 02/ 1.2e − 02	**1.06e − 02/3.3e − 03**	3.80e − 02/ 3.7e − 03	**9.80e − 03/1.1e − 03**
HFSP_30_4	**5.82e − 01/** 8.2e − 02	6.23e − 01/ **4.6e − 02**	1.16e − 01/ 1.3e − 02	**8.59e − 03/1.8e − 03**	4.34e − 02/ 3.1e − 03	**6.57e − 03/1.2e − 03**
HFSP_30_6	**5.86e − 01/ 5.3e − 02**	1.10e + 00/ 6.7e − 02	1.69e − 01/ 2.4e − 02	**8.33e − 03/3.6e − 03**	5.24e − 02/ 5.1e − 03	**9.79e − 03/2.7e − 03**
HFSP_40_4	**6.01e − 01/** 5.8e − 02	7.31e − 01/ **5.5e − 02**	1.23e − 01/ 1.4e − 02	**7.28e − 03/1.7e − 03**	3.80e − 02/ 1.4e − 03	**6.63e − 03/1.2e − 03**
HFSP_40_6	**5.86e − 01/** 5.5e − 02	7.02e − 01/ **5.4e − 02**	1.16e − 01/ 1.6e − 02	**8.61e − 03/2.2e − 03**	4.36e − 02/ 2.0e − 03	**6.86e − 03/1.0e − 03**
HFSP_50_4	**5.81e − 01/ 5.4e − 02**	8.18e − 01/ 7.1e − 02	1.11e − 01/ 1.7e − 02	**7.30e − 03/2.2e − 03**	3.84e − 02/ 1.8e − 03	**7.47e − 03/1.3e − 03**
HFSP_50_6	**5.82e − 01/ 5.0e − 02**	6.67e − 01/ 5.3e − 02	9.96e − 02/ 1.2e − 02	**7.89e − 03/1.7e − 03**	4.10e − 02/ 2.0e − 03	**6.84e − 03/9.4e − 04**
HFSP_60_4	**5.92e − 01/** 8.7e − 02	5.97e − 01/ **3.9e − 02**	1.51e − 01/ 1.7e − 02	**8.32e − 03/2.0e − 03**	5.87e − 02/ 3.1e − 03	**7.91e − 03/1.1e − 03**
HFSP_60_6	**6.29e − 01/** 6.9e − 02	6.98e − 01/ **5.1e − 02**	1.86e − 01/ 2.7e − 02	**1.24e − 02/3.7e − 03**	7.31e − 02/ 4.0e − 03	**1.02e − 02/2.3e − 03**
HFSP_70_4	**6.09e − 01/ 5.8e − 02**	7.51e − 01/ 7.3e − 02	2.30e − 01/ 2.9e − 02	**9.33e − 03/3.0e − 03**	7.71e − 02/ 3.0e − 03	**7.95e − 03/1.5e − 03**
HFSP_70_6	**6.21e − 01/ 5.6e − 02**	6.74e − 01/ 6.1e − 02	2.20e − 01/ 3.4e − 02	**9.54e − 03/3.8e − 03**	6.91e − 02/ 4.6e − 03	**1.00e − 02/2.4e − 03**
HFSP_80_4	**6.11e − 01/ 4.2e − 02**	9.20e − 01/ 1.0e − 01	2.80e − 01/ 3.3e − 02	**1.19e − 02/3.7e − 03**	1.11e − 01/ 5.0e − 03	**1.16e − 02/2.0e − 03**
HFSP_80_6	**6.44e − 01/ 6.1e − 02**	8.14e − 01/ 1.1e − 01	3.61e − 01/ 4.6e − 02	**1.37e − 02/6.3e − 03**	1.14e − 01/ 3.4e − 03	**1.27e − 02/3.0e − 03**

(continued)

Table 4.4 (continued)

Problems	Spread (Mean/std)		GD (Mean/std)		IGD (Mean/std)	
	$MOCGWO_{no}$	MOCGWO	$MOCGWO_{no}$	MOCGWO	$MOCGWO_{no}$	MOCGWO
HFSP_90_4	**5.99e − 01/** 5.2e − 02	6.19e − 01/ **4.2e − 02**	1.35e − 01/ 2.1e − 02	**8.60e − 03/1.7e − 03**	5.14e − 02/ 1.9e − 03	**7.65e − 03/1.1e − 03**
HFSP_90_6	**6.35e − 01/ 3.5e − 02**	9.85e − 01/ 7.7e − 02	3.24e − 01/ 7.5e − 02	**1.84e − 02/8.9e − 03**	1.14e − 01/ 7.0e − 03	**1.87e − 02/2.7e − 03**
HFSP_100_4	**6.34e − 01/** 5.2e − 02	6.51e − 01/ **4.7e − 02**	3.04e − 01/ 5.2e − 02	**1.10e − 02/3.4e − 03**	9.04e − 02/ 3.8e − 03	**9.49e − 03/2.1e − 03**
HFSP_100_6	6.66e − 01/ **4.5e − 02**	**6.31e − 01/** 6.4e − 02	3.16e − 01/ 4.0e − 02	**1.33e − 02/4.8e − 03**	9.01e − 02/ 4.1e − 03	**1.11e − 02/2.6e − 03**
Mj10c5a2	**5.32e − 01/ 5.5e − 02**	5.75e − 01/ 6.0e − 02	2.09e − 01/ 3.1e − 02	**1.16e − 01/1.2e − 03**	5.78e − 03/ 4.8e − 04	**1.35e − 03/2.4e − 04**
Mj10c5a3	**5.04e − 01/ 5.0e − 02**	6.33e − 01/ 6.5e − 02	1.32e − 01/ 1.2e − 02	**2.15e − 02/8.6e − 03**	5.34e − 02/ 4.0e − 03	**8.67e − 03/2.2e − 03**
Mj10c5a4	**5.78e − 01/** 8.2e − 02	7.14e − 01/ **7.2e − 02**	3.78e − 01/ 3.9e − 02	**3.21e − 01/4.6e − 03**	1.25e − 01/ 3.2e − 03	**1.09e − 02/2.2e − 03**
Mj10c5a5	**6.16e − 01/ 6.1e − 02**	7.58e − 01/ 7.7e − 02	2.45e − 01/ 3.7e − 02	**1.30e − 02/3.2e − 03**	6.68e − 02/ 4.5e − 03	**1.21e − 02/2.3e − 03**
Mj10c5a6	**6.24e − 01/ 6.2e − 02**	8.10e − 01/ 1.0e − 01	3.17e − 01/ 4.5e − 02	**1.65e − 02/7.4e − 03**	7.52e − 02/ 3.9e − 03	**7.68e − 03/1.7e − 03**
mj15c5a1	**5.91e − 01/ 7.4e − 02**	8.02e − 01/ 8.2e − 02	3.73e − 01/ 4.1e − 02	**2.32e − 02/3.3e − 03**	1.19e − 01/ 6.3e − 03	**1.09e − 02/2.5e − 03**
mj15c5a2	**6.37e − 01/ 6.3e − 02**	8.32e − 01/ 1.2e − 01	3.24e − 01/ 5.0e − 02	**2.84e − 01/4.2e − 03**	4.34e − 02/ 3.4e − 03	**6.45e − 03/8.3e − 04**
mj15c5a3	**6.25e − 01/** 5.3e − 02	7.23e − 01/ **4.6e − 02**	2.64e − 01/ 4.2e − 02	**1.37e − 02/3.5e − 03**	5.46e − 02/ 6.6e − 03	**7.84e − 03/5.8e − 03**
mj15c5a4	**6.55e − 01/ 4.2e − 02**	8.25e − 01/ 1.3e − 01	3.21e − 01/ 3.8e − 02	**2.46e − 02/5.6e − 03**	5.30e − 02/ 5.3e − 03	**7.53e − 03/2.4e − 03**
mj15c5a5	**5.89e − 01/** 5.6e − 02	6.24e − 01/ **4.7e − 02**	2.16e − 01/ 3.3e − 02	**8.69e − 03/4.7e − 03**	8.76e − 02/ 4.5e − 03	**8.03e − 03/2.6e − 03**
mj15c5a6	**5.89e − 01/ 5.6e − 02**	7.41e − 01/ 6.8e − 02	4.02e − 01/ 4.1e − 02	**1.88e − 03/3.9e − 03**	6.31e − 02/ 6.4e − 03	**1.21e − 02/4.8e − 03**

Table 4.5 Wilcoxon signed-rank test results for each instance (a level of significant $\alpha = 0.05$)

Problems	MOCGWO vs. $MOGWO_{no}$								
	Spread			GD			IGD		
	R^+	R^-	*p*-value/win	R^+	R^-	*p*-value/win	R^+	R^-	*p*-value/win
HFSP_10_4	32	178	6.42e − 03/−	210	0	**8.86e − 05/ +**	210	0	**8.86e − 05/ +**
HFSP_10_6	5	205	1.89e − 04/−	210	0	**8.86e − 05/ +**	210	0	**8.86e − 05/ +**
HFSP_20_4	11	199	4.49e − 04/−	210	0	**8.86e − 05/ +**	210	0	**8.86e − 05/ +**
HFSP_20_6	10	200	3.90e − 04/−	210	0	**8.86e − 05/ +**	210	0	**8.86e − 05/ +**
HFSP_30_4	55	155	6.20e − 02/−	210	0	**8.86e − 05/ +**	210	0	**8.86e − 05/ +**
HFSP_30_6	0	210	8.86e − 05/−	210	0	**8.86e − 05/ +**	210	0	**8.86e − 05/ +**
HFSP_40_4	0	210	8.86e − 05/−	210	0	**8.86e − 05/ +**	210	0	**8.86e − 05/ +**
HFSP_40_6	2	208	1.20e − 04/−	210	0	**8.86e − 05/ +**	210	0	**8.86e − 05/ +**
HFSP_50_4	0	210	8.86e − 05/−	210	0	**8.86e − 05/ +**	210	0	**8.86e − 05/ +**
HFSP_50_6	13	197	5.93e − 04/-	210	0	**8.86e − 05/ +**	210	0	**8.86e − 05/ +**
HFSP_60_4	100	110	8.52e − 01/ =	210	0	**8.86e − 05/ +**	210	0	**8.86e − 05/ +**
HFSP_60_6	29	181	4.54e − 03/−	210	0	**8.86e − 05/ +**	210	0	**8.86e − 05/ +**
HFSP_70_4	1	209	1.03e − 04/−	210	0	**8.86e − 05/ +**	210	0	**8.86e − 05/ +**
HFSP_70_6	35	175	8.96e − 03/−	210	0	**8.86e − 05/ +**	210	0	**8.86e − 05/ +**

(continued)

Table 4.5 (continued)

Problems	MOCGWO vs. $MOGWO_{no}$								
	Spread			GD			IGD		
	R^+	R^-	*p*-value/win	R^+	R^-	*p*-value/win	R^+	R^-	*p*-value/win
HFSP_80_4	0	210	8.86e − 05/−	210	0	**8.86e − 05/ +**	210	0	**8.86e − 05/ +**
HFSP_80_6	8	202	2.93e − 04/−	210	0	**8.86e − 05/ +**	210	0	**8.86e − 05/ +**
HFSP_90_4	74	136	2.47e − 01/ =	210	0	**8.86e − 05/ +**	210	0	**8.86e − 05/ +**
HFSP_90_6	0	210	8.86e − 05/−	210	0	**8.86e − 05/ +**	210	0	**8.86e − 05/ +**
HFSP_100_4	80	130	3.50e − 01/ =	210	0	**8.86e − 05/ +**	210	0	**8.86e − 05/ +**
HFSP_100_6	159	51	**4.38e − 02/ +**	210	0	**8.86e − 05/ +**	210	0	**8.86e − 05/ +**
mj10c5a2	100	110	8.52e − 01/ =	210	0	**8.86e − 05/ +**	210	0	**8.86e − 05/ +**
mj10c5a3	0	210	8.86e − 05/−	210	0	**8.86e − 05/ +**	210	0	**8.86e − 05/ +**
mj10c5a4	0	210	8.86e − 05/−	210	0	**8.86e − 05/ +**	210	0	**8.86e − 05/ +**
mj10c5a5	0	210	8.86e − 05/−	210	0	**8.86e − 05/ +**	210	0	**8.86e − 05/ +**
mj10c5a6	0	210	8.86e − 05/−	210	0	**8.86e − 05/ +**	210	0	**8.86e − 05/ +**
mj15c5a1	68	142	1.67e − 01/ =	210	0	**8.86e − 05/ +**	210	0	**8.86e − 05/ +**
mj15c5a2	0	210	8.86e − 05/−	210	0	**8.86e − 05/ +**	210	0	**8.86e − 05/ +**
mj15c5a3	1	209	1.03e − 04/−	210	0	**8.86e − 05/ +**	210	0	**8.86e − 05/ +**
mj15c5a4	0	210	8.86e − 05/−	210	0	**8.86e − 05/ +**	210	0	**8.86e − 05/ +**
mj15c5a5	0	210	8.86e − 05/−	210	0	**8.86e − 05/ +**	210	0	**8.86e − 05/ +**
mj15c5a6	0	210	8.86e − 05/−	210	0	**8.86e − 05/ +**	210	0	**8.86e − 05/ +**
+ / = /−	1/5/25			31/0/0			31/0/0		

one problem. The cause of this phenomena is that the VNS strategy places more emphasis on taking advantage of superior solutions nearby than it does on exploring uncharted territory. In this instance, the search diversity suffers to some extent as a result of this method. Overall, however, it is clear that MOCGWO totally surpasses $\mathrm{MOCGWO_{no}}$ on all problems pertaining to the IGD metric. Therefore, it can be said that the VNS has a more favorable impact on the MOCGWO's overall performance, notably for the convergence performance. It is the VNS that makes the comparison experiment successful. If the objectives were slightly improved in this MOP, it would be able to outperform other solutions. The VNS can thus be advantageous to MOCGWO. $\mathrm{MOCGWO_{no}}$, in contrast, omits this VNS tactic that can cause the derived solutions to converge on the PF*.

We can infer from the aforementioned comparative experiment that the MOCGWO performs significantly better than $\mathrm{MOCGWO_{no}}$ on the majority of problems. This supports the viability of the VNS approach.

4.4.6 Comparison of MOCGWO with Other Algorithms

To assess the algorithm's comprehensive performance, the proposed method is contrasted with the already available MOEAs such as NSGA-II [15], SPEA2 [13], and MOEA/D [16]. As far as we know, there is no MOEA for the HFSP with noise pollution criterion. As a result, we adjusted the comparable MOEAs taken into consideration in this study to adapt the problem's characteristics. The parameter settings for all the MOEAs are the same in the previous Sect. 4.4.2. Each algorithm was conducted 20 run times independently on each problem. For multiple comparisons involving more than two methods, a pairwise comparison test, such as the Wilcoxon signed-rank test, is not appropriate. To identify variations between a collection of algorithms' performance that are significant at the 0.05 level of significance, a multiple comparison test (also known as the Friedman test) is performed.

The mean and standard deviation values for the GD metric are shown in Table 4.6. These findings in Table 4.6 show that, in the majority of cases, the proposed MOCGWO algorithm performs noticeably better than the other MOEAs. According to the GD metric, Table 4.7 shows the overall rankings of these MOEAs across all of the Friedman test questions. According to the results, the MOCGWO algorithm outperforms all other MOEAs in terms of the GD metric. In a similar vein, Table 4.8 makes it extremely evident that the proposed MOCGWO can achieve highly promising outcomes in terms of the Spread metric. On 27 of the 31 problems, the MOCGWO generates the best mean results, with the exception of "HFSP_20_4", "HFSP_30_4", "mj10c5a2" and "mj10c5a3". On 17 out of the 31 problems, it can produce the best deviation outcomes, demonstrating the robustness of the proposed approach. Additionally, it can be observed from the Table 4.9 that the MOCGWO outperforms its rivals on nearly all issues in terms of the Spread metric. The statistical findings for these MOEAs regarding the IGD metric are shown in Tables 4.10 and 4.11. Table 4.10 shows that, when compared to other MOEAs, the proposal can produce 24 best mean IGD values and 10 best standard deviation values. According to the IGD metric, Table 4.11 displays the overall rankings of these MOEAs after applying the Friedman test. The suggested MOCGWO clearly performs better than the other MOEAs in terms of the IGD metric. These experimental findings suggest that the proposed MOCGWO can successfully address this scheduling problem.

Figures 4.4, 4.6, and 4.8 show the estimated PFs discovered by these algorithms for three example issues, namely "HSP_40_6" (small scale problem), "HFSP_60_6" (middle scale problem), and "HFSP_80_4". These final approximations PFs were found in a single run using the best IGD measure for each MOEA during 20 independent run times. According to expectations, the suggested MOCGWO algorithm can offer superior non-dominated solutions than other MOEAs when compared in terms of solution quality and distribution. In addition, Figs. 4.5, 4.7, and 4.9 illustrate the Gantt charts of the extreme makespan criterion (i.e., point A in Pareto fronts). These figures support our findings from the experiment with numerical comparisons mentioned above. Therefore, it can be concluded that for solving such a scheduling problem, the suggested MOCGWO greatly beats the previous algorithms. The proposal's advantage can be due to the MOCGWO's integration of the benefits of CA for diversification and VNS for intensification, which maintains a healthy balance between exploration and exploitation.

4.5 Conclusion

In this article, from the perspective of noise pollution, a hybrid flow shop scheduling problem (HFSP) with strong application background is studied. This article is an important addition to the existing literature, in which most studies on HFSP have emphasized the importance of production efficiency while ignoring the green environmental issues represented by noise pollution. We view this new HFSP as a multi-objective optimization problem with the goal of simultaneously minimizing completion time, noise pollution, and total energy consumption. We develop a new multi-objective HFSP mathematical model. Meanwhile, to achieve the goal of green manufacturing, one energy conservation/noise reduction strategy is added into this model. To address this problem, a multi-objective cellular grey wolf optimizer (MOCGWO) is developed for the first time. In this algorithm, it combines the diversification of cellular automata (CA) and the reinforcement of variable neighborhood search (VNS) and maintains a good balance between exploration and development. We conducted comparative experiments on 20 test instances to determine the reliability of the proposal to address multi-objective HFSPs. The empirical results show that MOCGWO is more reliable than other MOEA in such problems. In addition, in order to determine the feasibility of CA mechanism and VNS strategy, we conducted many numerical experiments to compare and verify.

Table 4.6 Mean and standard deviation of GD metric by NSGA-II, SPEA2, MOEA/D, and MOCGWO

problems	NSGA-II (Mean/std)	SPEA2 (Mean/std)	MOEA/D (Mean/std)	MOCGWO (Mean/std)
HFSP_10_4	5.70e − 03/2.3e − 03	3.68e − 03/1.6e − 03	3.75e − 03/1.1e − 03	**9.86e − 04/ 2.6e − 04**
HFSP_10_6	5.11e − 03/1.3e − 03	3.57e − 03/**7.2e − 04**	3.31e − 03/9.5e − 04	**1.46e − 03/** 9.6e − 04
HFSP_20_4	9.73e − 03/2.7e − 03	7.60e − 03/1.5e − 03	6.75e − 03/1.7e − 03	**4.30e − 03/ 1.4e − 03**
HFSP_20_6	1.03e − 02/1.6e − 03	8.13e − 03/1.8e − 03	7.51e − 03/1.7e − 03	**4.54e − 03/ 1.4e − 03**
HFSP_30_4	8.64e − 03/1.7e − 03	7.30e − 03/2.1e − 03	6.42e − 03/1.9e − 03	**3.79e − 03/ 1.4e − 03**
HFSP_30_6	2.12e − 02/1.6e − 02	2.19e − 02/1.6e − 02	1.69e − 02/1.8e − 02	**1.45e − 02/ 1.7e − 02**
HFSP_40_4	1.08e − 02/1.9e − 03	8.34e − 03/1.6e − 03	7.40e − 03/1.5e − 03	**3.41e − 03/ 9.1e − 04**
HFSP_40_6	1.06e − 02/2.5e − 03	6.72e − 03/1.5e − 03	6.94e − 03/1.5e − 03	**3.06e − 03/ 9.7e − 04**
HFSP_50_4	9.92e − 03/1.7e − 03	8.62e − 03/1.7e − 03	6.19e − 03/**1.4e − 03**	**4.89e − 03/** 1.8e − 03
HFSP_50_6	9.24e − 03/1.8e − 03	8.11e − 03/1.7e − 03	5.56e − 03/**1.1e − 03**	**3.01e − 03/** 1.3e − 03
HFSP_60_4	1.00e − 02/1.7e − 03	6.14e − 03/1.4e − 03	6.81e − 03/1.2e − 03	**3.14e − 03/ 8.8e − 04**
HFSP_60_6	1.64e − 02/3.7e − 03	1.32e − 02/3.0e − 03	**9.39e − 03/2.0e − 03**	9.50e − 03/ 2.7e − 03
HFSP_70_4	1.38e − 02/2.6e − 03	9.74e − 03/2.6e − 03	7.21e − 03/**1.9e − 03**	**4.69e − 03/** 2.8e − 03
HFSP_70_6	1.28e − 02/2.1e − 03	9.84e − 03/2.6e − 03	8.13e − 03/2.2e − 03	**4.10e − 03/ 1.7e − 03**
HFSP_80_4	1.65e − 02/5.4e − 03	1.01e − 02/3.4e − 03	**7.54e − 03/2.7e − 03**	1.03e − 02/ 3.4e − 03
HFSP_80_6	1.35e − 02/5.0e − 03	9.50e − 03/3.0e − 03	6.92e − 03/2.3e − 03	**4.49e − 03/ 1.9e − 03**
HFSP_90_4	1.40e − 02/2.4e − 03	8.50e − 03/2.0e − 03	7.15e − 03/1.9e − 03	**3.70e − 03/ 1.5e − 03**
HFSP_90_6	1.18e − 02/3.1e − 03	1.14e − 02/2.9e − 03	1.03e − 02/5.3e − 03	**1.03e − 02/ 4.2e − 03**
HFSP_100_4	1.21e − 02/2.8e − 03	1.25e − 02/3.6e − 03	9.57e − 03/3.4e − 03	**5.19e − 03/ 2.7e − 03**
HFSP_100_6	1.86e − 02/4.4e − 03	1.38e − 02/3.1e − 03	**1.08e − 02/2.8e − 03**	1.46e − 02/ 4.3e − 03

(continued)

Table 4.6 (continued)

problems	NSGA-II (Mean/std)	SPEA2 (Mean/std)	MOEA/D (Mean/std)	MOCGWO (Mean/std)
mjec5a2	4.30e – 03/3.2e – 03	4.58e – 03/4.1e – 03	4.05e – 03/3.4e – 03	**8.39e – 04/ 2.5e – 03**
mjec5a3	7.23e – 03/6.5e – 03	5.72e – 03/4.6e – 03	4.89e – 03/2.7e – 03	**3.73e – 03/ 1.3e – 03**
mjec5a4	6.17e – 03/2.3e – 03	5.34e – 03/6.2e – 03	3.34e – 03/7.8e – 04	**2.43e – 03/ 6.6e – 04**
mjec5a5	8.10e – 03/3.5e – 03	8.36e – 03/3.7e – 03	6.34e – 03/3.4e – 03	**5.68e – 03/ 2.7e – 03**
mjec5a6	9.03e – 03/2.2e – 03	7.79e – 03/1.8e – 03	5.90e – 03/2.5e – 03	**4.25e – 03/ 8.9e – 04**
mj15c5a1	1.54e – 02/1.6e – 03	5.84e – 03/4.4e – 03	5.69e – 03/5.2e – 03	**4.07e – 03/ 9.0e – 04**
mj15c5a2	1.84e – 02/2.4e – 03	7.04e – 03/3.6e – 03	7.15e – 03/3.4e – 03	**4.37e – 03/ 2.6e – 03**
mj15c5a3	1.05e – 02/4.5e – 03	8.53e – 03/3.4e – 03	5.94e – 03/5.3e – 03	**4.53e – 03/ 2.3e – 03**
mj15c5a4	2.30e – 02/2.6e – 02	5.02e – 02/3.6e – 02	4.46e – 02/3.8e – 02	**2.53e – 02/ 2.4e – 02**
mj15c5a5	3.75e – 02/3.3e – 02	4.14e – 02/3.2e – 02	3.66e – 02/**2.7e – 02**	**3.07e – 02/** 4.1e – 02
mj15c5a6	5.49e – 02/5.1e – 02	5.83e – 02/5.9e – 02	7.48e – 02/6.3e – 02	**4.56e – 02/ 4.8e – 02**

Table 4.7 Overall ranks through the Friedman test of IGD metric among different variants (a level of significant $\alpha = 0.05$)

Algorithms	Rank	p-value
NSGA-II	3.95	3.86e–10
SPEA2	3.00	
MOEA/D	1.85	
MOCGWO	1.20	

Table 4.8 Mean and standard deviation of Spread metric obtained by NSGA-II, SPEA2, MOEA/D, and MOCGWO

Problems	NSGA-II (Mean/std)	SPEA2 (Mean/std)	MOEA/D (Mean/std)	MOCGWO (Mean/std)
HFSP_10_4	8.85e − 01/4.1e − 02	6.33e − 01/**2.0e − 02**	1.79e + 00/2.2e − 02	**6.15e − 01/** 3.4e − 02
HFSP_10_6	9.30e − 01/4.3e − 02	6.80e − 01/3.5e − 02	1.80e + 00/**2.4e − 02**	**6.66e − 01/** 3.3e − 02
HFSP_20_4	8.12e − 01/5.0e − 02	**6.37e − 01**/5.2e − 02	1.74e + 00/**3.0e − 02**	6.38e − 01/ 4.4e − 02
HFSP_20_6	8.75e − 01/6.9e − 02	6.75e − 01/7.0e − 02	1.76e + 00/**5.6e − 02**	**6.52e − 01/** 6.2e − 02
HFSP_30_4	7.51e − 01/4.8e − 02	**6.41e − 01**/4.7e − 02	1.65e + 00/4.1e − 02	6.46e − 01/ **4.0e − 02**
HFSP_30_6	1.20e + 00/5.8e − 02	1.17e + 00/4.9e − 02	1.61e + 00/6.2e − 02	**1.09e + 00/ 1.1e − 02**
HFSP_40_4	8.37e − 01/6.4e − 02	9.08e − 01/5.6e − 02	1.62e + 00/**3.3e − 02**	**6.98e − 01/ 3.7e − 02**
HFSP_40_6	7.58e − 01/4.3e − 02	7.73e − 01/6.1e − 02	1.63e + 00/4.3e − 02	**6.76e − 01/ 3.9e − 02**
HFSP_50_4	9.05e − 01/6.5e − 02	9.04e − 01/8.3e − 02	1.62e + 00/6.1e − 02	**7.64e − 01/ 5.9e − 02**
HFSP_50_6	7.62e − 01/**3.4e − 02**	7.03e − 01/5.3e − 02	1.60e + 00/4.2e − 02	**6.77e − 01/** 4.6e − 02
HFSP_60_4	7.35e − 01/5.3e − 02	6.15e − 01/5.1e − 02	1.60e + 00/5.1e − 02	**6.07e − 01/ 4.2e − 02**
HFSP_60_6	8.12e − 01/6.1e − 02	7.22e − 01/5.8e − 02	1.59e + 00/**3.7e − 02**	**7.08e − 01/** 4.8e − 02
HFSP_70_4	8.06e − 01/6.1e − 02	8.29e − 01/7.5e − 02	1.60e + 00/**3.9e − 02**	**6.71e − 01/** 4.7e − 02
HFSP_70_6	7.99e − 01/6.5e − 02	7.02e − 01/6.7e − 02	1.62e + 00/**4.3e − 02**	**6.67e − 01/** 5.5e − 02
HFSP_80_4	9.11e − 01/**6.0e − 02**	9.59e − 01/6.9e − 02	1.58e + 00/6.9e − 02	**8.60e − 01/** 9.1e − 02
HFSP_80_6	9.08e − 01/7.7e − 02	9.52e − 01/1.1e − 01	1.51e + 00/5.2e − 02	**7.56e − 01/ 5.1e − 02**
HFSP_90_4	6.83e − 01/4.4e − 02	6.18e − 01/5.2e − 02	1.60e + 00/3.9e − 02	**6.14e − 01/ 3.8e − 02**
HFSP_90_6	1.15e + 00/**8.0e − 02**	1.16e + 00/1.1e − 01	1.44e + 00/1.5e − 01	**1.02e + 00/** 8.8e − 02
HFSP_100_4	7.87e − 01/6.2e − 02	7.14e − 01/4.3e − 02	1.54e + 00/7.0e − 02	**6.32e − 01/ 3.3e − 02**
HFSP_100_6	7.86e − 01/5.2e − 02	6.51e − 01/6.2e − 02	1.48e + 00/4.9e − 02	**6.51e − 01/ 4.6e − 02**

(continued)

Table 4.8 (continued)

Problems	NSGA-II (Mean/std)	SPEA2 (Mean/std)	MOEA/D (Mean/std)	MOCGWO (Mean/std)
mj10c5a2	6.52e − 01/5.8e − 02	**6.04e − 01/4.9e − 02**	8.65e − 01/6.5e − 02	6.75e − 01/ 5.8e − 02
mj10c5a3	8.07e − 01/7.6e − 02	**6.58e − 01/6.5e − 02**	7.94e − 01/7.3e − 02	6.59e − 01/ 6.7e − 02
mj10c5a4	7.73e − 01/5.4e − 02	7.21e − 01/5.4e − 02	8.03e − 01/5.6e − 02	**6.63e − 01/ 4.7e − 02**
mj10c5a5	7.84e − 01/6.3e − 02	7.14e − 01/6.1e − 02	7.54e − 01/5.7e − 02	**5.87e − 01/ 5.1e − 02**
mj10c5a6	8.20e − 01/6.2e − 02	8.54e − 01/**5.9e − 02**	8.15e − 01/6.1e − 02	**7.49e − 01/** 6.0e − 02
mj15c5a1	8.68e − 01/7.3e − 02	7.96e − 01/6.8e − 02	8.72e − 01/7.4e − 02	**7.03e − 01/ 5.3e − 02**
mj15c5a2	7.93e − 01/6.3e − 02	7.05e − 01/5.3e − 02	8.98e − 01/7.8e − 02	**6.47e − 01/ 4.6e − 02**
mj15c5a3	6.96e − 01/6.7e − 02	6.76e − 01/5.2e − 02	8.69e − 01/6.5e − 02	**6.08e − 01/ 4.1e − 02**
mj15c5a4	9.07e − 01/7.3e − 02	9.24e − 01/**7.0e − 02**	9.78e − 01/7.3e − 02	**8.43e − 01/** 7.5e − 02
mj15c5a5	8.98e − 01/7.5e − 02	8.68e − 01/8.8e − 02	9.35e − 01/7.9e − 02	**7.86e − 01/ 6.0e − 02**
mj15c5a6	7.85e − 01/8.0e − 02	8.06e − 01/7.5e − 02	8.29e − 01/6.9e − 02	**7.32e − 01/ 5.8e − 02**

Table 4.9 Overall ranks through the Friedman test of Spread metric among different variants (a level of significant $\alpha = 0.05$)

Algorithms	Rank	*p*-value
NSGA-II	2.56	1.45e−10
SPEA2	2.00	
MOEA/D	4.38	
MOCGWO	1.06	

Table 4.10 Mean and standard deviation of IGD metric obtained by NSGA-II, SPEA2, MOEA/D, and MOCGWO

Problems	NSGA-II (Mean/std)	SPEA2 (Mean/std)	MOEA/D (Mean/std)	MOCGWO (Mean/std)
HFSP_10_4	**3.80e − 04/5.0e − 05**	3.91e − 04/7.2e − 05	1.52e − 03/2.8e − 04	5.46e − 04/ 8.9e − 05
HFSP_10_6	**3.75e − 04/6.0e − 05**	4.84e − 04/6.3e − 05	1.43e − 03/1.2e − 04	6.10e − 04/ 8.8e − 05
HFSP_20_4	**3.66e − 03**/1.2e − 03	5.25e − 03/1.4e − 03	4.96e − 03/4.1e − 04	4.68e − 03/ **1.0e − 03**
HFSP_20_6	8.43e − 03/1.7e − 03	8.99e − 03/1.9e − 03	**6.07e − 03/7.1e − 04**	7.61e − 03/ 1.4e − 03
HFSP_30_4	**4.02e − 03**/6.3e − 04	5.06e − 03/9.8e − 04	1.28e − 02/1.2e − 03	4.73e − 03/ **4.5e − 04**
HFSP_30_6	8.88e − 03/2.0e − 03	9.03e − 03/1.9e − 03	7.94e − 03/**1.6e − 03**	**7.65e − 03/** 1.7e − 03
HFSP_40_4	7.23e − 03/1.9e − 03	8.49e − 03/1.7e − 03	1.52e − 02/**9.8e − 04**	**6.74e − 03/** 1.4e − 03
HFSP_40_6	5.56e − 03/1.9e − 03	7.15e − 03/1.7e − 03	7.84e − 03/**4.8e − 04**	**5.32e − 03/** 1.2e − 03
HFSP_50_4	7.10e − 03/2.1e − 03	6.79e − 03/1.5e − 03	9.45e − 03/**1.0e − 03**	**5.18e − 03/** 1.3e − 03
HFSP_50_6	5.30e − 03/1.6e − 03	6.74e − 03/2.0e − 03	5.11e − 03/**3.9e − 04**	**4.60e − 03/** 1.4e − 03
HFSP_60_4	8.26e − 03/1.0e − 03	9.36e − 03/1.6e − 03	7.21e − 03/**5.2e − 04**	**7.19e − 03/** 2.0e − 03
HFSP_60_6	1.31e − 02/2.5e − 03	1.31e − 02/2.5e − 03	1.47e − 02/**1.1e − 03**	**1.04e − 02/** 2.1e − 03
HFSP_70_4	8.53e − 03/2.1e − 03	9.54e − 03/2.3e − 03	9.74e − 03/**9.8e − 04**	**7.74e − 03/** 2.1e − 03
HFSP_70_6	8.30e − 03/1.9e − 03	8.97e − 03/2.3e − 03	8.97e − 03/**6.3e − 04**	**6.58e − 03/** 2.1e − 03
HFSP_80_4	1.73e − 02/2.5e − 03	1.85e − 02/2.7e − 03	1.95e − 02/**1.1e − 03**	**1.46e − 02/** 3.2e − 03
HFSP_80_6	1.73e − 02/3.4e − 03	1.72e − 02/3.4e − 03	1.85e − 02/**1.3e − 03**	**1.43e − 02/** 2.8e − 03
HFSP_90_4	7.24e − 03/1.9e − 03	8.04e − 03/2.0e − 03	7.27e − 03/**5.8e − 04**	**5.74e − 03/** 1.0e − 03
HFSP_90_6	1.76e − 02/1.8e − 03	**1.40e − 02**/2.4e − 03	2.87e − 02/**1.8e − 03**	1.42e − 02/ 1.9e − 03
HFSP_100_4	1.05e − 02/1.8e − 03	9.85e − 03/1.9e − 03	**7.75e − 03/6.1e − 04**	9.26e − 03/ 1.7e − 03
HFSP_100_6	1.57e − 02/3.4e − 03	1.63e − 02/2.7e − 03	2.82e − 02/**2.1e − 03**	**1.24e − 02/** 2.6e − 03

(continued)

Table 4.10 (continued)

Problems	NSGA-II (Mean/std)	SPEA2 (Mean/std)	MOEA/D (Mean/std)	MOCGWO (Mean/std)
mj10c5a2	7.46e − 03/3.7e − 03	7.86e − 03/3.9e − 03	8.77e − 03/**2.1e − 04**	**7.04e − 03/** 2.4e − 03
mj10c5a3	8.03e − 03/2.4e − 03	8.23e − 03/2.3e − 03	6.95e − 03/2.6e − 03	**5.72e − 03/ 1.6e − 03**
mj10c5a4	6.32e − 03/4.8e − 03	6.43e − 03/2.7e − 03	1.05e − 02/5.8e − 03	**4.68e − 03/ 2.4e − 03**
mj10c5a5	6.35e − 03/2.9e − 03	7.07e − 03/5.3e − 03	8.23e − 03/3.3e − 03	**5.47e − 03/ 2.0e − 03**
mj10c5a6	4.68e − 03/5.3e − 03	5.47e − 03/4.8e − 03	1.09e − 02/1.5e − 03	**4.01e − 03/ 5.6e − 04**
mj15c5a1	4.47e − 03/2.1e − 03	6.89e − 03/2.2e − 03	7.44e − 03/**1.3e − 03**	**3.54e − 03/** 2.2e − 03
mj15c5a2	5.64e − 03/2.6e − 03	7.01e − 03/2.7e − 03	6.35e − 03/2.0e − 03	**4.07e − 03/ 1.5e − 03**
mj15c5a3	3.26e − 03/1.8e − 03	4.74e − 03/2.1e − 03	4.86e − 03/2.8e − 03	**2.74e − 03/ 1.1e − 03**
mj15c5a4	2.84e − 02/1.9e − 03	2.40e − 02/2.4e − 03	2.31e − 02/**1.7e − 03**	**1.42e − 02/** 2.2e − 03
mj15c5a5	1.76e − 02/1.8e − 03	2.15e − 02/2.6e − 03	2.05e − 02/1.9e − 03	**8.91e − 03/ 1.6e − 03**
mj15c5a6	2.35e − 02/3.2e − 03	5.31e − 02/4.2e − 03	4.63e − 02/4.4e − 03	**9.07e − 03/ 2.3e − 03**

Table 4.11 Overall ranks through the Friedman test of GD metric among different variants (a level of significant $\alpha = 0.05$)

Algorithms	Rank	p-value
NSGA-II	2.13	3.13e−06
SPEA2	3.02	
MOEA/D	3.60	
MOCGWO	1.23	

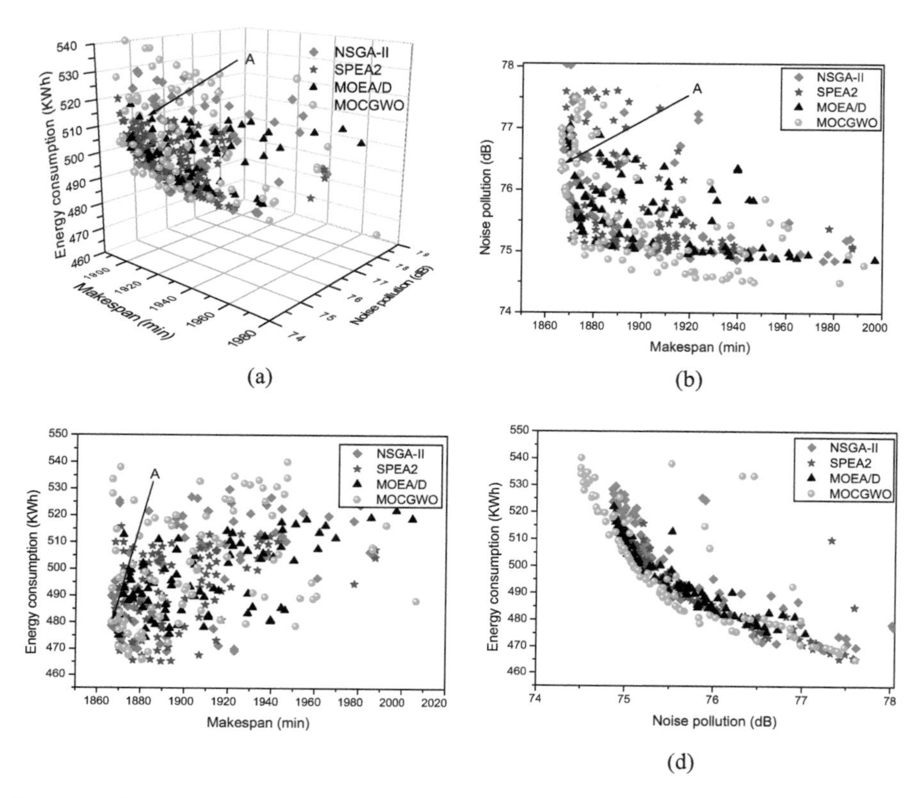

Fig. 4.4 PFs obtained by NSGA-II, SPEA2, MOEA/D, and MOCGWO on HFSP_40_6. **a** PFs by different MOEAs with three criteria, **b** PFs with makespan and noise, **c** PFs with makespan and energy consumption, **d** PFs with noise and energy consumption

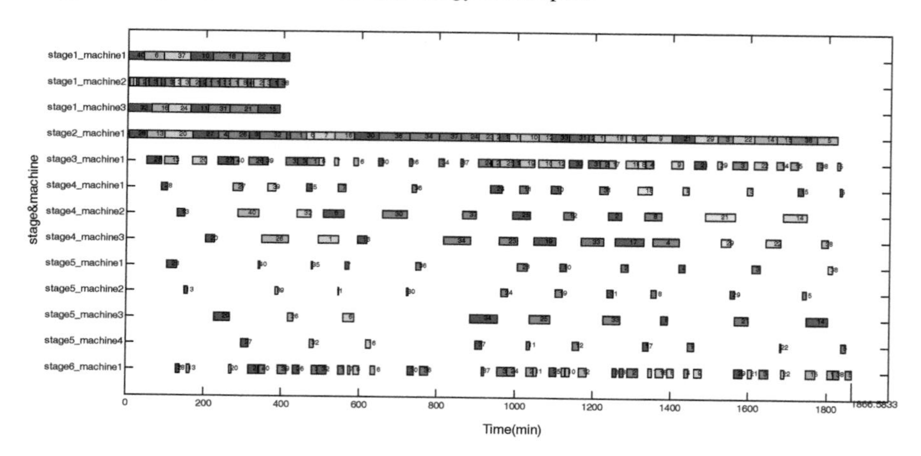

Fig. 4.5 Gantt chart in the corresponding solution A on HFSP_40_6

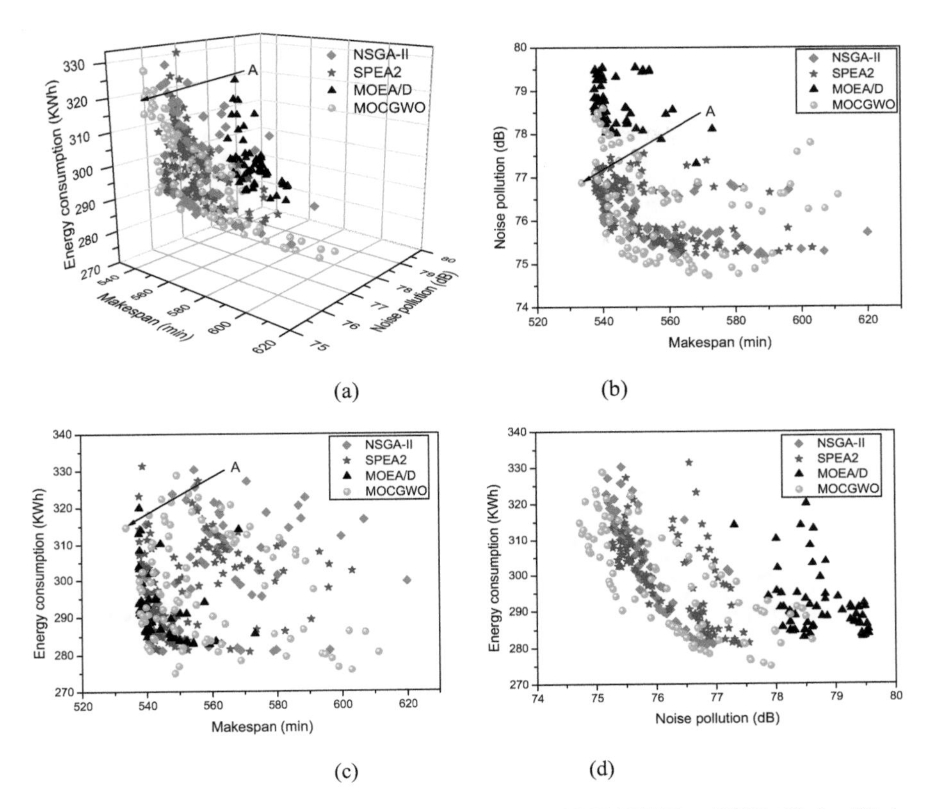

Fig. 4.6 PFs obtained by NSGA-II, SPEA2, MOEA/D, and MOCGWO on HFSP_60_6. **a** PFs by different MOEAs with three criteria, **b** PFs with makespan and noise, **c** PFs with makespan and energy consumption, **d** PFs with noise and energy consumption

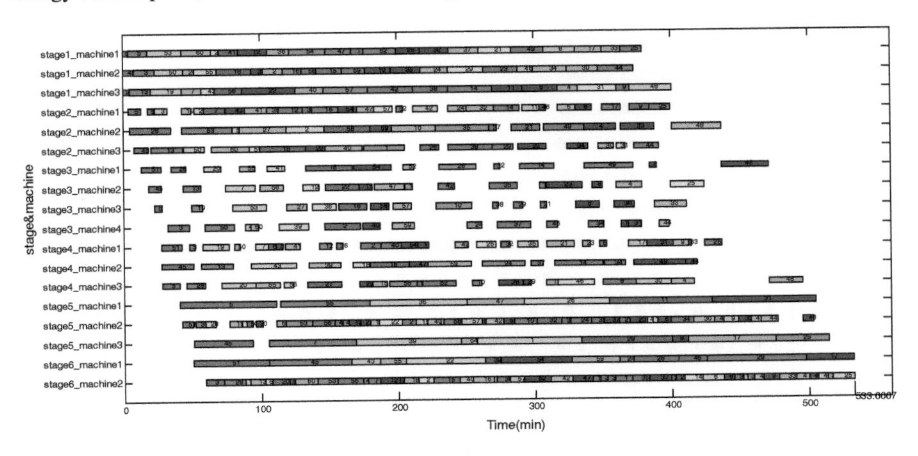

Fig. 4.7 Gantt chart of the solution A on HFSP_60_6

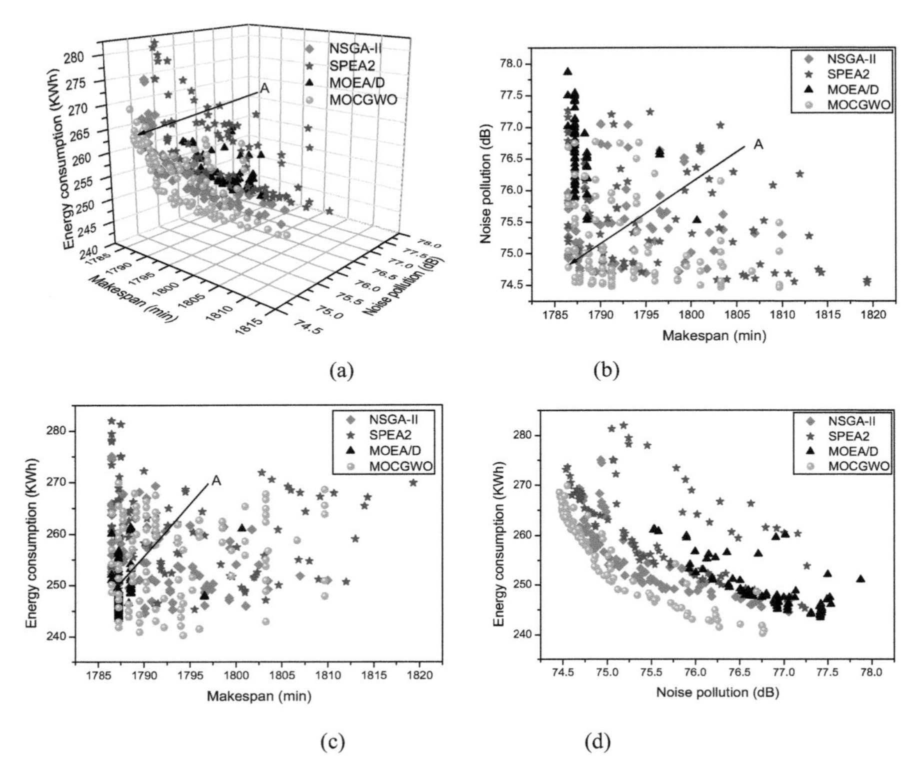

Fig. 4.8 PFs obtained by NSGA-II, SPEA2, MOEA/D, and MOCGWO on HFSP_80_4. **a** PFs by different MOEAs with three criteria, **b** PFs with makespan and noise, **c** PFs with makespan and energy consumption, **d** PFs with noise and energy consumption

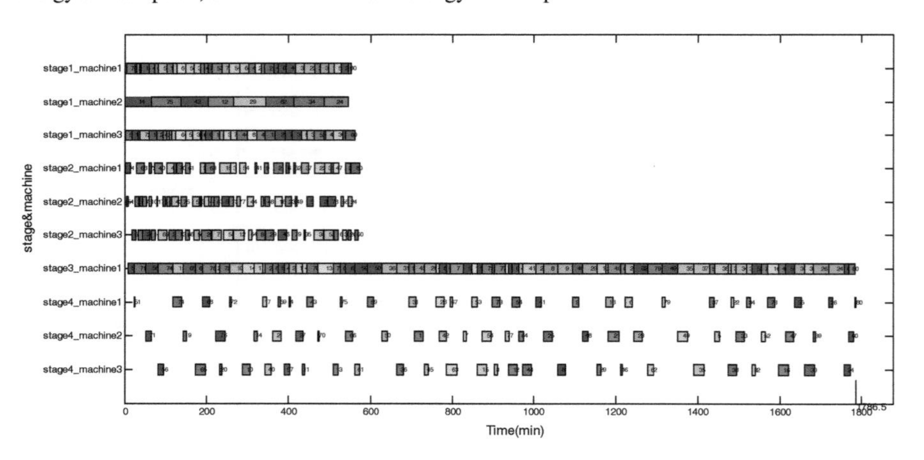

Fig. 4.9 Gantt chart of the solution A on HFSP_80_4

References

1. Yin, L., Li, X., Gao, L., et al.: A novel mathematical model and multi-objective method for the low-carbon flexible job shop scheduling problem. Sustain. Comput. Inf. Syst. **13**, 15–30 (2017)
2. Lu, C., Gao, L., Li, X., et al.: Energy-efficient permutation flow shop scheduling problem using a hybrid multi-objective backtracking search algorithm. J. Clean. Prod. **144**, 228–238 (2017)
3. Heidari, A.A., Pahlavani, P.: An efficient modified grey wolf optimizer with lévy flight for optimization tasks. Appl. Soft Comput.. Soft Comput. **60**, 115–134 (2017)
4. Qais, M.H., Hasanien, H.M., Alghuwainem, S.: Augmented grey wolf optimizer for grid-connected PMSG-based wind energy conversion systems. Appl. Soft Comput. S1668392745 (2018)
5. Gao, L., Pan Q.-K., et al.: A shuffled multi-swarm micro-migrating birds optimizer for a multi-resource-constrained flexible job shop scheduling problem. Inf. Sci. (2016)
6. Zitzler, E., Laumanns, M., Thiele, L.: SPEA2: improving the strength pareto evolutionary algorithm for multiobjective optimization: evolutionary methods for design, optimization and control with applications to industrial problems. In: Proceedings of the EUROGEN'2001. Athens. Greece, 19–21 Sept 2001
7. Pan, Q., Gao, L., Li, X., et al.: Effective metaheuristics for scheduling a hybrid flowshop with sequence-dependent setup times. Appl. Math. Comput.Comput. **303**, 89–112 (2017)
8. Golchin, M., Liew, A.: Parallel biclustering detection using strength pareto front evolutionary algorithm. Inf. Sci. **s 415–416**, 283–297 (2017)
9. Mladenovi N, Hansen P. Variable Neighborhood Search[J]. Computers & Operations Research. 1997, 24(11): e97–1e0.
10. Zhao, F., Yang, L., Yi, Z., et al.: A hybrid harmony search algorithm with efficient job sequence scheme and variable neighborhood search for the permutation flow shop scheduling problems. Eng. Appl. Artif. Intell.Artif. Intell. **65**, 178–199 (2017)
11. Carlier, J., Neron, E.: An exact method for solving the multi-processor flow-shop. RAIRO Oper. Res. **34**(1), 1–25 (2000)
12. Zitzler, E., Deb, K., Thiele, L.: Comparison of multiobjective evolutionary algorithms: empirical results. Evol. Comput. (2000)
13. Zitzler, E., Thiele, L.: Multiobjective evolutionary algorithms: a comparative case study and the strength Pareto approach. IEEE Trans. Evol. Comput.Evol. Comput. **3**(4), 257–271 (1999)
14. Derrac, J., García, S., Molina, D., Herrera, F.: A practical tutorial on the use of nonparametric statistical tests as a methodology for comparing evolutionary and swarm intelligence algorithms. Swarm Evol. Comput.Evol. Comput. **1**(1), 3–18 (2011)
15. Deb, K., Pratap, A., Agarwal, S., et al.: A fast and elitist multiobjective genetic algorithm: NSGA-II. IEEE Trans. Evol. Comput.Evol. Comput. **6**(2), 182–197 (2002)
16. Hui, L., Zhang, Q.: Multiobjective optimization problems with complicated Pareto sets, MOEA/D and NSGA-II. IEEE Trans. Evol. Comput.Evol. Comput. **13**(2), 284–302 (2009)

Chapter 5
Green Scheduling in Job Shop Environment

5.1 Brief Introduction

Due of the greenhouse effect and global warming, green manufacturing is one of the hottest subjects in both academia and business today. Job-shop scheduling is a crucial component of production systems. This article examines the green job-shop scheduling problem with variable machining speeds (JSPVMS), which aims to reduce the makespan as well as total energy consumption (TEC). The first step in developing this green JSPVMS is to create a new mixed-integer linear programming model. A novel knowledge-based multi-objective memetic algorithm (MOMA) is then created to deal with this issue. To better achieve trade-off solutions between makespan and TEC in our MOMA, an unique decoding approach based on the problem property is properly constructed. Furthermore, by learning about the specifics of the problem, a novel local search is proposed to search for plausible non-dominated solutions. In order to balance exploration and exploitation, the suggested MOMA also makes use of the advantages of the genetic operator and local search. By comparing experiments across several MOMA variations, the efficacy of each enhancement component (decoding method and local search) in our MOMA is confirmed. On JSPVMS instances, we compare our MOMA with a number of well-known multi-objective optimization techniques, including NSGA-II, SPEA2, and MOEA/D. According to experimental findings, our MOMA performs much better than the other algorithms in the majority of the problems.

C. Lu et al., *Intelligence Optimization for Green Scheduling in Manufacturing Systems*, Engineering Applications of Computational Methods 18,
https://doi.org/10.1007/978-981-99-6987-6_5

5.2 Problem Statement and Modeling

5.2.1 Problem Statement

Formally, this multi-objective JSPVMS with makespan and TEC can be stated briefly as follows: There are n jobs ($J = \{J_1, ..., J_i, ..., J_n\}$) and m machines ($M = \{M_1, ..., M_k, ..., M_m\}$) available. Each job i contains a chain of n_i operations $\{O_{i,1}, ..., O_{i,j}, ..., O_{i,n_i}\}$. Each operation $O_{i,j}$ of the job i should be completed on one given machine in advance. Each job can be handled on only one given machine at a time, and each machine also can handle only one job at a time. The main difference between JSPVMS and traditional JSP is that the machining speeds for different operations in JSPVMS are variable (i.e., JSP with the speed-scaling framework). In other words, the actual processing time for each operation varies on the machine's processing speed. The actual processing time of operation $O_{i,j}$ on this designated machine at a speed $v_{i,j,r}$ shall be marked as $p_{i,j}/v_{i,j,r}$ if the standard processing time of operation $O_{i,j}$ on one specified machine is signified by the symbol $p_{i,j}$. Consider the possibility that choosing a machine with a higher speed level could result in faster processing but more energy use. As a result, there could be a disagreement between the makespan and the TEC. Finding some acceptable trade-off solutions between the makespan and TEC is the goal of this JSPVMS. In addition, the JSPVMS is predicated on the following in this study.

(1) Interruptions are not permitted while in run mode.
(2) Job opportunities exist at time zero.
(3) Operation setup time is disregarded
(4) Machines cannot be completely turned off until they have completed the operations that have been assigned to them.
(5) When processing a single operation, the machine's processing speed cannot be changed again.

There are some notions about JSPVMS:

(1) Parameters

n: Total number of jobs.
m: Total number of machines.
i, i': Job index, i, $i' = 1,...,n$.
k, k': Machine index, k, $k' = 1,...,m$.
n_i: Total number of operations of job i.
$O_{i,j}$: jth operation of job i on a given machine in advance, $j = 1,..., n_i$.
CO: Set of critical operations.
s: Total number of available speeds of the given machine.
V: Finite and discrete set of available speeds of the given machine.
$p_{i,j}$: Standard processing time of $O_{i,j}$.
$p_{i,j,r}$: Actual processing time of $O_{i,j}$ on the given machine at a speed $v_{i,j,r}$, i.e., $p_{i,j,r} = p_{i,j}/v_{i,j,r}$.

$p^{w}_{k,i,j,r}$: Energy consumption per unit time of a machine k at a speed $v_{i,j,r}$ during the working mode.

p^{w}_{k}: Energy consumption per unit time of a machine k during the stand-by mode.

C_k: Completion time of all operations assigned on the machine k.

$C_{i,j}$: Completion time of $O_{i,j}$.

$C_{\max}$: Makespan of a schedule.

L: Infinite large positive number.

EC_w: Total energy consumption during the working mode.

(2) Decision variables

$S_{i,j}$: Starting time of $O_{i,j}$.

$v_{i,j,r}$: rth speed in a set of machine speeds of $O_{i,j}$ on the given machine, i.e., $v_{i,j,r} \in V$.

$$x_{i,j,j'} = \begin{cases} 1, \text{ if } O_{i,j} \text{ precedes } O_{i',j'} \text{ for processing job } i \\ 0, \quad \text{otherwise.} \end{cases}$$

$$y_{i,j,i',j'} = \begin{cases} 1, \text{ if } O_{i,j} \text{ precedes } O_{i',j'} \text{ on the same machine} \\ 0, \quad \text{otherwise} \end{cases}$$

$$z_{i,j,r} = \begin{cases} 1, \text{ if } O_{i,j} \text{ is handled on the given machine with speed } v_{i,j,r} \\ 0, \quad \text{otherwise.} \end{cases}$$

5.2.2 Mathematical Modeling

An MILP model of this green JSPVMS is formulated in this section. The first objective is to optimize $C_{\max}$ which is formulated briefly as

$$\min f_1 = C_{max} = \max_{1 \le i \le n, 1 \le j \le n_i} \{C_{i,j}\} \tag{5.1}$$

The second objective is to minimize the TEC throughout a schedule during the production system, and the function of TEC is defined as [1]

$$\begin{aligned} \text{Min } f_2 = TEC &= \mathrm{EC}_w + \mathrm{EC}_s \\ &= \sum_{k=1}^{m} \sum_{i=1}^{n} \sum_{j=1}^{n_i} \sum_{v_{i,j,r} \in \mathcal{V}} z_{i,j,r} \cdot P^{w}_{k,i,j,r} \cdot p_{i,j} / v_{i,j,r} \\ &+ \sum_{k=1}^{m} P^{s}_{k} \left(C_k - \sum_{i=1}^{n} \sum_{j=1}^{n_i} \sum_{v_{i,j,r} \in \mathcal{V}} z_{i,j,r} \cdot \frac{p_{i,j}}{v_{i,j,r}} \right) \end{aligned} \tag{5.2}$$

where the energy-consumption function $\mathrm{EC_w}$ during the woing mode can be expressed by $\sum_{k=1}^{m} \sum_{i=1}^{n} \sum_{j=1}^{n_i} \sum_{v_{i,j,r} \in \mathcal{V}} z_{i,j,r} \cdot P^{w}_{k,i,j,r} \cdot p_{i,j} / v_{i,j,r}$.

Similarly, EC_s function during the stand-by mode can be formulated by $\sum_{k=1}^{m} P_k^s\left(C_k - \sum_{i=1}^{n}\sum_{j=1}^{n_i}\sum_{v_{i,j,r}\in\mathcal{V}} z_{i,j,r}\cdot\frac{p_{i,j}}{v_{i,j,r}}\right)$.

In short, an MILP model of this green JSPVMS is given as follows:

$$\text{objectives}:\begin{cases}\min: f_1 = C_{\max}\\ \min: f_2 = \text{TEC}\end{cases} \tag{5.3}$$

Subjects to:

$$\sum_{v_{i,j,r}\in V} z_{i,j,r} = 1 \forall i = 1,\cdots,n;\ j = 1,\cdots,n_i \tag{5.4}$$

$$\begin{gathered} S_{i,j} \geq S_{i,j'} + \sum_{v_{i',r}\in\mathcal{V}} z_{i,j',r}\cdot\frac{p_{i,j'}}{v_{i',j'r}} - L\cdot y_{i,j,i,j'} \\ \forall O_{i,j}, O_{i',j'}, r = 1,\ldots,s \end{gathered} \tag{5.5}$$

$$\begin{gathered} S_{i'j'} \geq S_{i,j} + \sum_{v_{i,j,r}\in\mathcal{V}} z_{i,j,r}\cdot\frac{p_{i,j}}{v_{i,j,r}} - L\cdot\left(1 - y_{i,j,i,j'}\right) \\ \forall O_{i,j}, O_{i,j'}, r = 1,\ldots,s \end{gathered} \tag{5.6}$$

$$\begin{gathered} S_{i,j} \geq S_{i,j'} + \sum_{v_{i,j',r}\in\mathcal{V}} z_{i,j',r}\cdot\frac{p_{i,j'}}{v_{i,j',r}} - L\cdot x_{x,j,j'} \\ \forall O_{i,j}, O_{i,j'}, r = 1,\ldots,s \end{gathered} \tag{5.7}$$

$$\begin{gathered} S_{i,j'} \geq S_{i,j} + \sum_{v_{i,j,r}\in\mathcal{V}} z_{i,j,r}\cdot\frac{p_{i,j}}{v_{i,j,r}} - L\cdot\left(1 - x_{i,j,j'}\right) \\ \forall O_{i,j}, O_{i,j'}, r = 1,\ldots,s \end{gathered} \tag{5.8}$$

$$x_{i,j,j'} \in \{0,1\} \forall i = 1,\ldots,n;\ j, j' = 1,\ldots,m \tag{5.9}$$

$$y_{i,j,i'j'} \in \{0,1\} \forall i, i' = 1,\ldots,n;\ j, j' = 1,\ldots,m \tag{5.10}$$

$$z_{i,j,\mathrm{r}} \in \{0,1\} \forall i = 1,\ldots,n;\ j = 1,\ldots,m; \mathrm{r} = 1,\ldots., \mathrm{s} \tag{5.11}$$

Equation (5.3) denotes that the objective function of this JSPVMS is to optimize the makespan and TEC, simultaneously. Constraint (5.4) ensures that each operation is handled on one given machine at only one speed selected from a set of available speeds. Constraints (5.5)–(5.7) define the order priority of different operations to be handled on the same machine. Constraints (5.7)–(5.8) impose the operation order priority of the same job. Constraints (5.9)–(5.11) show that these three decision variable parameters are binary values. To address this energy-efficient JSPVMS, a knowledge-based multi-objective metaheuristic algorithm should be proposed.

5.3 Proposed Algorithm

The proposed multi-objective memetic algorithm (MOMA) is a Pareto-based optimization algorithm and its framework is depicted in Algorithm 5.1. In MOMA, a fast non-dominated sort technique, crowding distance assignment, and elitist strategy are the same with NSGA-II. In this section, the main improvement procedures of MOMA are elaborated.

Algorithm 5.1: Framework of the Proposed MOMA

1: **Input**: Parameter settings such as PS, Maximum number of function evaluations (MaxNFEs), crossover probability (P) and mutation probability (Pm)
2: **Output**: Non-dominated solutions and Pareto Front
3: P0 $\leftarrow$ Initialize population (PS)
4: t $\leftarrow$ 0
5: **while** t $\leq$ MaxNFEs **do**
6: S_t $\leftarrow$ Parent_selection (P_t)
7: $S_t{\prime}$ $\leftarrow$ Crossover (S_t, P_c)
8: $S_t^{''}$ $\leftarrow$ Mutation ($S_t{\prime}$, Pm)
9: Qt $\leftarrow$ Local Search ($S_t^{''}$)
10: $C_t \leftarrow Q_t \cup P_t$
11: $\{F_1, F_2, \cdots\}$ $\leftarrow$ Fast non-dominated sort (C_t)
12: $P_{t+1} \leftarrow \emptyset$
13: $i \leftarrow 1$
14: **while** $|P_{t+1}| + |F_i| \leq PS$ **do**
15: Crowding distance assignment (F_i)
16: $Pt + 1 \leftarrow Pt + 1 \cup F$
17: $i++$
18: **end while**
19: P_{t+1} $\leftarrow$ Elitist strategy (P_{t+1})
20: Sort (F_i)
21: $t = t + 1$
22: **end while**

5.3.1 Encoding and Decoding

(1) *Encoding:* The encoding scheme was critical in improving the performance of the metaheuristic for scheduling problems. As previously stated, JSPVMS encoding consists of two parts: (1) the operation sequence on each machine (denoted as O) and (2) the machine processing speed level assigned to each operation (denoted as V). As a result, a two-part encoding scheme is well-suited to representing a solution in this article. More specifically, the first part O is made up of a chain of operations in which job i occurs n_i times. The second component V is a vector of integer numbers that represent the machine processing speed levels of the operation $O_{i,j}$. It is worth noting that the length

of each part is the same and equals $\sum_{i=1}^{n} n_i$. To demonstrate this encoding mechanism, Table 5.1 depicts a JSPVMS with four jobs and four machines, with jobs 1 and 4 having three operations and jobs 2 and 3 having two operations. To ensure the discrete nature of the processing times and to facilitate calculation, the actual processing time is set arbitrarily in this example rather than strictly by the equation $P_{i,j}/V_{i,j,r}$. Figure 5.1 depicts a JSPVMS solution encoding. The length of each solution component is ten. The first section O contains a set of job indices (operations) denoted as 1–4–2–1–3–4–3–4–2–1, which denotes a chain of operations $O_{11} \prec O_{41} \prec O_{21} \prec O_{12} \prec O_{31} \prec O_{42} \prec O_{32} \prec O_{43} \prec O_{22} \prec O_{13}$. The operation O_{11} possesses the highest priority and is handled at first; then, O_{41} has the second highest priority, and so on. The second part V records a list of machine processing speed levels of the corresponding operations shown as 1–2–3–3–1–2–2–3–1–1. In this part, the first number "1" in this section represents that the operation O_{11} is handled on the given machine at the first speed level $v_{1,1,1}$ and the second number "2" represents that the operation O_{41} is handled on the given machine at the second speed level $v_{4,1,2}$ and so on. The initial population is generated at random using this encoding scheme.

(2) *Decoding:* Decoding schedules are broadly classified into three types: non-delay decoding schedules, semi-active decoding schedules, and active decoding schedules. The active decoding schedule has been demonstrated to be optimal in terms of regular criteria such as makespan and tardiness [2]. Nonetheless, the active decoding schedule is advantageous for makespan rather than TEC.

To address the aforementioned issue, we devise a more effective decoding scheme that makes use of problem-specific knowledge. The JSPVMS's detailed properties are as follows.

Table 5.1 Example of JSPVMS with four jobs and four machines

Job	Operation $O_{i,j}$		
	Machine M/Speed V/Processing time p		
J_1	M_1	M_2	M_3
	$V = [1,2,3]$	$V = [1,2]$	$V = [1,2]$
	$P = [14,7,6]$	$P = [7,6]$	$P = [8,7]$
J_2	M_2	M_4	–
	$V = [1,2,3]$	V = [1,2]	–
	$P = [7,5,6]$	$P = [4,3]$	–
J_3	M_1	M_4	–
	$V = [1,2]$	$V = [1,2,3]$	–
	$P = [5,4]$	$P = [13,12,8]$	–
J_4	M_3	M_4	M_1
	$V = [1,2]$	$V = [1,2]$	$V = [1,2,3]$
	$P = [9,6]$	$P = [6,3]$	$P = [5,3,2]$

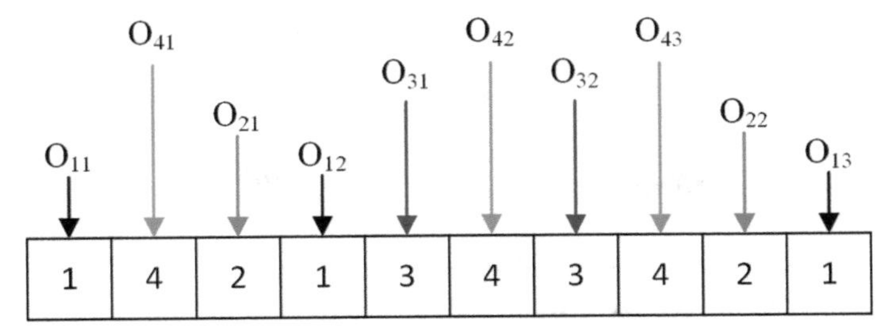

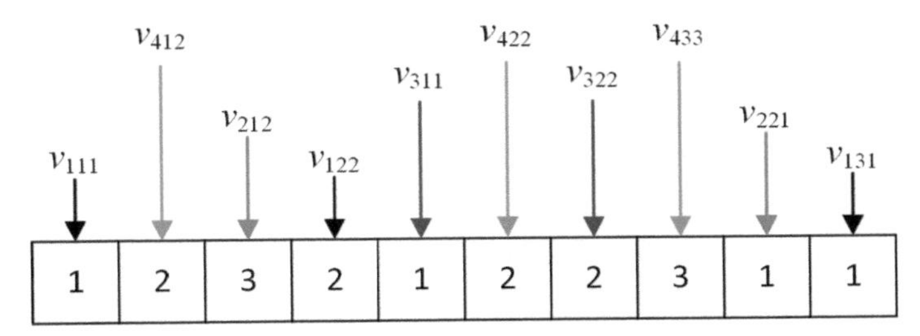

Fig. 5.1 Encoding of a solution in JSPVMS

Property 1 After decoding one solution into an active schedule, it is likely to reduce the TEC by slowing down the processing speed of operations as much as possible without affecting other operations. Finally, TEC and makespan can be reduced at the same time.

Proof

Suppose that x' is a new solution obtained by performing our proposed decoding scheme (active decoding and reduce speed without postponing any other operations) on the solution x. The detailed procedure of the proof is as follows: For the first objective $C_{\max}$, the makespan of solutions x and x' is defined as (1). It is clear that $C_{\max}(x') \leq C_{\max}(x)$. Because the solution x' is obtained by inserting each operation from right to left on each machine as compact as possible on the solution x, i.e., $C_k(x') \leq C_{\max}(x), \forall k = 1, \ldots, m$. For the second objective TEC, it is the sum of EC_w and EC_s. In the solution x, the energy consumption during the processing period $EC_w(x) = \sum_{k=1}^{m} \sum_{i=1}^{n} \sum_{j=1}^{n_i} \sum_{v_{i,j,r} \in V} z_{i,j,r} * P_{k,i,j,r}^{w} * p_{i,j}/v_{i,j,r}$ and the energy consumption during the stand-by period $EC_s(x) = \sum_{k=1}^{m} p_k^s * (C_k(x) - \sum_{i=1}^{n} \sum_{j=1}^{n_i} \sum_{v_{i,j,r} \in V} z_{i,j,r} * P_{k,i,j,r}^{w} * p_{i,j}/v_{i,j,r})$. Similarly, the solution x' is plugged into (5.2). Because a lower processing speed leads to less energy consumption during the work period, we can get $EC_w(x') < EC_w(x)$ based on $V_{i,j,r'} < V_{i,j,r}$. Meanwhile, because the EC_s is proportional to the size of the stand-by interval, increasing the processing time can fill the gap. By reducing machine processing, the stand-by interval time speed without interfering with other operations (if it can). Consequently, we can get $\mathrm{EC}_s(x') < \mathrm{EC}_s(x)$. In other words, $\mathrm{TEC}(x') < \mathrm{TEC}(x)$. According to the previous deduction, the first objective $C_{\max}(x') \leq C_{\max}(x)$ and the second objective $\mathrm{TEC}(x') < \mathrm{TEC}(x)$. Thus, we can draw a conclusion that x' dominates x (denoted as $x' \prec x$).

The procedure of the proposed decoding scheme for JSPVMS is described as follows.

Step 1: One solution can be decoded into an active schedule.

Step 2: Each operation can be decelerated until its completing time affects its job or machine successor. That is, the processing speed of the operation O_{ij} can be further reduced, if it meets the following condition:

$$\begin{cases} S_{i,j} + p_{i,j,r}/v_{i,j,r-1} \leq \min\{S_{\mathrm{SM}(i,j)}, S_{SJ(i,j)}\} \\ \text{if } r \geq 2 \text{ and } (\mathrm{SM}(i,\ j)! = 0 \text{ or } SJ(i,\ j)! = 0) \\ S_{i,j} + p_{i,j,r}/v_{i,j,r-1} \leq C_{\max} \\ \text{if } r \geq 2 \text{ and } \mathrm{SM}(i,\ j) = \mathrm{SJ}(i,\ j) = 0 \end{cases} \tag{5.12}$$

where SM(i, j) is the immediate machine successor of O_{ij}, SJ(i, j) is the immediate job successor of O_{ij}, $S_{\mathrm{SM}}(i, j)$ and $S_{\mathrm{SJ}}(i, j)$ represent the starting time of SM(i, j) and SJ(i, j), respectively. SM(i, j) = SJ(i, j) = 0 denotes that O_{ij} is its last operation on the given machine and is also the last operation of job i (i.e., $j = n_i$). If operation O_{ij} can be decelerated without affecting its job or machine successor, the completing time of O_{ij} is as close as possible to $\min\{S_{\mathrm{SM}}(i, j), S_{\mathrm{SJ}}(i, j)\}$ or $C_{\max}$.

Figure 5.2 gives a simple decoding illustration of a solution that has been appeared in Fig. 5.1. Suppose that the working power of machine k at a speed $v_{i,j,r}$ is $P^{w}_{k,i,j,r} = 2 * V^2_{i,j,r}$ and the stand-by power of machine k is $P^k_s = 1$. Figure 5.2a shows a semi-active decoding schedule, where the makespan is 35 and TEC is 443. Meanwhile, Fig. 5.2b presents our proposed decoding scheme, where the makespan is 31 and TEC is 335. In this example, we can observe that the proposed decoding scheme can find a better solution than the other decoding rule in terms of the makespan and TEC criterion. Additionally, Algorithm 5.2 also provides the pseudocode of the proposed decoding scheme.

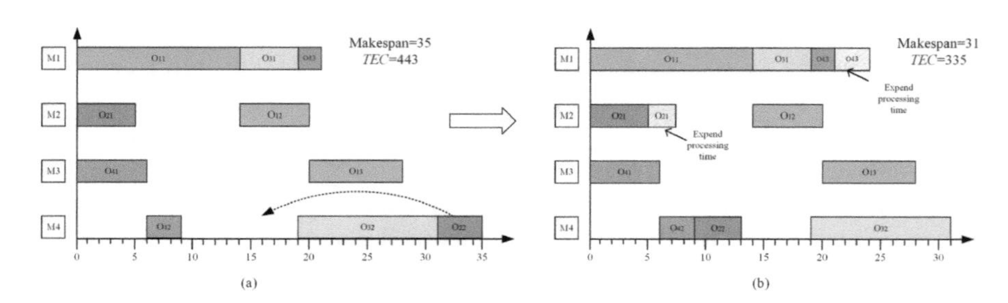

Fig. 5.2 New decoding scheme of a solution in JSPVMS

Algorithm 5.2: Proposed Decoding Scheme

1: **Input**: the current solution or schedule $\boldsymbol{x} = [O, V]$, *SM* (i, j) is the immediate machine successor of $O_{i,j}$ and *SJ* (i, j) is the immediate job successor of $O_{i,j}$
2: **Output**: the improved solution or schedule $\boldsymbol{x}' = [O', V']$
3: An active decoding schedule is performed on the solution $\boldsymbol{x}$
4:: **for** $i = 1$ to n **do**
5:: /*n is the total number of jobs*/
6:: **for** $j = 1$ to n_i **do**
7:: /*n_i is the total number of operations in job i*/
8:: **if** *SM* $(i, j) == 0$ and *SJ* $(i, j) == 0$ **then**
9: **while** $r > 1$ and $S_{i,j} + p_{i,j}/v_{i,j,r-1} < C_{\max}(\boldsymbol{x})$ **do**
10:: $vi,j,r \leftarrow vi,j,r - 1$
11:: **end while**
12: **else if** *SM* $(i, j)! = 0$ and *SJ* $(i, j) == 0$ **then**
13: **while** $r > 1$ and $S_{i,j} + p_{i,j}/v_{i,j,r-1} < S_{SM(i,j)}$ **do**
14: /*$S_{SM(i,j)}$ is the starting time of *SM* (i, j)*/
15: $vi,j,r \leftarrow vi,j,r - 1$
16: **end while**
17: **else if** *SM* $(i, j) == 0$ and *SJ* $(i, j)! = 0$ **then**
18: **while** $r > 1$ and $S_{i,j} + p_{i,j}/v_{i,j,r-1} < S_{SJ(i,j)}$ **do**
19: /*$S_{SJ(i,j)}$ is the starting time of *SJ* (i, j)*/
20:: $vi,j,r \leftarrow vi,j,r - 1$
21:: **end while**
22: **else**
23:: $Smin = \min\{S_{SM}(i, j), S_{SJ}(i,j)\}$
24: **while** $r > 1$ and $S_{i,j} + p_{i,j}/v_{i,j,r-1} < S_{\min}$ **do**
25: /*$S_{SJ(i,j)}$ is the starting time of *SJ* (i, j)*/
26: $vi,j,r \leftarrow vi,j,r - 1$
27: **end while**
28: **end if**
29: $Ci,j = Si,j + pi,j/vi,j,r - 1$
30: **end for**
31: **end for**

5.3.2 Crossover and Mutation

The crossover's goal is to use parental information to explore more promising unknown solution space. Two-point crossover has a higher search capacity than one-point crossover and has been successfully applied to shop scheduling problems. Thus, a two-point crossover operator is used in our proposed MOMA, and the specific steps are as follows.

Step 1: Choose two solutions from a binary tournament selection as two parents labeled Parent 1 and Parent 2, respectively.

Step 2: Choose two different crossover points of operation (i.e., O) parts at random. Copy the genes from Parent 1 between two points into the corresponding area of a new solution marked with Child 2. Similarly, copy all genes in Parent 2 between two points into the corresponding area of a new solution Child 1.

Step 3: To ensure the legitimacy of the newly generated solution, scan each gene of Parent 2 from left to right and move genes from Parent 2 into Child 1 until job I occurs n_i times (n_i denotes the total number of operations of job i). Obtain a new speed sequence for the speed (i.e., V) part based on the mapping relation of newly generated operations.

The mutation operator only makes changes to one solution. The mutation operation is divided into two parts, which will be discussed further below.

Step 4: For the first part (i.e., O) of the chosen solution, choose two mutation points at random and swap the corresponding operations on the two points.

Step 5: Select a gene at random for the second part (i.e., V) of the selected solution and adjust the value in its range based on the mapping relation of newly generated operations.

5.3.3 Local Search Based on Problem Property

Local search is a useful heuristic for resolving scheduling issues [2–10]. However, because JSPVMS has two objectives that conflict with each other, improving the first objective value will degrade the second objective value. In this case, we should use problem-specific knowledge to create a strong local search heuristic for exploiting more promising trade-off solutions between two objectives. We give the two properties of this problem to extract useful problem-specific knowledge.

Property 2 The makespan criterion can only be reduced by moving critical operations.

Proof

Assume that x is a schedule or solution x's neighbor. Let P represent a critical path in a schedule x that consists of a set of critical operations. $\boldsymbol{CO} = \{O_1^c \rightarrow \cdots \rightarrow O_l^c\}$. According to the reverse proof, if $O_{i,j}$/CO, moving the $O_{i,j}$ will have three conditions:

(1) move to the location between two adjacent operations on one given machine;
(2) move to the first location of all operations on one given machine;
(3) move to the rear location of all operations on one given machine.

In the first condition, suppose that $O_{i,j}$ is shifted to the location between O_x and O_y on one given machine. It is clear that if there is not g $\in \{1, 2, \ldots, l-1\}$, $O_g^c = O_x$ and $O_{g+1}^c = O_y$, then P would not be affected and still exits in x'. Therefore, $C_{\max}$(x) $\leq C_{\max}$(x'). Otherwise $O_c^1 \rightarrow O_g^c \rightarrow O_{i,j} \rightarrow O_{g+1}^c \rightarrow O_l^c$ would exit in x' and $C_{\max}(x) \leq C_{\max}(x')$. Thus, we conclude that $C_{\max}(x) \leq C_{\max}(x')$. Similarly, this proof is the same for the other two conditions. In other words, only some critical operations are moved or adjusted to reduce the makespan.

Property 3 Makespan and TEC criteria may be simultaneously reduced by accelerating critical operation and decelerating noncritical operation.

Proof

Suppose that x' is a neighbor of a schedule or solution x, and there is only one critical path for solution x. Let OT_g^c be the processing time of one critical operation $O_g^c = CO$ in this schedule x. The makespan of this schedule is $C_{\max}(x) = \sum_{g=1}^{l} OT_g^c$. If O_g^c is accelerated, its processing time will become shorter (denoted as $OT_g^{c'}$). It is obvious that $OT_g^{c'} < OT_g^c$ and the critical path would change, i.e., $C_{\max}(x') \leq C_{\max}(x)$. However, this will lead to the additional energy consumption ΔE_1. We can decelerate one noncritical operation $O_{i,j} \notin CO$ to save energy consumption ΔE_2, which compensates the increase in energy consumption. More precisely, if $\Delta E_2 \geq \Delta E_1$, then TEC($x'$) < TEC($x$). In this case, we can conclude $C_{\max}(x') \leq C_{\max}(x)$ and TEC(x') < TEC(x). That is to say, x' will dominate x if one critical operation is accelerated to make room for the slowdown of profitable noncritical operations.

Unlike previous literature [11], the proposed local search in this article consists of two procedures: (1) move the critical operations of blocks; and (2) accelerate the critical operation while decelerating the noncritical operation.

This local search is depicted in Figs. 5.3 and 5.4. Assume that the working power of machine k at speed $v_{i,j,r}$ is $P_{k,i,j,r}^{w} = 2 * V_{i,j,r}^{2}$ and the stand-by power is $P_s^k = 1$. This schedule's critical path is shown in Fig. 5.3a as $\{O_{3,1} \rightarrow O_{1,1} \rightarrow O_{1,2} \rightarrow O_{1,3}\}$; its length is 33; and its TEC is 339. We perform the swap move near the borderline of each critical block and then use the proposed decoding scheme in Fig. 5.3 to generate a new schedule (b). In Fig. 5.3b, this critical path is denoted as $\{O_{1,1} \rightarrow O_{3,1} \rightarrow O_{3,2}\}$, with a length of 31 and a TEC of 335. This solution is improved by swapping critical block operations. Figure 5.4 shows a Gantt chart comparison of a solution before and after the speed adjustment. The makespan of this solution in Fig. 5.4a cannot be reduced further by swapping critical operations of blocks, where the makespan is 31. In this case, we can reduce makespan by increasing the machine processing speed of one critical operation (e.g., $O_{3,1}$), where the makespan can be reduced to 30. It will, however, result in an increase in TEC. To reduce TEC and achieve a better solution, we will slow down one or more noncritical operations (e.g., $O_{4,3}$). To further complicate matters, we assume that the critical operation $O_{3,1}$'s processing speed is increased from 1 to 2, while the non-critical operation $O_{4,3}$'s processing speed is decreased from 3 to 1. The extra energy consumed by accelerating operation $O_{3,1}$ is 22, while the extra energy consumed by slowing down operation $O_{4,3}$ is 27. The total amount saved in terms of energy consumption is 5. As a result, Fig. 5.4 shows how to achieve the dual benefits of improving makespan and lowering TEC(b). Algorithm 5.3 also displays the pseudocode for this local search.

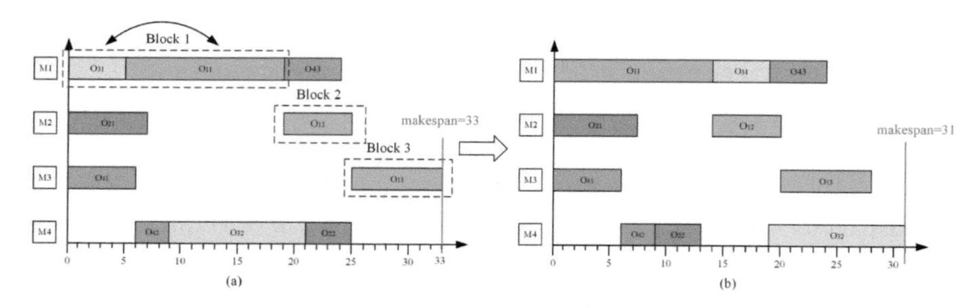

Fig. 5.3 Move the critical operations of blocks

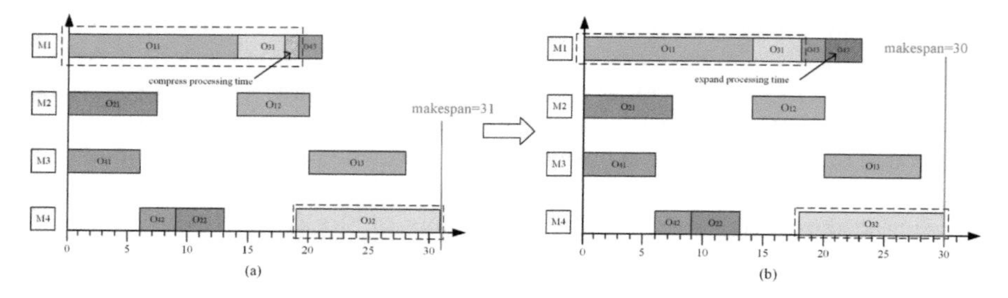

Fig. 5.4 Accelerate the critical operation and decelerate the non-critical operation

Algorithm 5.3: Proposed Local Search

1: **Input**: the current solution or schedule $\boldsymbol{x} = [\text{O}, V]$
2: **Output**: the improved solution or schedule $\boldsymbol{x}' = [\text{O}', V']$
3: $\boldsymbol{CO} = \{O_1^c, \ldots, O_g^c, \ldots, O_l^c\}$of in $\boldsymbol{x}$, where O_g^c is the gth critical operation, and critical blocks, i.e.,
$\boldsymbol{block} = \{block_1, \ldots, block_h, \ldots, block_H\}$, where H is the total number of critical blocks, $block_h$ denotes the hth block
4: flag $\leftarrow 0$
5: **for** $h = 1$ to H **do**
6: generate a new solution $\boldsymbol{x}^{\text{J}}$ by swapping the two adjacent critical operations in the $block_h$
7: **if** $\boldsymbol{x}^{\text{J}} \prec \boldsymbol{x}$ **then**
8: $\boldsymbol{x}^{\text{J}} \leftarrow \boldsymbol{x}$
9: flag $\leftarrow 1$
10: break
11: **end if**
12: **end for**

(continued)

(continued)

Algorithm 5.3: Proposed Local Search

13: **if** *flag* $= =$ 0 **then**
14: **for** each critical operation $O_g^c \in \boldsymbol{CO}$ **do**
15: $vg,r \leftarrow vg,r + 1$ and compute the increased energy consumption ΔE_1
16: **for** $i = 1$ to n **do**
17: **for** $j = 1$ to n_i **do**
18: **if** $O_{i,j} \in / \boldsymbol{CO}$ **then**
19: $vi,j,r \leftarrow vi,j,r - 1$ and compute the increased energy consumption ΔE_2
20: **end if**
21: **if** $\Delta E_1 \leq \Delta E_2$ **then**
22: update the operations followed by O_g^c and $O_{i,j}$
23: *break;*
24: **end if**
25: **end for**
26: **end for**
27: **end for**
28: **end if**

5.4 Experiments

5.4.1 Test Instances

There is no established benchmark for JSPVMS because the JSPVMS problem is an extensive version of JSP. We converted some traditional JSP benchmarks into JSPVMS in this article.by adding variable processing speeds to benchmarks. It should be noted that the processing times of the initial operations in JSP benchmarks can be compared to the processing times of the operations in JSPVMS instances with the lowest speed level. The machine processing speed level $v_{i,j,r}$ for operation $O_{i,j}$ has four available speed levels, where $i \in \{1, 2,\ldots,n\}, j \in \{1, 2,\ldots,n_i\}$. Thus, the actual processing time of each operation is $p_{i,j,r} = [p_{i,j}/v_{i,j,r}]$, $v_{i,j,r} = \{1, 2, 3, 4\}$ where the symbol "[]" represents that we can take the integer part from a real number by using a round function. Suppose that the working power of machine k at a speed $v_{i,j,r}$ is $P^w_{k,i,j,r} = 2 * V^2_{i,j,r}$, and the stand-by power of machine k is $P^k_s = 1$. Each instance is named as "mlaX." For example, "mlae" represents the modified classic benchmark "lae" by adding different machine processing speed levels.

5.4.2 Parameter Settings

Metaheuristics typically respond differently depending on how their parameter configurations are set up. In order to determine the ideal parameter configuration for MOMA, the Taguchi method is used. Population size (P_S), crossover (P_c), and mutation probability are the three primary MOMA parameters (P_m). Each parameter's level is as follows: $P_S = \{20, 50, 80, 100\}$, $P_c = \{0.6, 0.7, 0.8, 0.9\}$, $P_m = \{0.1, 0.2, 0.3, 0.4\}$.To assess the effects of various parameter configurations on all instances, we use a comprehensive performance metric called IGD. Twenty independent runs of this experiment are conducted for each parameter setting. The main effect plot of three parameters is displayed in Fig. 5.5. This figure shows that the ideal MOMA parameter configuration is $P_S = 100$, $P_c = 0.8$, and $P_m = 0.4$. We also follow the same steps to obtain the optimal parameter configurations for the compared competitors. However, due to space limitation, we only provide their final parameter configurations as follows.

(1) The crossover and swap mutation: Two-point crossover and swap mutation operators are applied in all considered algorithms. In this article, the values of crossover and mutation probability are 0.9 and 0.4 for NSGA-II and SPEA2, 0.8 and 0.4 for MOEA/D.
(2) The external archive and population size (or other parameters): The external archive size of SPEA2 is set to 60. The population size is 80 for MOEA/D and e0 for other MOEAs. The neighborhood size is 20 for MOEA/D.
(3) The stop criterion: The maximum number of function evaluations (MaxNFEs) is considered as the stop criterion. The number of MaxNFEs depends on the

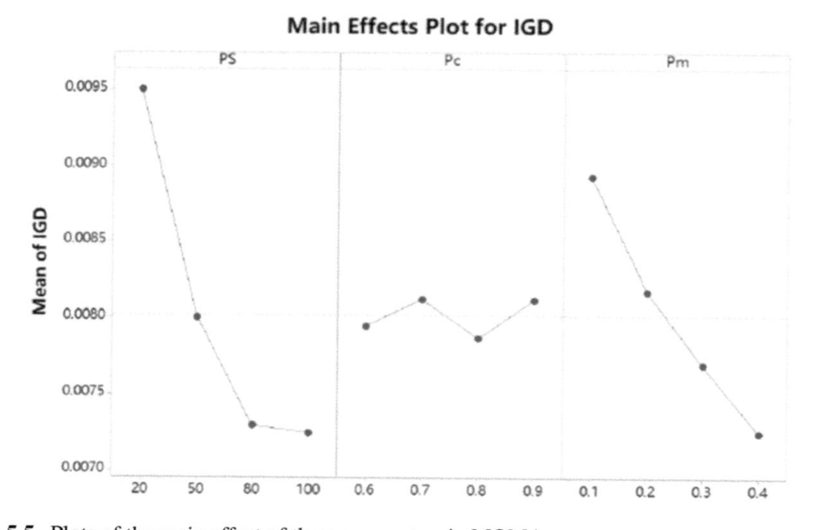

Fig. 5.5 Plots of the main effect of three parameters in MOMA

problem scale. Thus, MaxNFEs is set to 40 000 for mla01-mla20, 60 000 for mla21-mla40.

(4) Run times: Each algorithm is executed 20 times independently for each instance. A Wilcoxon signed-rank test is adopted to check whether there are significant differences at a 0.05 significance level between the algorithms. The symbol " + " ("−") represents that our MOMA is significantly better (worse) than the second-best (best) rival, whereas " = " indicates no significant differences exist between them.

5.4.3 *Performance Metrics*

To evaluate the behavior of these different strategies or algorithms for solving JSPVMS instances, three performance metrics such as Spread (Δ) [12], generational distance (GD) [13], and inverted generational distance (IGD) [14] are utilized. Δ metric is used to test the diversity of Pareto approximations, and GD can assess the convergence behavior of a certain algorithm. IGD reflects a comprehensive performance regarding diversity and convergence.

5.4.4 *Effectiveness of the Proposed Decoding Scheme*

To prove the validity of the proposed decoding scheme, we executed a comparison experiment between the MOMA with and without the proposed decoding scheme. MOMA represents the MOMA with all the proposed components, and MOMA-ND denotes the MOMA with a traditional active decoding scheme. The parameter settings for MOMA and MOMA-ND are the same as the Sect. 5.4.2. Experimental statistical results (average and standard deviation values of three metrics) are provided in Table 5.2, and the optimal statistical results are marked in bold font. It should be pointed out that the proposed decoding scheme is also an energy-saving strategy for JSPVMS since it can slow down the processing speed to reduce the energy consumption.

It is observed from Table 5.2 that MOMA outperforms MOMAND on most instances concerning three metrics, especially for GD metric. Meanwhile, MOMA has more " + " counts than MOMA-ND on most instances, which denotes that MOMA is superior to MOMA-ND by a significant margin. These further suggest that MOMA has an overwhelming advantage compared to MOMA-ND for solving these instances. The reason behind this is that when a solution is decoded into an active schedule, operation sequences on machines and makespan are fixed. Then one minimization energy consumption problem will be addressed if makespan is fixed. At this time, the TEC can be reduced by slowing down the processing speeds of some operations without postponing any other operations. A better trade-off solution between makespan and TEC may be generated by using our proposed decoding

Table 5.2 Statistical values on all the metrics between MOMA and MOMA-ND

Instances	Spread(mean/std)		GD(mean/std)		IGD (mean/std)	
	MOMA	MOMA-ND	MOMA	MOMA-ND	MOMA	MOMA-ND
mla01	6.83e−1/ 7.20e−2 +	7.31e−1/ 1.00e−1	5.60e−3/ 1.40e−3 +	9.29e−3/ 2.00e−3	6.64e−3/ 1.60e−3 +	8.79e−3/ 1.20e−3
mla02	7.20e−1/ 7.90e−2 +	8.26e−1/ 1.20e−1	4.98e−3/ 1.90e−3 +	1.10e−2/ 1.70e−3	7.00e−3/ 2.20e−3 +	8.31e−3/ 1.10e−3
mla03	7.32e−1/ 7.80e−2 +	8.12e−1/ 1.20e−1	5.79e−3/ 1.20e−3 +	9.04e−3/ 1.50e−3	9.42e−3/ 2.90e−3 +	1.13e−2/ 2.80e−3
mla04	7.21e−1/ 4.70e−2 +	7.68e−1/ 1.10e−1	5.72e−3/ 1.50e−3 +	9.26e−3/ 2.10e−3	7.36e−3/ 1.80e−3 +	8.93e−3/ 1.80e−3
mla05	6.68e−1/ 8.80e−2 +	6.99e−1/ 1.20e−1	4.84e−3/ 1.30e−3 +	7.52e−3/ 1.30e−3	6.81e−3/ 2.10e−3 +	7.61e−3/ 1.50e−3
mla06	6.76e−1/ 6.10e−2 =	6.88e−1/ 8.00e−2	4.97e−3/ 1.30e−3 +	7.17e−3/ 1.50e−3	6.94e−3/ 1.90e−3 +	8.77e−3/ 2.00e−3
mla07	6.81e−1/ 6.80e−2 +	7.22e−1/ 7.30e−2	5.59e−3/ 1.70e−3 +	1.07e−2/ 2.00e−3	7.56e−3/ 1.90e−3 +	9.49e−3/ 1.90e−3
mla08	6.88e−1/ 5.80e−2-	6.67e−1/ 6.90e−2	5.01e−3/ 1.40e−3 +	7.55e−3/ 1.60e−3	7.57e−3/ 1.90e−3 +	8.64e−3/ 1.70e−3
mla09	6.00e−1/ 7.00e−2 +	6.54e−1/ 1.10e−1	4.91e−3/ 1.70e−3 +	7.24e−3/ 2.00e−3	5.70e−3/ 1.10e−3 +	7.88e−3/ 1.90e−3
mla10	6.29e−1/ 5.20e−2-	6.67e−1/ 7.90e−2	5.28e−3/ 1.40e−3 +	8.69e−3/ 1.90e−3	5.71e−3/ 1.70e−3 +	8.78e−3/ 2.10e−3
mla11	5.80e−1/ 7.30e−2-	5.39e−1/ 8.20e−2	5.06e−3/ 1.50e−3 +	8.58e−3/ 2.00e−3	5.19e−3/ 1.70e−3 +	7.07e−3/ 1.50e−3
mla12	5.68e−1/ 6.10e−2 =	5.44e−1/ 4.90e−2	4.42e−3/ 1.30e−3 +	7.37e−3/ 1.80e−3	5.38e−3/ 1.80e−3 +	6.68e−3/ 9.10e−4
mla13	6.17e−1/ 8.20e−2-	5.89e−1/ 8.40e−2	4.84e−3/ 1.30e−3 +	7.14e−3/ 2.00e−3	7.18e−3/ 1.80e−3 +	7.66e−3/ 1.20e−3
mla14	6.37e−1/ 6.30e−2-	5.66e−1/ 8.40e−2	4.63e−3/ 9.60e−4 +	7.48e−3/ 1.70e−3	6.67e−3/ 1.40e−3 +	7.49e−3/ 1.80e−3
mla15	6.59e−1/ 6.60e−2 =	6.35e−1/ 8.40e−2	5.73e−3/ 1.20e−3 +	9.33e−3/ 1.80e−3	6.36e−3/ 1.40e−3 +	8.27e−3/ 1.10e−3
mla16	9.11e−1/ 7.90e−2 =	9.30e−1/ 1.20e−1	9.19e−3/ 2.20e−3 +	1.52e−2/ 3.20e−3	1.25e−2/ 3.00e−3 +	1.51e−2/ 2.60e−3
mla17	8.90e−1/ 8.90e−2 =	9.13e−1/ 1.10e−1	8.25e−3/ 2.60e−3 +	9.57e−3/ 3.00e−3	1.21e−2/ 2.70e−3 =	1.30e−2/ 2.10e−2
mla18	8.78e−1/ 9.60e−2 +	9.66e−1/ 1.00e−1	9.85e−3/ 2.90e−3 +	1.42e−2/ 3.60e−3	1.22e−2/ 3.50e−3 +	1.46e−2/ 2.90e−3
mla19	9.18e−1/ 9.00e−2 +	9.63e−1/ 9.50e−2	1.01e−2/ 3.00e−3 +	1.68e−2/ 3.50e−3	1.05e−2/ 1.90e−3 +	1.34e−2/ 2.00e−3
mla20	8.85e−1/ 8.30e−2-	8.55e−1/ 1.10e−1	9.53e−3/ 3.20e−3 +	1.47e−2/ 3.00e−3	1.30e−2/ 3.40e−3 =	1.37e−2/ 2.40e−3

(continued)

Table 5.2 (continued)

Instances	Spread(mean/std)		GD(mean/std)		IGD (mean/std)	
	MOMA	MOMA-ND	MOMA	MOMA-ND	MOMA	MOMA-ND
mla21	1.21e + 0/ 7.50e−2 =	1.21e + 0/ 7.80e−2	7.75e−3/ 1.70e−3 +	9.51e−3/ 2.00e−3	8.61e−3/ 1.20e−3 +	9.70e−3/ 1.10e−3
mla22	1.14e + 0/ 6.90e−2 =	1.21e + 0/ 6.60e−2	8.99e−3/ 1.80e−3 +	9.16e−3/ 2.80e−3	8.35e−3/ 1.00e−3 +	9.30e−3/ 1.60e−3
mla23	1.14e + 0/ 6.90e−2 =	1.20e + 0/ 6.60e−2	5.80e−3/ 1.40e−3 +	7.49e−3/ 1.80e−3	5.53e−3/ 1.00e−3 +	6.90e−3/ 1.30e−3
mla24	1.17e + 0/ 7.70e−2 =	1.17e + 0/ 5.80e−2	5.92e−3/ 1.80e−3 +	6.95e−3/ 1.70e−3	7.68e−3/ 1.60e−3 =	7.69e−3/ 1.10e−3
mla25	1.16e + 0/ 7.40e−2 =	1.20e + 0/ 7.30e−2	6.48e−3/ 1.90e−3 +	8.40e−3/ 2.20e−3	6.26e−3/ 1.40e−3 +	7.29e−3/ 1.70e−3
mla26	1.16e + 0/ 8.70e−2 =	1.17e + 0/ 1.10e−1	7.97e−3/ 2.20e−3 +	8.20e−3/ 2.60e−3	7.49e−3/ 1.20e−3 +	8.05e−3/ 2.10e−3
mla27	1.15e + 0/ 5.10e−2 =	1.14e + 0/ 6.90e−2	5.98e−3/ 1.90e−3 +	6.14e−3/ 2.00e−3	7.27e−3/ 1.20e−3 =	7.34e−3/ 1.00e−3
mla28	1.17e + 0/ 8.90e−2 =	1.17e + 0/ 8.10e−2	7.21e−3/ 2.50e−3 =	7.39e−3/ 2.00e−3	8.22e−3/ 1.60e−3-	7.70e−3/ 2.10e−3
mla29	1.13e + 0/ 7.10e−2 =	1.15e + 0/ 7.50e−2	4.70e−3/ 1.70e−3 +	5.90e−3/ 1.10e−3	6.52e−3/ 1.70e−3-	6.17e−3/ 1.40e−3
mla30	1.14e + 0/ 5.60e−2 =	1.16e + 0/ 7.10e−2	6.35e−3/ 2.20e−3 +	6.84e−3/ 2.20e−3	7.08e−3/ 1.90e−3-	6.72e−3/ 1.60e−3
mla31	1.05e + 0/ 9.00e−2 =	1.08e + 0/ 1.00e−1	6.26e−3/ 2.00e−3-	5.75e−3/ 1.90e−3	5.34e−3/ 1.00e−3 =	5.12e−3/ 1.00e−3
mla32	1.05e + 0/ 8.60e−2 =	1.08e + 0/ 1.00e−1	6.22e−3/ 1.70e−3 =	6.39e−3/ 2.10e−3	6.27e−3/ 1.40e−3 =	6.14e−3/ 1.20e−3
mla33	1.01e + 0/ 8.10e−2 =	1.03e + 0/ 5.60e−2	5.91e−3/ 1.40e−3 =	5.96e−3/ 2.00e−3	6.19e−3/ 1.10e−3 +	6.43e−3/ 1.00e−3
mla34	9.98e−1/ 7.50e−2 =	1.02e + 0/ 7.40e−2	4.91e−3/ 1.90e−3 +	6.48e−3/ 1.90e−3	5.97e−3/ 9.80e−4 +	6.62e−3/ 9.40e−4
mla35	1.05e + 0/ 8.00e−2 +	1.12e + 0/ 1.10e−1	7.47e−3/ 3.00e−3 +	1.09e−2/ 2.80e−3	6.50e−3/ 1.90e−3 +	8.00e−3/ 1.60e−3
mla36	1.26e + 0/ 7.20e−2 =	1.29e + 0/ 6.00e−2	7.32e−3/ 1.80e−3 +	8.33e−3/ 1.50e−3	8.03e−3/ 1.80e−3 +	8.95e−3/ 2.00e−3
mla37	1.30e + 0/ 6.20e−2 =	1.30e + 0/ 7.40e−2	1.06e−2/ 3.20e−3 =	1.10e−2/ 3.00e−3	8.85e−3/ 2.20e−3 =	8.86e−3/ 1.60e−3
mla38	1.23e + 0/ 6.20e−2 =	1.25e + 0/ 5.90e−2	6.97e−3/ 1.90e−3 +	7.27e−3/ 2.10e−3	8.65e−3/ 2.20e−3 +	9.04e−3/ 1.90e−3
mla39	1.25e + 0/ 7.60e−2 =	1.29e + 0/ 6.10e−2	8.83e−3/ 3.00e−3 +	1.13e−2/ 3.00e−3	8.71e−3/ 1.20e−3 +	9.04e−3/ 1.40e−3

(continued)

Table 5.2 (continued)

Instances	Spread(mean/std)		GD(mean/std)		IGD (mean/std)	
	MOMA	MOMA-ND	MOMA	MOMA-ND	MOMA	MOMA-ND
mla40	1.25e + 0/ 4.50e−2 =	1.29e + 0/ 5.60e−2	8.52e−3/ 2.50e−3 =	8.71e−3/ 2.50e−3	7.68e−3/ 1.30e−3 =	7.77e−3/ 1.60e−3
'+' counts	10	–	34	–	29	–
'=' counts	24	–	5	–	8	–
'−' counts	6	–	1	–	3	–

scheme. Thus, the proposed decoding scheme has positive impacts on the algorithm and can further improve the algorithm's convergence performance.

5.4.5 Effectiveness of the Proposed Local Search

To verify the effectiveness of the proposed local search strategy of MOMA, we conduct a comparison experiment between the MOMA with and without the proposed local search mechanism (denotes as MOMA and MOMA-NL). Table 5.3 reports the statistical values of three performance metrics based on 20 independent runs on each instance for each algorithm.

Numerical results in Table 5.3 demonstrate that MOMA is significantly better than MOMA-NL on all instances concerning the GD metric. This suggests that the proposed local search heuristic can significantly boost the convergence performance of the optimization algorithm.

Concerning the Spread metric, MOMA significantly outperforms MOMA-NL on 19 out of the total 40 instances. On the contrary, MOMA-NL remarkably outperforms MOMA on only one instance (mla30). Neither of the two algorithms has an overwhelming advantage on the rest instances regarding the Spread metric. In other words, MOMA shows a slightly better diversity performance compared to MOMA-NL. This implies that the proposed local search heuristic lays emphasis on exploiting the more promising solutions. Thus, MOMA can integrate the merits of a genetic search for diversification and local search for intensification. On a whole, MOMA completely overwhelms MOMA-NL since MOMA is significantly superior to MOMA-NL on most instances concerning the comprehensive metric IGD.

Accordingly, the proposed local search rule can enhance the overall behavior of MOMA, especially for the GD metric.

Table 5.3 Statistical values on all the metrics between MOMA and MOMA-NL

Instances	Spread (mean/std)		GD (mean/std)		IGD (mean/std)	
	MOMA	MOMA-NL	MOMA	MOMA-NL	MOMA	MOMA-NL
mla01	9.40e−1/ 9.7e−2 +	1.12e + 0/ 8.0e−2	3.33e−3/ 1.0e−3 +	5.90e−3/ 1.1e−3	4.83e−3/ 1.6e−3 +	6.21e−3/ 1.1e−3
mla02	1.10e + 0/ 9.1e−2 =	1.03e + 0/ 1.4e−1	3.72e−3/ 7.4e−4 +	7.06e−3/ 1.6e−3	6.53e−3/ 2.0e−3 =	6.61e−3/ 1.2e−3
mla03	9.93e−1/ 9.0e−2 +	1.09e + 0/ 1.0e−1	3.64e−3/ 7.2e−4 +	5.76e−3/ 1.3e−3	5.56e−3/ 2.2e−3 =	6.40e−3/ 1.8e−3
mla04	9.24e−1/ 6.6e−2 +	1.07e + 0/ 1.1e−1	2.97e−3/ 8.7e−4 +	5.17e−3/ 9.9e−4	4.25e−3/ 6.7e−4 +	5.01e−3/ 9.0e−4
mla05	9.06e−1/ 8.5e−2 +	9.89e−1/ 5.8e−2	2.55e−3/ 7.4e−4 +	4.17e−3/ 9.1e−4	5.33e−3/ 1.8e−3 =	5.41e−3/ 1.2e−3
mla06	8.75e−1/ 6.1e−2 =	9.06e−1/ 8.9e−2	2.87e−3/ 8.1e−4 +	4.79e−3/ 9.8e−4	4.91e−3/ 1.2e−3 =	4.69e−3/ 8.8e−4
mla07	1.00e + 0/ 7.5e−2 =	1.02e + 0/ 7.5e−2	4.50e−3/ 1.2e−3 +	7.91e−3/ 1.3e−3	5.59e−3/ 1.4e−3 =	6.11e−3/ 8.8e−4
mla08	8.42e−1/ 1.1e−1 +	9.21e−1/ 9.4e−2	3.49e−3/ 1.2e−3 +	7.20e−3/ 1.2e−3	3.88e−3/ 1.0e−3 +	5.72e−3/ 9.1e−4
mla09	8.07e−1/ 7.0e−2 +	8.75e−1/ 8.2e−2	2.58e−3/ 6.9e−4 +	4.90e−3/ 1.1e−3	4.14e−3/ 1.3e−3 +	4.69e−3/ 1.0e−3
mla10	8.32e−1/ 4.9e−2 =	8.82e−1/ 9.3e−2	2.42e−3/ 7.3e−4 +	4.65e−3/ 9.3e−4	3.82e−3/ 7.6e−4 +	4.54e−3/ 7.0e−4
mla11	7.71e−1/ 7.1e−2 =	7.88e−1/ 6.8e−2	2.66e−3/ 6.7e−4 +	4.12e−3/ 9.2e−4	4.37e−3/ 1.5e−3 =	5.00e−3/ 1.4e−3
mla12	7.56e−1/ 5.2e−2 =	7.82e−1/ 9.0e−2	2.34e−3/ 5.5e−4 +	4.47e−3/ 9.4e−4	4.23e−3/ 8.9e−4 +	5.22e−3/ 9.7e−4
mla13	7.71e−1/ 9.6e−2 =	7.96e−1/ 9.3e−2	3.09e−3/ 9.2e−4 +	5.44e−3/ 1.3e−3	5.32e−3/ 1.5e−3 =	5.34e−3/ 9.8e−4
mla14	7.52e−1/ 5.5e−2 +	8.52e−1/ 7.4e−2	3.45e−3/ 8.4e−4 +	5.43e−3/ 1.3e−3	4.76e−3/ 8.6e−4 +	5.91e−3/ 9.9e−4
mla15	8.44e−1/ 1.1e−1 +	9.49e−1/ 1.1e−1	3.90e−3/ 1.2e−3 +	6.91e−3/ 1.4e−3	4.54e−3/ 1.3e−3 +	5.75e−3/ 7.6e−4
mla16	1.11e + 0/ 7.2e−2 +	1.25e + 0/ 6.4e−2	5.31e−3/ 1.5e−3 +	7.15e−3/ 1.4e−3	6.81e−3/ 1.6e−3 +	8.44e−3/ 1.7e−3
mla17	1.13e + 0/ 7.0e−2 +	1.22e + 0/ 6.8e−2	6.31e−3/ 2.1e−3 +	1.14e−2/ 2.3e−3	6.80e−3/ 2.1e−3 +	1.03e−2/ 1.5e−3
mla18	1.12e + 0/ 6.1e−2 +	1.18e + 0/ 7.5e−2	5.72e−3/ 2.0e−3 +	9.30e−3/ 1.8e−3	6.94e−3/ 1.6e−3 +	8.77e−3/ 9.7e−4
mla19	1.16e + 0/ 5.9e−2 +	1.21e + 0/ 6.4e−2	5.19e−3/ 1.1e−3 +	9.79e−3/ 2.1e−3	5.92e−3/ 1.4e−3 +	8.02e−3/ 1.4e−3
mla20	1.07e + 0/ 5.8e−2 +	1.17e + 0/ 6.6e−2	4.82e−3/ 1.3e−3 +	7.78e−3/ 1.9e−3	7.13e−3/ 1.3e−3 +	8.86e−3/ 1.6e−3

(continued)

Table 5.3 (continued)

Instances	Spread (mean/std)		GD (mean/std)		IGD (mean/std)	
	MOMA	MOMA-NL	MOMA	MOMA-NL	MOMA	MOMA-NL
mla21	1.14e + 0/ 7.2e−2 +	1.21e + 0/ 8.8e−2	5.39e−3/ 2.0e−3 +	1.04e−2/ 2.7e−3	6.90e−3/ 1.4e−3 +	8.31e−3/ 1.5e−3
mla22	1.11e + 0/ 5.4e−2 +	1.18e + 0/ 7.2e−2	4.28e−3/ 1.3e−3 +	8.70e−3/ 2.3e−3	8.46e−3/ 1.4e−3 +	1.02e−2/ 1.6e−3
mla23	1.18e + 0/ 8.9e−2 +	1.24e + 0/ 8.0e−2	8.04e−3/ 2.6e−3 +	1.45e−2/ 2.3e−3	8.72e−3/ 1.9e−3 +	1.30e−2/ 1.8e−3
mla24	1.09e + 0/ 8.4e−2 +	1.18e + 0/ 6.8e−2	6.61e−3/ 2.6e−3 +	1.32e−2/ 1.9e−3	6.58e−3/ 1.4e−3 +	1.13e−2/ 1.3e−3
mla25	1.13e + 0/ 8.1e−2 =	1.17e + 0/ 7.9e−2	5.68e−3/ 1.7e−3 +	1.12e−2/ 1.9e−3	7.10e−3/ 1.4e−3 +	9.41e−3/ 1.4e−3
mla26	1.12e + 0/ 9.7e−2 =	1.13e + 0/ 1.0e−1	4.48e−3/ 1.7e−3 +	9.15e−3/ 1.9e−3	7.29e−3/ 1.6e−3 +	8.78e−3/ 1.2e−3
mla27	1.12e + 0/ 8.2e−2 =	1.13e + 0/ 7.3e−2	7.35e−3/ 1.6e−3 +	1.22e−2/ 2.5e−3	8.38e−3/ 1.4e−3 +	1.07e−2/ 1.6e−3
mla28	1.17e + 0/ 6.3e−2 =	1.17e + 0/ 7.4e−2	6.52e−3/ 2.1e−3 +	1.27e−2/ 2.1e−3	7.31e−3/ 1.9e−3 +	1.14e−2/ 2.3e−3
mla29	1.13e + 0/ 8.8e−2 +	1.18e + 0/ 8.0e−2	6.32e−3/ 2.2e−3 +	1.30e−2/ 2.4e−3	6.91e−3/ 1.4e−3 +	1.07e−2/ 1.6e−3
mla30	1.18e + 0/ 9.8e−2 =	1.16e + 0/ 8.5e−2	7.73e−3/ 3.1e−3 +	1.56e−2/ 2.3e−3	6.44e−3/ 1.5e−3 +	1.21e−2/ 1.7e−3
mla31	1.09e + 0/ 8.0e−2 =	1.07e + 0/ 8.7e−2	5.79e−3/ 2.6e−3 +	1.01e−2/ 2.1e−3	5.86e−3/ 1.7e−3 +	8.44e−3/ 1.6e−3
mla32	1.15e + 0/ 8.0e−2-	1.10e + 0/ 7.3e−2	4.76e−3/ 1.6e−3 +	8.96e−3/ 1.8e−3	5.76e−3/ 1.3e−3 +	7.26e−3/ 9.7e−4
mla33	1.10e + 0/ 6.6e−2 =	1.04e + 0/ 1.0e−1	5.46e−3/ 2.1e−3 +	1.05e−2/ 1.8e−3	6.72e−3/ 1.1e−3 +	9.08e−3/ 1.1e−3
mla34	1.08e + 0/ 6.5e−2 =	1.09e + 0/ 6.3e−2	6.42e−3/ 2.5e−3 +	1.13e−2/ 2.0e−3	8.31e−3/ 1.7e−3 +	1.10e−2/ 1.6e−3
mla35	1.15e + 0/ 9.7e−2 =	1.10e + 0/ 9.1e−2	6.85e−3/ 2.0e−3 +	1.16e−2/ 2.6e−3	7.49e−3/ 1.4e−3 +	1.03e−2/ 1.9e−3
mla36	1.25e + 0/ 6.4e−2 =	1.28e + 0/ 5.0e−2	8.18e−3/ 2.5e−3 +	1.56e−2/ 3.1e−3	8.14e−3/ 1.8e−3 +	1.26e−2/ 2.5e−3
mla37	1.25e + 0/ 6.1e−2 =	1.29e + 0/ 7.3e−2	7.96e−3/ 3.2e−3 +	1.35e−2/ 3.1e−3	1.01e−2/ 2.7e−3 +	1.25e−2/ 1.7e−3
mla38	1.22e + 0/ 7.0e−2 =	1.27e + 0/ 5.6e−2	7.98e−3/ 2.8e−3 +	1.36e−2/ 3.5e−3	9.20e−3/ 1.7e−3 +	1.34e−2/ 3.4e−3
mla39	1.28e + 0/ 5.4e−2 =	1.30e + 0/ 3.9e−2	7.60e−3/ 2.6e−3 +	1.40e−2/ 3.5e−3	8.01e−3/ 1.9e−3 +	1.11e−2/ 2.1e−3

(continued)

Table 5.3 (continued)

Instances	Spread (mean/std)		GD (mean/std)		IGD (mean/std)	
	MOMA	MOMA-NL	MOMA	MOMA-NL	MOMA	MOMA-NL
mla40	1.20e + 0/ 5.3e−2 +	1.29e + 0/ 5.7e−2	6.73e−3/ 1.7e−3 +	1.05e−2/ 3.0e−3	6.78e−3/ 9.7e−4 +	9.19e−3/ 7.7e−4
'+'counts	19	–	40	–	33	–
'='counts	20	–	0	–	7	–
'−' counts	1	–	0	–	0	–

5.4.6 Comparison Between MOMA and Other MOEAs

To further measure the behavior of our proposed MOMA, the MOMA is also compared to some classic MOEAs including NSGA-II [12], SPEA2 [14], and MOEA/D [14] on 40 JSPVMS instances. All the compared algorithms adopt the same termination criterion. In addition, they also use the aforementioned encoding and decoding mechanism, crossover, and mutation (if they exist), and energy-saving strategy. The parameters of all MOEAs are set as stated in Sect. 5.4.2.

Figure 5.6 shows box-and-whisker plots of all algorithms on all instances in terms of three metrics (i.e., IGD, GD, and Spread). Box-and-whisker plots clearly show that the distribution of solutions, the minimum, first quartile, median, third quartile, and maximum values are marked. The dots that are close to the median positions represent the average values. We can see that both the average and median values of the proposal are lower than those of its competitors regarding three metrics. It verifies the effectiveness of the proposal.

Thus, on a whole, MOMA is better than its competitors regarding the comprehensive metric on most instances. These results further confirm the effectiveness of our MOMA in handling such scheduling problems. The reasons for the good behavior of the MOMA are stated as follows. First, the memetic hybrid mechanism is able to balance exploration and exploitation. By contrast, the other compared algorithms do not utilize this hybrid mechanism. Second, to further improve the quality of solutions, we fully utilize the problem-specific knowledge to design a powerful local search

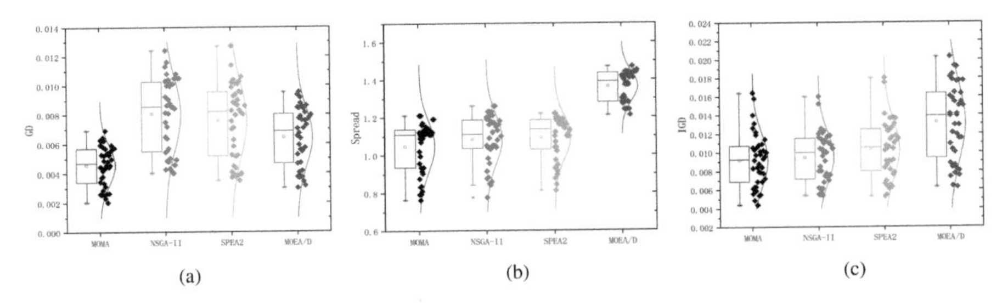

Fig. 5.6 Box-and-whisker plots. **a** GD. **b** SPREAD. **c** IGD

heuristic for exploiting trade-off solutions between makespan and TEC. Although the proposed local search can help a little bit improvement of the solution quality, any little improvement will make this solution dominate many other solutions in MOPs. Therefore, it is concluded that the proposed MOMA can effectively deal with JSPVMS instances.

5.5 Conclusion

In this article, we examine a green JSPVMS, which has both significant academic value and practical engineering applications. The goals of JSPVMS, a multi-objective optimization problem, are to simultaneously minimize the makespan and the TEC. We suggest a novel MOMA based on the JSPVMS's properties to solve this issue. On 40 different problem instances, comparison experiments have been done to show the efficacy of the proposed MOMA. According to experimental findings, the proposed algorithm performs better than the other MOEAs on the majority of problems. Additionally, we have shown through comparison experiments on various MOMA variants the efficacy of each improvement component of MOMA. We will research more challenging scheduling issues in the future, such as distributed and dynamic scheduling issues. To solve the JSPVMS problem, improved metaheuristics will also be created.

References

1. Yin, L.J., et al.: A novel mathematical model and multi-objective method for the low-carbon flexible job shop scheduling problem. Sustain. Comput. Inf. Syst. **13**, 15–30 (2017)
2. Lu, C., et al.: An effective multi-objective discrete virus optimization algorithm for flexible job-shop scheduling problem with controllable processing times. Comput. Ind. Eng.. Ind. Eng. **104**, 156–174 (2017)
3. Shao, W.S., Pi, D.C., Shao, Z.S.: Local search methods for a distributed assembly no-idle flow shop scheduling problem. IEEE Syst. J. **13**(2), 1945–1956 (2019)
4. Lu, C., et al.: Energy-efficient scheduling of distributed flow shop with heterogeneous factories: a real-world case from automobile industry in China. IEEE Trans. Industr. Inf.Industr. Inf. **17**(10), 6687–6696 (2021)
5. Lu, C., et al.: Sustainable scheduling of distributed permutation flow-shop with non-identical factory using a knowledge-based multi-objective memetic optimization algorithm. Swarm Evol. Comput.Evol. Comput. **60**, 100803 (2021)
6. Dai, M., et al.: Energy-efficient scheduling for a flexible flow shop using an improved genetic-simulated annealing algorithm. Robot. Comput.-Integr. Manuf. **29**(5), 418–429 (2013)
7. Lu, C., et al.: A multi-objective cellular grey wolf optimizer for hybrid flowshop scheduling problem considering noise pollution. Appl. Soft Comput.Comput. **75**, 728–749 (2019)
8. Li, X.Y., et al.: An effective multiobjective algorithm for energy-efficient scheduling in a real-life welding shop. IEEE Trans. Industr. Inf.Industr. Inf. **14**(12), 5400–5409 (2018)
9. Lu, C., et al.: A multi-objective approach to welding shop scheduling for makespan, noise pollution and energy consumption. J. Clean. Prod. **196**, 773–787 (2018)

10. Shao, W.S., Pi, D.C., Shao, Z.S.: A Pareto-based estimation of distribution algorithm for solving multiobjective distributed no-wait flow-shop scheduling problem with sequence-dependent setup time. IEEE Trans. Autom. Sci. Eng.Autom. Sci. Eng. **16**(3), 1344–1360 (2019)
11. Gao, J., Sun, L.Y., Gen, M.S.: A hybrid genetic and variable neighborhood descent algorithm for flexible job shop scheduling problems. Comput. Oper. Res.. Oper. Res. **35**(9), 2892–2907 (2008)
12. Deb, K., et al.: A fast and elitist multiobjective genetic algorithm: NSGA-II. IEEE Trans. Evol. Comput.Evol. Comput. **6**(2), 182–197 (2002)
13. Zitzler, E., Deb, K., Thiele, L.: Comparison of multiobjective evolutionary algorithms: empirical results. Evol. Comput.. Comput. **8**(2), 173–195 (2000)
14. Zitzler, E., Thiele, L.: Multiobjective evolutionary algorithms: a comparative case study and the Strength Pareto approach. IEEE Trans. Evol. Comput.Evol. Comput. **3**(4), 257–271 (1999)

Chapter 6
Green Scheduling in Flexible Job Shop Environment

6.1 Brief Introduction

For most traditional scheduling problems, the main goal of optimization is to improve production efficiency, cost, and quality. However, with the rise in energy prices and environmental pollutants, "low-carbon scheduling" is gaining attention from scholars and engineers. For the flexible job shop scheduling problem, a novel low-carbon mathematical model is proposed to optimize productivity, energy efficiency, and noise reduction. The model allows flexibility in production time, power, and noise of the machining spindle speed as a separate variable for decision making. Evaluation methods for production efficiency, energy consumption, and noise are proposed. To successfully solve this mixed-integer programming model, a multi-objective genetic algorithm based on a simplex lattice design is proposed. The corresponding encoding/decoding methods, fitness functions, and crossover/variance operators are consistent with this problem. The effectiveness of the method is demonstrated and evaluated by three problem examples of different sizes and an engineering case study. The results illustrate the feasibility of the proposed model and method for the low-carbon job shop scheduling problem.

6.2 Problem Statement and Modeling

6.2.1 Problem Statement

Productivity, energy consumption, and noise emission are three optimization objectives in this chapter. The evaluation models for these three objectives are first proposed. Based on the proposed evaluation model, a low-carbon FJSP mixed-integer programming model is given.

C. Lu et al., *Intelligence Optimization for Green Scheduling in Manufacturing Systems*, Engineering Applications of Computational Methods 18,
https://doi.org/10.1007/978-981-99-6987-6_6

The mathematical model of the scheduling problem can be explained as follows. A set of m machines can handle a set of n jobs. Each job can be done on a different machine. Each machine has a different processing time, energy consumption, and processing noise [1, 2]. To optimize the scheduling objectives, the mathematical model defines on which machine each operation of each job is executed and the order in which the jobs are processed.

To solve this problem, the following premises are established.

(1) The jobs are independent. There is no space for jobs to grab, and each machine can only process one job at a time.
(2) A job cannot be processed on multiple machines at the same time.
(3) At time zero, all jobs and machines are available at the same time.
(4) After a job is completed on one machine, it is transported directly to the next machine. That is, the transfer time is negligible.
(5) The setup time for operations on a machine is independent of the sequence of operations and is included in the processing time.

6.2.2 Mathematical Modeling

(1) **Evaluation model of energy consumption**

When a machine is not actively engaged in machining operations, its entire input power is used for idling: $p^l = p^u$; when it is machining jobs, some power is required to maintain idling while the rest is used for machining, and some of the power is lost to the load conditions of the machine. This lost power is called load power loss.

$$p^l = p^u + p^c + p^a \tag{6.1}$$

where

p^α—Input power of the machine.
p^μ—Idling power of the machine.
p^c—Output power of the machine (machining power).
p^α—Loading power loss.

Given that the processing time for job j on machine i is t, as per Eq. (6.1), the energy consumption can be represented as e_{ij}:

$$e_{ij} = \int_0^{t_{ij}} p_{ij}^l(t) \cdot \mathrm{d}t = \int_0^{t_{ij}} p_{ij}^u(t) \cdot \mathrm{d}t + \int_0^{t_{ij}} p_{ij}^c(t) \cdot \mathrm{d}t + \int_0^{t_{ij}} p_{ij}^a(t) \cdot \mathrm{d}t \tag{6.2}$$

In practical applications, the idling power consumption of a machine is frequently substituted for the output power consumption of machining in order to simplify the calculation and real operation, and a matrix of energy consumption may be derived via practice [3, 4].

The loading power loss is caused by the loading condition of the machine tool and is proportional to the output power p^c within the allowable range. The machining power balance equation is as follows:

$$p^l = p^u + (1+\alpha)p^c \tag{6.3}$$

where α represents the machine's loading loss power factor.

Equation (6.2) can be transformed into the following equation:

$$e_{ij} = \int_0^{t_{ij}} p_{ij}^l(t)\cdot \mathrm{d}t = \int_0^{t_{ij}} p_{ij}^u(t)\cdot \mathrm{d}t + \int_0^{t_{ij}} (1+\alpha)p_{ij}^c(t)\cdot \mathrm{d}t \tag{6.4}$$

When the process parameters are identical, the energy consumption for cutting identical components is roughly equivalent, and the increase in energy usage relative to idle is moderate. In addition, the effect of a change in α is small in comparison to the energy spent during the machining process. Consequently, the majority of the energy spent throughout the machining process is due to idling, and Eq. (6.4) may be simplified as follows:

$$e_{ij} \approx \int_0^{t_{ij}} p_{ij}^u(t)\cdot \mathrm{d}t \tag{6.5}$$

(2) **Evaluation model of noise emission**

Before analyzing the noise, evaluation indicators must be determined. There are several physical indicators that can be used to evaluate sound, such as sound pressure, sound pressure level, sound intensity, sound intensity level. The most common metric is the sound pressure level, which is measured in the well-known decibels (dB). When all machines in a machine shop are running at the same processing speed, the noise emitted is broadband and stationary. A weighting method is used to correct this noise. Therefore, in this work, the A-level sound pressure is used to analyze the radiation of machine noise.

Typical machining noise includes structural noise and cutting noise. Typically, noise from idling machines is used to measure structural noise; cutting noise is caused by the dynamics of cutting. Cao [4, 5] observed that cutting power consumption, dynamic cutting forces, and cutting noise in a machining system can be considered nearly equal if the same criteria for process parameters are met (e.g., fixtures). The main source of machining noise is structural noise, which depends on the machine frame.

Based on the previous analysis, this research makes a noise coefficient matrix for the scheduling model based on the idle noise of the spindle during machining. The idling noise of a machine is a parameter related to the spindle speed in the matrix $l_{ij} = l_i\,(n)$.

After building the noise coefficient matrix in the scheduling model, we were able to collect all noise maps for each processing job; the job scheduling consisted of several processing stages. This work evaluated the noise performance in various scheduling environments based on comparable sound levels A [6, 7].

To interpret the equivalent sound of class A, typical noise varies throughout time rather than remaining constant and fixed. In general, the average sound energy over a period of time is used to determine the average energy of the fluctuating and varying class A sounds that occur over a specific period of time, and the continuously stable class A sound over an equivalent period is chosen to represent the noise loudness. This continuously stable class A sound is referred to as the equivalent continuous sound class of unstable noise, denoted by leq; it is similar to the fluctuating sound in the real world, but exists continuously during this period. This is also referred to as the equivalent continuous sound class A or equivalent sound class A. Defined by the equation:

$$l_{\text{eq}} = 10 \cdot \lg \frac{1}{T} \int_0^T 10^{0.1 \cdot l_i} \mathrm{d}t \tag{6.6}$$

where l_i—the instantaneous value of sound class A in the exposure time.

T—the exposure time of noise.

In the workshop analyzed in this study, the radiated noise value in each period is stable while the machine performs the task at a fixed spindle speed; so, Eq. (6.6) may be simplified as follows:

$$l_{\text{eq}} = 10 \cdot \lg \frac{\sum_i 10^{0.1 \cdot l_i} t_i}{\sum_i t_i} \tag{6.7}$$

where l_i—the sound class A measured in "i" paragraphed time.

t_i—"i" paragraphed time.

Based on this, productivity, energy consumption, and noise emission assessment models were developed and analyzed.

(3) **Mixed-integer programming model of low-carbon FJSP**

The mixed-integer programming model is made in the way outlined below, taking into account the goal of low-carbon scheduling and the constraints:

The notation used to explain the model is described below:

n: the total number of jobs.

m: the total number of machines.

Ω: the total set of machines.

i,e: the number of a machine, $i,e = 1,2,\ldots,m$

j,k: the number of a job, $j,k = 1,2,\ldots,n$

h_j: the total number of operations of the jth job.

l: the number of the operation, $l = 1,2\ldots,h$

Ω_{jh}: the alternative machine set of the hth operation of the jth job.

m_{jh}: the alternative machine number of the hth operation of the jth job.

O_{jh}: the hth operation of the jth job.

p_{ijh}: the processing time of O_{jh} on machine i.

s_{jh}: the starting time of processing O_{jh}.
c_{jh}: the ending time of processing O_{jh}.
q_{ijh}: the processing efficiency of O_{jh} on machine i.
n_{ijh}: the processing noise of O_{jh} on machine i.
L: a sufficiently large positive number

$$x_{ijh} = \begin{cases} 1, & \text{if } O_{jh} \text{ is processed on machine } i \\ 0, & \text{otherwise} \end{cases}$$

$$y_{ijhkl} = \begin{cases} 1, & \text{if } O_{ijh} \text{ is processed prior to } O_{ikl} \\ 0, & \text{otherwise} \end{cases}$$

Objectives:
Minimizing the makespan:

$$\min T = \max_{1 \le j \le n} \left\{ \max_{1 \le h \le h_j} c_{jh} \right\} \tag{6.8}$$

Minimizing the total energy consumption:

$$\min E = \sum_{j=1}^{n} \sum_{h=1}^{h_j} q \cdot p \cdot xi \in \{1, 2, ..., m\}; j \in \{1, 2, ..., n\} \tag{6.9}$$

Minimizing the noise emission:

$$\min N = 10 \cdot \lg \frac{\sum_{i=1}^{m} \sum_{j=1}^{n} \sum_{h=1}^{h_j} 10^{0.1 \cdot n_{ijh}} \cdot p_{ijh} \cdot x_{ijh}}{\sum_{i=1}^{m} \sum_{j=1}^{n} \sum_{h=1}^{h_j} p_{ijh} \cdot x_{ijh}} i \in \{1, 2, ..., m\}; j \in \{1, 2, ..., n\} \tag{6.10}$$

Subject to:

$$s_{jh} + p_{ijh} \cdot x_{ijh} = c_{jh} i \in \{1, 2, ..., m\}; j \in \{1, 2, ..., n\} \tag{6.11}$$

$$c_{jh} \le s_{j(h+1)} j \in \{1, 2, ..., n\} \tag{6.12}$$

$$s_{jh} + p_{ijh} \le s_{kl} + L \cdot (1 - y_{ijhkl}) i \in \{1, 2, ..., m\}; j, k \in \{1, 2, ..., n\}; l \in \{1, 2, ..., h\} \tag{6.13}$$

$$c_{kl} \le s_{j(h+1)} + L \cdot (1 - y_{iklk(h+1)}) s_{jh} + p_{ijh} \le s_{kl} + L \cdot (1 - y_{ijhkl}) i \in \{1, 2, ..., m\}; j, k \in \{1, 2, ..., n\}; l \in \{1, 2, ..., h\} \tag{6.14}$$

$$\sum_{i=1}^{m_{jh}} x_{ijh} = 1 s_{jh} + p_{ijh} \leq s_{kl} + L \cdot \left(1 - y_{ijhkl}\right) i \in \{1, 2, ..., m\};\ j, k \in \{1, 2, ..., n\} \tag{6.15}$$

Equations (6.8) to (6.10) define the objective functions. Due to constraint (6.12), the various actions of one job cannot be completed at the same time. Each machine can only process one job at a time due to constraints (6.13) on the system. Jobs cannot be preempted due to the constraint (6.14) in place. Just one spindle speed may be chosen for each operation, due to constraint (6.15). It is possible to use mixed-integer programming techniques to solve this low-carbon FJSP model. Low-carbon FJSP is likewise an NP-complete issue, and FJSP has been shown to be one of the most challenging NP-complete problems [6]. Finding optimal solutions quickly for huge issues is challenging, and the model suggested in this research is a multi-objective optimization problem, which is quite challenging [8]. This work suggests a multi-objective genetic algorithm (MOGA) based on a simplex lattice design to solve the model and get a set of solutions distributed uniformly over the Pareto border in order to successfully address this problem.

6.3 Proposed Algorithm

6.3.1 Simplex Lattice Design

The earliest mixing formula designs featured a simplex lattice structure. Its objective is to uniformly distribute each weight of the mixing formula in the design space, distribute the weights of the individual components in a reasonable manner, and then conduct independent tests for each weight of the distribution to determine the optimal production formulation [9, 10].

The second-order lattice point set, denoted as {3, 2}, is formed when each side of an equilateral triangle in Fig. 6.1a is divided in half, as shown in Fig. 6.1b The three original vertices and three midpoints on each side of the triangle are known as the second-order lattice point set, where "3" stands for the number of vertices of the regular simplex and "2" stands for the corresponding division number on either side for the case when $m = 3$, making the order "d" = 2.

If each side of the triangle is divided into three equal portions, as shown in Fig. 6.1c. The vertices of the smaller equilateral triangles are referred to as the third-order lattice point set, marked as the {3, 3} simplex lattice design, and form a lattice of multiple smaller equilateral triangles on the original equilateral triangle.

This roughly similar approach may be used to obtain more lattice point sets. This test design would be referred to as a simplex lattice point design if the points on the standard simplex lattice were chosen as the test points. This can guarantee a uniform distribution of the test points, as is seen from the simplex points' regular distribution in the design space. The computation also gets easier and more accurate.

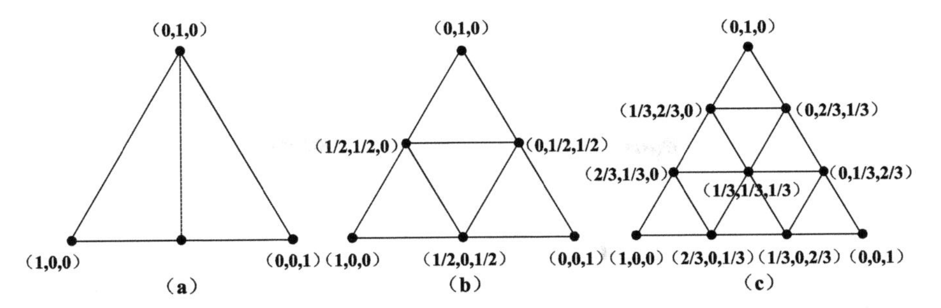

Fig. 6.1 Simplex lattice design

It does this by applying the simplex lattice point design to the different target weight evaluations, getting a set of weights that are all about the same size through the simplex lattice point design, and then putting everything together. The current algorithm guides the solution in various directions of the solution space until it arrives at the uniformly distributed Pareto set.

There are 6 points in the point set $\{3, 2\}$ and e points in the point set $\{3, 3\}$ in the simplex lattice design. As proven, there are $\frac{(m+d-1)!}{d!(m-1)!}$ points in point set $\{m, d\}$ and m groups of d-order lattice points. This paper's solution has three target functions, therefore, $m = 3$. The algorithm has a variable called order "d". While the number of points grows significantly as "d" increases, a greater value of "d" specifies more points in the lattice point set $\{m, d\}$ and a more accurate solution. As a result, the solution must take into account the actual problems and be logically constructed with a "d" number that strikes a balance between calculation accuracy and efficiency.

6.3.2 Encoding and Decoding

Appropriate encoding and decoding techniques may greatly simplify the algorithm's operation and enhance computation performance.

According to Fig. 6.2, each chromosome in the scheduling population is divided into two equal-length segments. The scheduling string appears first. The operation-based representation used in this study is the scheduling encoding made up of the job numbers [11]. Each job number occurs S_i times in the chromosome in this form, which employs an unpartitioned permutation with S_i repeats of the job numbers (S_i is the total number of operations of job i). The fth occurrence of a job number, while reading a chromosome from left to right, corresponds to the fth operation of this job. The key characteristic of this format is the ability to decode any chromosomal permutation into a workable solution. It is assumed that there are n jobs. The scheduling plan string therefore has a length of $\sum s_i$. The machine string makes up the second half of a chromosome. This string, which indicates the chosen machine set for the relevant

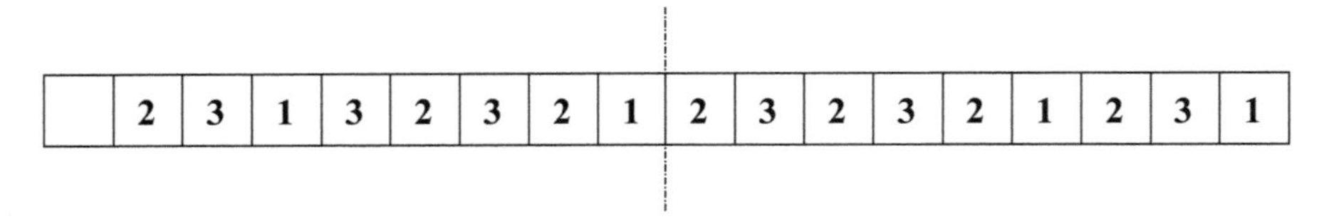

Fig. 6.2 Chromosome of the problem

operations for each task in the scheduling string, is the same length as the scheduling string.

Figure 6.2 depicts a chromosome in action. In this case, $n = 3$ and $S_1 = S_2 = S_3 = 3$. Hence, there are 9 items in both the scheduling string and the machine string. As there are 3 operations for job 1, there is 1 in each of the 3 parts of the scheduling string. It is possible to interpret the other chromosomal components in a similar manner.

Schedules that are semi-active, active, non-delay, and hybrid can be derived from the permutations. In this article, the active schedule is utilized. The following is the decoding process: first scan the scheduling string from left to right, read the machine information of each operation corresponding to the gene in the scheduling string (processing time, energy consumption and noise emission can be obtained from the speed information). Then insert the operation sequence and sequence of the scanned chromosome into the earliest idle time corresponding to the machine that will process the operation. Therefore, each operation is set at the best accessible site. At the same time, the energy requirements of all process sequences are summarized, and class A sound is used to eliminate the sound.

6.3.3 *Initial Population and Fitness Evaluation*

A chromosomal scheduling string's encoding concept uses an operation-based format. The ability of this format to decode any chromosomal permutation into a workable timetable is its defining feature. It is unable to overcome the restrictions imposed by the order of operations.

The makespan, energy use, and noise emission should all be included in the fitness function because this is a multi-objective optimization issue. Given a scheduling plan, its maximum makespan is $T(x)$, its total energy consumption is $E(x)$, and the equivalent sound class A is $N(x)$. The fitness function may thus be stated as follows:

$$F(x) = w_1 \cdot T(x) + w_2 \cdot E(x) + w_3 \cdot N(x) \quad (6.16)$$

where w_1, w_2, and w_3 are the weight coefficients of each target, and

$$w_1 + w_2 + w_3 = 1 \quad (6.17)$$

The sum cannot be calculated directly due to the inconsistent dimensions of the goal functions, and the substantial variations in the values of the objectives make it simple for one of them to dominate the fitness function. To achieve consistency, each goal should be recast as dimensionless and normalized. Equation (6.17) is modified as follows as a result of the fitness functions' normalizations and dimensionality removals:

$$F(x) = w_1 \frac{T(x) - T_{\min}}{T_{\max} - T_{\min}} + w_2 \frac{E(X) - E_{\min}}{E_{\max} - E_{\min}} + w_3 \frac{N(x) - N_{\min}}{N_{\max} - N_{\min}} \tag{6.18}$$

where, e.g., $T_{\max}$ and $T_{\min}$ denote the maximum and minimum, respectively, that could be achieved for objective T. The additional parameters are similar. The range of each target will be between 0 and 1 if Eq. (6.18) is solved. Because of Eq. (6.17), the fitness function also remains between 0 and 1, and its value decreases for groups of the same weight, indicating that the scheduling plan is doing better.

The weight coefficients w_1, w_2, and w_3 of the fitness function are not the same for every case. This means that there are many different ways to set up a schedule. The simplex lattice point design discussed in Sect. 6.3.1 produces their values.

6.3.4 Genetic Operations

(1) **Selection operation**

The selector uses the scale selection method. Select the appropriate individuals using the introduction of the proportion of individual fitness level. If $\sum_{j=1}^{i=1} f_j / \sum_{j=1}^{\mathrm{Pop_{size}}} f_j \mathrm{rand} \leq \sum_{j=1}^{i} f_j / \sum_{j=1}^{\mathrm{Pop_{size}}} f_j$, so, select individual "j", where f_j represents the fitness level of the individual.

(2) **Crossover operation**

Single-point crossing method is adopted for cross operation. The procedure is as follows (Fig. 6.3 provides an example):

Step 1: Randomly choose two people from the population to serve as the parents, A and B, and then create a random integer x that follows the length l of the front half of the chromosomal string, where $1 < x < l$.

Step 2: Exchange the front and rear gene sites of parents A and B before gene X, and generate children A and B of parents A and B.

Step 3: Check the superfluous and missing genes in the front segment of children's chromosomes A and B, modify the superfluous genes with the missing genes, and then adjust the modified genes according to the machines of individuals before crossing to ensure the legitimacy of chromosomes.

(3) **Mutation operation**

There are two phases to the mutation process. The specific stages are as follows (Fig. 6.4 provides an example):

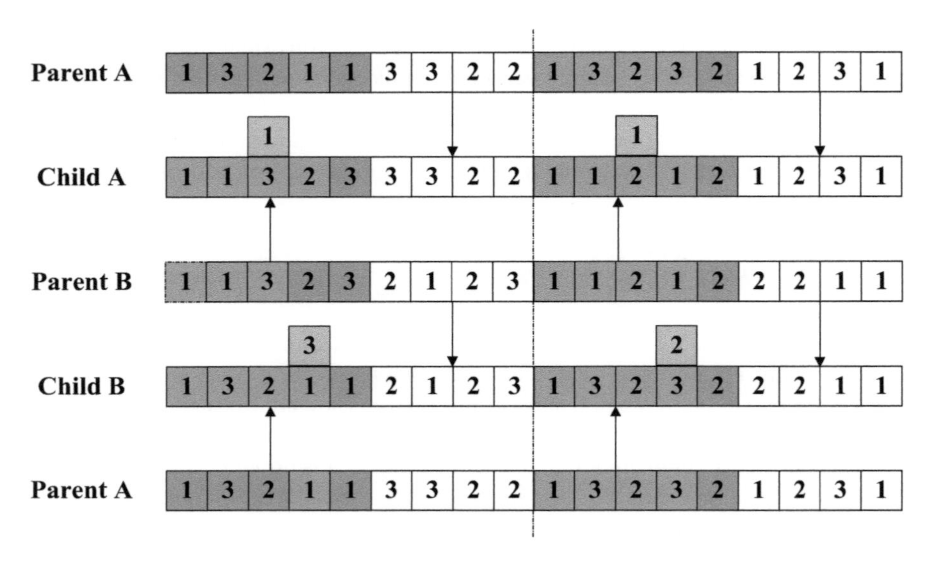

Fig. 6.3 Crossover operation

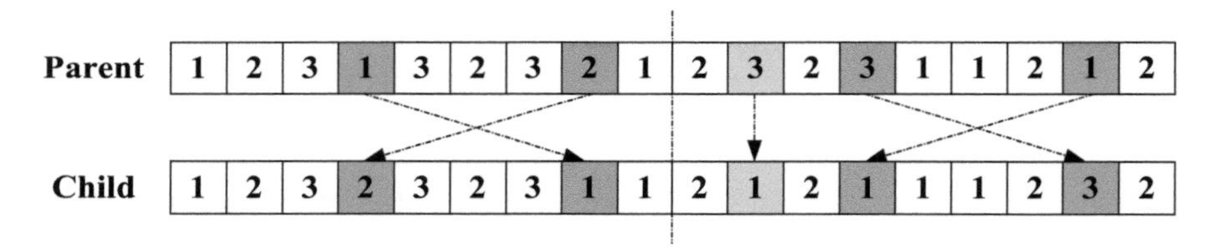

Fig. 6.4 Mutation operation

Step 1: Randomly choose two places on the parent chromosomes and exchange the sequences at those two sites. Exchange the genes in the chromosome's back region as well, as these genes each encode machine information.

Step 2: For the rear sections of the parent chromosome, randomly pick a gene representing the machine information if it has multiple machine selections, and randomly select the others.

Figure 6.4 depicts a simple schematic of the mutation process.

6.3.5 Elite Strategy

The quality protection method is typically included in the algorithm to preserve population size or species diversity while ensuring quick convergence on the best solution. During the evolution of each generation of the population, the most fit individual is chosen from the parent generation G in terms of proportions; subsequently, the individuals of the remaining proportion (I-G) are produced by the crossover mutation operation and enter the next generation of the population directly. This guarantees that both the parent population's and the offspring population's individuals are the same.

6.3.6 Framework of the Proposed MOGA

Figure 6.5 shows a flowchart of how the algorithm method works. These steps are based on the key operations listed above:

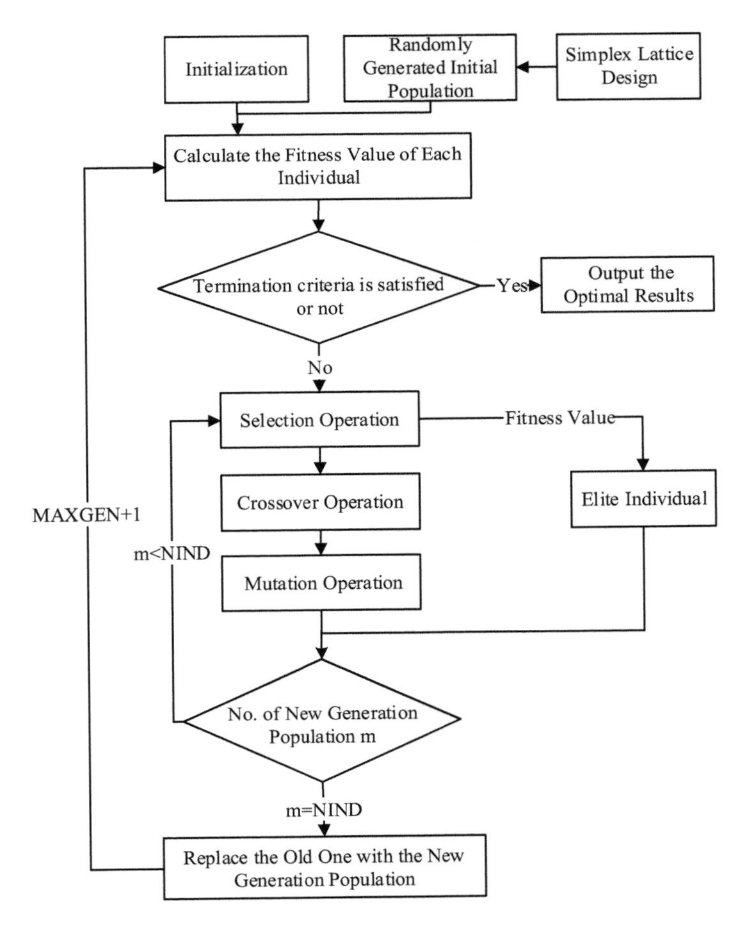

Fig. 6.5 Flowchart of MOGA

Step 1: Initialization: Randomly create the initial population and establish the settings. Add the following information: the simplex lattice design order d, the population size NIND, the maximum number of iterations MAXGEN, the quality protection ratio G, the crossover probability XOVR, and the mutation probability MUTR.

Step 2: Calculate the fitness value of each member of the population to see if the algorithm's end condition is met. If so, achieve the best outcomes. If not, move on to step 3.

Step 3: Compare the fitness value of each individual in the population, and conduct quality protection operation, and select the better individual according to the fitness ratio G. Directly enter the next generation of population.

Step 4: Conduct the population's selection operation; choose two parent individuals with a predetermined probability in accordance with the individuals' fitness values.

Step 5: Use the crossover procedure to create two intermediate individuals from the chosen parent individuals using the crossover probability XOVR.

Step 6: MUTR runs the mutation operation on the intermediate individuals based on the mutation probability. Create two children and add them to the population of the following generation.

Step 7: Verify the number of individuals in the new generation population m. If $m <$ NIND, return to step 4; if $m =$ NIND, replace the old algebra with the new generation of algebra plus 1, and perform step 2.

6.4 Experiments

Use a computer with a 2.0GHz Core 2 Duo CPU to implement the proposed algorithm, which was developed in MATLAB. Three cases of different scales are designed to evaluate the performance of the strategy.

6.4.1 Experiment 1

Table 6.1 shows the data of experiment 1. The first operation of Job 1 is 1–1, while 2–2 is the second operation of Job 2. "Processing time (min)/Unload power (kW)/Energy consumption (kWh)/Unload noise (dB)". For instance, "9/8/1.20/86.30" indicates that the first operation of Job 1 took 9 min to process, used 8 kW of power, 1.20 kWh of energy, and generated 86.30 dB of noise. For the flexible job shop scheduling problem, an operation of a task can be processed by different machines, albeit perhaps not all the machines. For instance, machines M3 and M5 can handle the second operation of job 1, while M1, M2, and M4 cannot for 1–2.

The maximum makespan, total processing energy consumption, and equivalent sound class A with regard to a single objective are initially taken into account in

Table 6.1 Data of instance 1 (4 jobs and 5 machines)

Job	Process	Processing time (min)/Unload power (kW)/Energy consumption (kWh)/ Unload noise (dB)				
		M1	M2	M3	M4	M5
1	1–1	9/8/1.20/86.30	6/10/1.00/ 88.60	–	12/5/1.00/ 80.50	–
	1–2	–	–	6/6/0.60/ 82.40	–	7/3/0.35/ 76.00
	1–3	–	2/10/0.33/ 88.60	4/6/0.40/ 82.40	–	–
2	2–1	5/8/0.67/86.30	–	–	7/5/0.58/ 80.50	9/3/0.45/ 76.00
	2–2	–	4/10/0.67/ 88.60	6/6/0.60/ 82.40	8/5/0.67/ 80.50	–
	2–3	–	–	4/6/0.40/ 82.40	–	8/3/0.40/ 76.00
3	3–1	3/8/0.40/86.30	–	–	–	–
	3–2	–	3/10/0.50/ 88.60	–	5/5/0.42/ 80.50	–
	3–3	8/8/1.07/86.30	–	10/6/1.00/ 82.40	–	13/3/0.65/ 76.00
4	4–1	6/8/0.80/86.30	–	9/6/0.90/ 82.40	–	10/3/0.50/ 76.00
	4–2	–	3/10/0.50/ 88.60	–	–	–
	4–3	2/8/0.27/86.30	–	–	4/5/0.33/ 80.50	–

the solution, and they are optimized to yield, respectively, $T_{\min} = 17$ and $T_{\max} = 47$, $E_{\min} = 5.87$ and $E_{\max} = 7.64$, $N_{\min} = 80.33$ and $N_{\max} = 86.82$. As a result, the algorithm's fitness function changes as follows:

$$F(x) = w_1 \cdot \frac{T(x) - 17}{30} + w_2 \cdot \frac{E(x) - 5.87}{1.77} + w_3 \cdot \frac{N(x) - 80.33}{6.49} \tag{6.19}$$

During the algorithm operation, set the relevant parameters as follows: the simplex lattice point design order $d = 5$, the number of population members NIND = 100, the evolution algebra order MAXGEN = 200, the quality protection ratio G = 10%, the crossover probability XOVR = 0.9, and the mutation probability MUTR = 0.8 should be set as the relevant parameters during the algorithm operation. Once the technique has been used several times, receive the issue solution set displayed in Table 6.2. The solution set's geographical distribution is shown in Fig. 6.6. The Gantt chart for scheduling scheme 17 is shown in Fig. 6.7.

Table 6.2 Scheduling schemes of case 1

No.	Weight of			T (min)	*E* (kWh)	*N* (dB)
	T	*E*	*N*			
1	0	0	1	47	6.07	80.33
2	0	0.2	0.8	47	6.07	80.33
3	0	0.4	0.6	47	6.00	80.50
4	0	0.6	0.4	47	5.94	80.78
5	0	0.8	0.2	47	5.87	81.21
6	0	1	0	39	5.87	81.59
7	0.2	0	0.8	32	6.32	81.09
8	0.2	0.2	0.6	39	6.00	80.88
9	0.2	0.4	0.4	39	6.00	80.88
10	0.2	0.6	0.2	39	5.87	81.59
11	0.2	0.8	0	30	6.00	83.25
12	0.4	0	0.6	32	6.32	81.09
13	0.4	0.2	0.4	32	6.32	81.09
14	0.4	0.4	0.2	25	6.25	83.54
15	0.4	0.6	0	30	6.00	83.25
16	0.6	0	0.4	22	6.85	82.96
17	0.6	0.2	0.2	20	6.48	84.14
18	0.6	0.4	0	19	6.50	84.91
19	0.8	0	0.2	17	6.86	85.44
20	0.8	0.2	0	17	6.86	85.44
21	1	0	0	17	6.86	85.44

In this case, the IBM ILOG CPLEX CP Optimizer tool 12.5 (CP optimizer) is exact methodology is compared with the suggested MOGA. When just the makespan conditions are taken into account, the precise algorithm's makespan is 20 min, compared to the MOGA is 20 min. The results show that the proposed MOGA is effective.

6.4.2 Experiment 2

Table 6.3 gives data of experiment 2. The maximum makespan, total processing energy consumption, and equivalent sound class A with regard to a single target are initially taken into account in the solution, and they are optimized to yield, respectively, $T_{\min} = 32$ and $T_{\max} = 105$, $E_{\min} = 15.27$ and $E_{\max} = 18.30$, and $N_{\min} = 79.94$ and $N_{\max} = 86.00$. As a result, the algorithm's fitness function changes as follows:

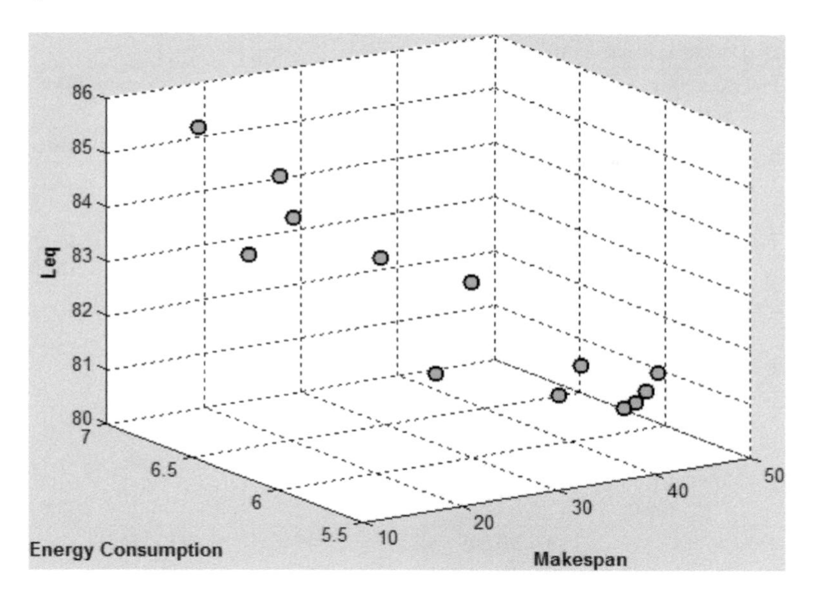

Fig. 6.6 Scatterplot of scheduling schemes of instance 1

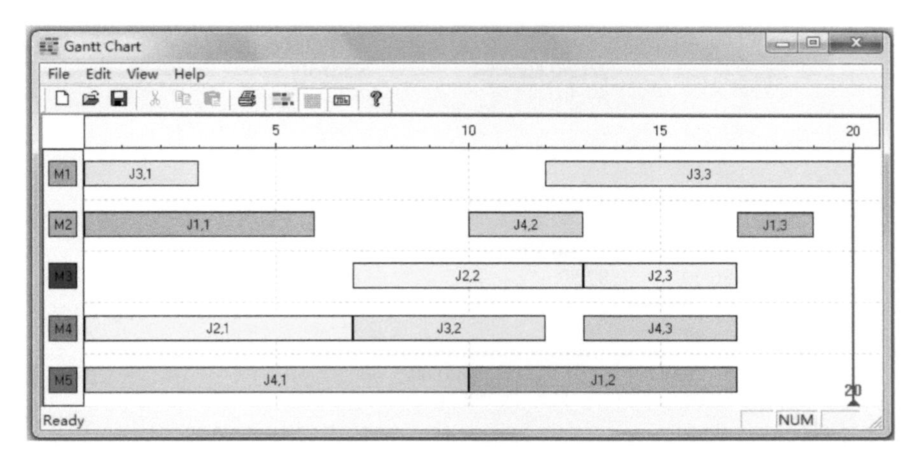

Fig. 6.7 Gantt chart of the 17th scheduling scheme of instance 1

$$F(x) = w_1 \cdot \frac{T(x) - 32}{73} + w_2 \cdot \frac{E(x) - 15.27}{3.03} + w_3 \cdot \frac{N(x) - 79.94}{6.06} \tag{6.20}$$

Set the following settings during algorithm operation: crossover probability XOVR = 0.9, mutation probability MUTR = 0.8, simplex lattice point design order $d = 5$, number of population individuals NIND = 500, evolution algebra order MAXGEN = 300. Once the method has been used a few times, get the set of

Table 6.3 Data of instance 2 (10 jobs and 6 machines)

Job	Process	Processing time (min)/Unload power (kW)/Energy consumption (kWh)/ Unload noise (dB)					
		M1	M2	M3	M4	M5	M6
1	1–1	7/10/1.17/ 86.00	5/12/1.00/ 89.50	9/8/1.20/ 82.50	–	–	–
	1–2	–	–	–	3/5/0.25/ 78.80	–	4/3/0.20/ 75.60
	1–3	–	–	2/8/0.27/ 82.50	–	8/1/0.13/ 72.60	–
	1–4	3/10/0.50/ 86.00	–	–	5/5/0.42/ 78.80	10/1/0.17/ 72.60	–
2	2–1	–	2/12/0.40/ 89.50	–	5/5/0.42/ 78.80	–	–
	2–2	2/10/0.33/ 86.00	–	3/8/0.40/ 82.50	–	–	6/3/0.30/ 75.60
	2–3	–	1/12/0.20/ 89.50	–	–	8/1/0.13/ 72.60	–
	2–4	–	–	4/8/0.53/ 82.50	6/5/0.50/ 78.80	–	9/3/0.45/ 75.60
3	3–1	4/10/0.67/ 86.00	–	–	–	–	8/3/0.40/ 75.60
	3–2	–	3/12/0.60/ 89.50	–	6/5/0.50/ 78.80	–	–
	3–3	–	–	–	–	6/1/0.10/ 72.60	–
	3–4	–	3/12/0.60/ 89.50	5/8/0.67/ 82.50	–	–	–
4	4–1	2/10/0.33/ 86.00	1/12/0.20/ 89.50	–	3/5/0.25/ 78.80	–	–
	4–2	–	–	–	–	6/1/0.10/ 72.60	3/3/0.15/ 75.60
	4–3	5/10/0.83/ 86.00	–	6/8/0.80/ 82.50	–	–	–
	4–4	–	–	–	2/5/0.17/ 78.80	8/1/0.13/ 72.60	–
5	5–1	3/10/0.50/ 86.00	–	–	–	12/1/0.20/ 72.60	7/3/0.35/ 75.60
	5–2	–	2/12/0.40/ 89.50	3/8/0.40/ 82.50	–	–	–
	5–3	–	–	–	5/5/0.42/ 78.80	–	8/3/0.40/ 75.60
	5–4	–	–	5/8/0.67/ 82.50	–	–	–

(continued)

Table 6.3 (continued)

Job	Process	Processing time (min)/Unload power (kW)/Energy consumption (kWh)/ Unload noise (dB)					
		M1	M2	M3	M4	M5	M6
6	6–1	–	3/12/0.60/ 89.50	–	–	–	10/3/0.50/ 75.60
	6–2	1/10/0.17/ 86.00	–	–	–	6/1/0.10/ 72.60	–
	6–3	–	–	–	–	–	5/3/0.25/ 75.60
	6–4	–	2/12/0.40/ 89.50	3/8/0.40/ 82.50	5/5/0.42/ 78.80	–	–
7	7–1	3/10/0.50/ 86.00	–	–	–	–	8/3/0.40/ 75.60
	7–2	–	4/12/0.80/ 89.50	–	–	–	–
	7–3	–	–	2/8/0.27/ 82.50	3/5/0.25/ 78.80	10/1/0.17/ 72.60	–
	7–4	1/10/0.17/ 86.00	–	–	–	8/1/0.13/ 72.60	–
8	8–1	–	–	–	4/5/0.33/ 78.80	12/1/0.20/ 72.60	6/3/0.30/ 75.60
	8–2	4/10/0.67/ 86.00	3/12/0.60/ 89.50	–	–	–	–
	8–3	–	–	4/8/0.53/ 82.50	6/5/0.50/ 78.80	–	–
	8–4	6/10/1.00/ 86.00	–	–	–	–	–
9	9–1	–	2/12/0.40/ 89.50	3/8/0.40/ 82.50	–	10/1/0.17/ 72.60	–
	9–2	–	–	–	4/5/0.50/ 78.80	–	–
	9–3	–	–	2/8/0.27/ 82.50	–	–	5/3/0.25/ 75.60
	9–4	1/10/0.17/ 86.00	–	–	–	7/1/0.12/ 72.60	–
10	10–1	–	–	4/8/0.53/ 82.50	–	–	9/3/0.45/ 75.60
	10–2	–	3/12/0.60/ 89.50	–	6/5/0.50/ 78.80	–	–

(continued)

Table 6.3 (continued)

Job	Process	Processing time (min)/Unload power (kW)/Energy consumption (kWh)/ Unload noise (dB)					
		M1	M2	M3	M4	M5	M6
	10–3	2/10/0.33/ 86.00	–	–	–	10/1/0.17/ 72.60	–
	10–4	–	3/12/0.60/ 89.50	–	–	–	8/3/0.40/ 75.60

problem-solving steps shown in Table 6.4. The solution set's geographical distribution is shown in Fig. 6.8. The Gantt chart for scheduling scheme 13 is seen in Fig. 6.9.

Table 6.4 Scheduling schemes of instance 2

No	Weight of			*T* (min)	*E* (kWh)	*N* (dB)
	T	*E*	*N*			
1	0	0	1	96	16.45	79.94
2	0	0.2	0.8	102	16.01	80.17
3	0	0.4	0.6	90	15.62	80.46
4	0	0.6	0.4	95	15.50	80.88
5	0	0.8	0.2	90	15.36	81.25
6	0	1	0	105	15.27	81.54
7	0.2	0	0.8	75	16.36	79.97
8	0.2	0.2	0.6	68	16.16	80.45
9	0.2	0.4	0.4	69	15.75	80.50
10	0.2	0.6	0.2	80	15.43	81.24
11	0.2	0.8	0	85	15.28	81.26
12	0.4	0	0.6	52	16.72	81.19
13	0.4	0.2	0.4	55	16.48	81.13
14	0.4	0.4	0.2	48	15.96	82.54
15	0.4	0.6	0	49	16.04	82.88
16	0.6	0	0.4	42	16.92	82.11
17	0.6	0.2	0.2	40	16.38	83.33
18	0.6	0.4	0	36	16.39	83.80
19	0.8	0	0.2	34	17.32	83.60
20	0.8	0.2	0	33	16.93	84.48
21	1	0	0	32	17.26	84.34

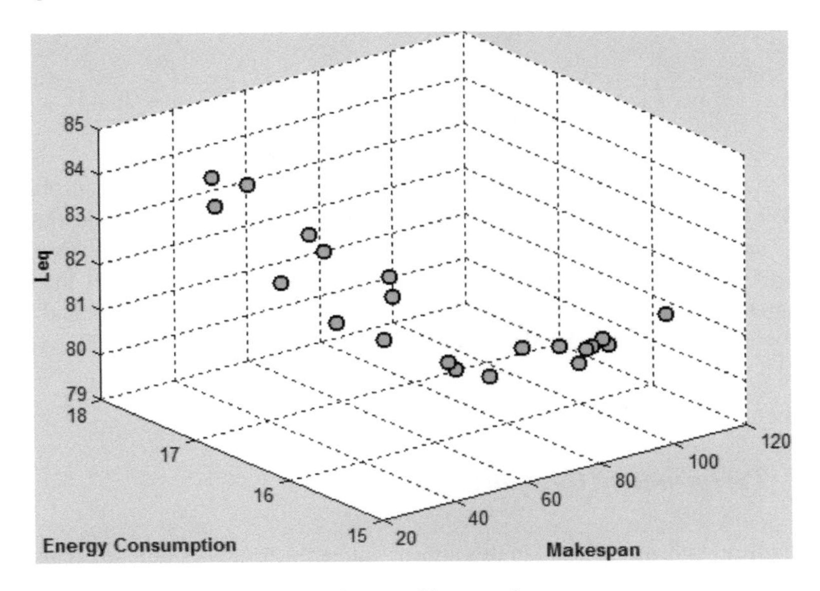

Fig. 6.8 Scatterplot of the scheduling schemes of instance 2

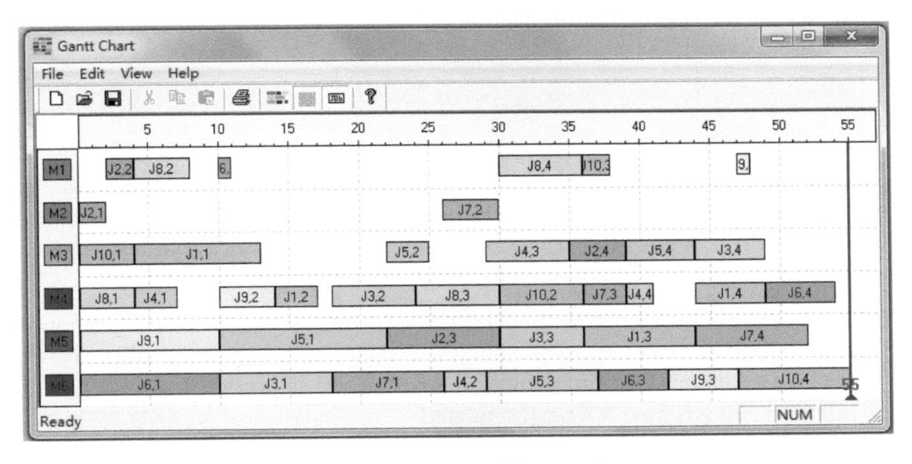

Fig. 6.9 Gantt chart of the 13th scheduling scheme of instance 2

6.4.3 Experiment 3

Table 6.5 gives data of experiment 3. The maximum makespan, total processing energy consumption, and equivalent sound class A with regard to a single target are initially taken into account in the solution, and they are optimized to yield, respectively, $T_{\min} = 63$ and $T_{\max} = 171$, $E_{\min} = 24.66$ and $E_{\max} = 28.51$, and $N_{\min} = 78.42$ and $N_{\max} = 82.26$. As a result, the algorithm's fitness function changes as follows:

$$F(x) = w_1 \cdot \frac{T(x) - 63}{108} + w_2 \cdot \frac{E(x) - 24.66}{3.85} + w_3 \cdot \frac{N(x) - 78.42}{3.84} \tag{6.21}$$

Set the following settings during algorithm operation: crossover probability XOVR = 0.9, mutation probability MUTR = 0.8, simplex lattice point design order $d = 5$, number of population individuals NIND = 500, evolution algebra order MAXGEN = 300. Once the method has been used a few times, get the set of problem-solving steps shown in Table 6.6. The solution set's geographical distribution is shown in Fig. 6.10. The Gantt chart for scheduling scheme 19 is seen in Fig. 6.11.

6.4.4 Performance Comparison

Three indicators are employed in this area to gauge the MOGA's efficacy. This work uses three indexes, each of which is explained as follows:

(1) The Pareto front (PF) obtained is separated from the PF* by the generational distance (GD) [12]. The formulation of this measure is as follows:

$$\mathrm{GD} = \frac{\sqrt{\sum_{i=1}^{n} D_i^2}}{n} \tag{6.22}$$

where n is the number of PF points found so far and D_i is the Euclidean distance between the ith member of the PF obtained and the nearest member of the optimal Pareto front (PF*). A low value of GD is desirable, which denotes a good convergence to PF*.

(2) The inverse convergence method for the GD method is called inverse generation convergence (IGD) [13]. The smallest distance between the uniform Pareto surface point and the non-dominated solution sets is averaged out using IGD. The following is how IGD is formulated:

$$\mathrm{IGD} = \frac{1}{N_{\mathrm{ture}}\sqrt{\sum_{i=1}^{N_{\mathrm{ture}}} d_i^2}} \tag{6.23}$$

where d_i is the distance between the vector in the real Pareto front and the vector in the target space that is the smallest. N_{ture} represents the number of elements in true Pareto front. Convergence is better when the IGD is lower. Moreover, IGD is utilized to assess universality and homogeneity.

(3) Spread (Δ) [12] is used to assess how evenly the Pareto front solution set is distributed, and it has the following definition:

Table 6.5 Data of instance 3 (20 jobs and 5 machines)

Job	Process	Processing time (min)/Unload power (kW)/Energy consumption (kWh)/ Unload noise (dB)				
		M1	M2	M3	M4	M5
1	1–1	5/10/0.83/ 86.00	6/8/0.80/82.50	–	–	–
	1–2	–	–	2/5/0.17/78.80	8/1/0.13/ 72.60	–
	1–3	3/10/0.50/ 86.00	–	–	12/1/0.20/ 72.60	7/3/0.35/ 75.60
2	2–1	–	4/8/0.53/82.50	6/5/0.50/78.80	–	–
	2–2	4/10/0.67/ 86.00	–	–	–	8/3/0.40/ 75.60
	2–3	–	–	6/5/0.50/78.80	–	–
3	3–1	7/10/1.17/ 86.00	9/8/1.20/82.50	–	–	–
	3–2	–	2/8/0.27/82.50	–	8/1/0.13/ 72.60	–
	3–3	3/10/0.50/ 86.00	–	5/5/0.42/78.80	10/1/0.17/ 72.60	–
4	4–1	–	–	–	8/1/0.13/ 72.60	–
	4–2	–	4/8/0.53/82.50	6/5/0.50/78.80	–	–
	4–3	4/10/0.67/ 86.00	–	–	–	8/3/0.40/ 75.60
5	5–1	–	–	5/5/0.42/78.80	–	8/3/0.40/ 75.60
	5–2	–	5/8/0.67/82.50	–	–	–
	5–3	–	–	4/5/0.33/78.80	12/1/0.20/ 72.60	6/3/0.30/ 75.60
6	6–1	–	4/8/0.53/82.50	6/5/0.50/78.80	–	–
	6–2	6/10/1.00/ 86.00	–	–	–	–
	6–3	–	3/8/0.40/82.50	–	10/1/0.17/ 72.60	–
7	7–1	1/10/0.17/ 86.00	–	–	7/1/0.12/ 72.60	–
	7–2	–	4/8/0.53/82.50	–	–	9/3/0.45/ 75.60
	7–3	–	–	6/5/0.50/78.80	–	–
8	8–1	2/10/0.33/ 86.00	–	–	10/1/0.17/ 72.60	–

(continued)

Table 6.5 (continued)

Job	Process	Processing time (min)/Unload power (kW)/Energy consumption (kWh)/ Unload noise (dB)				
		M1	M2	M3	M4	M5
	8–2	–	–	–	–	8/3/0.40/ 75.60
	8–3	3/10/0.50/ 86.00	–	–	12/1/0.20/ 72.60	7/3/0.35/ 75.60
9	9–1	–	3/8/0.40/82.50	5/5/0.42/78.80	–	–
	9–2	3/10/0.50/ 86.00	–	–	–	8/3/0.40/ 75.60
	9–3	–	–	–	8/1/0.13/ 72.60	–
10	10–1	–	–	3/5/0.25/78.80	–	4/3/0.20/ 75.60
	10–2	–	2/8/0.27/82.50	–	8/1/0.13/ 72.60	–
	10–3	3/10/0.50/ 86.00	–	5/5/0.42/78.80	10/1/0.17/ 72.60	–
11	11–1	–	4/8/0.53/82.50	–	–	9/3/0.45/ 75.60
	11–2	–	–	6/5/0.50/78.80	–	–
	11–3	–	–	–	–	8/3/0.40/ 75.60
12	12–1	–	4/8/0.53/82.50	6/5/0.50/78.80	–	–
	12–2	–	–	6/5/0.50/78.80	–	–
	12–3	4/10/0.67/ 86.00	–	–	–	8/3/0.40/ 75.60
13	13–1	7/10/1.17/ 86.00	9/8/1.20/82.50	–	–	–
	13–2	–	–	3/5/0.25/78.80	–	4/3/0.20/ 75.60
	13–3	–	2/8/0.27/82.50	–	8/1/0.13/ 72.60	–
14	14–1	3/10/0.50/ 86.00	–	–	12/1/0.20/ 72.60	7/3/0.35/ 75.60
	14–2	–	–	5/5/0.42/78.80	–	8/3/0.40/ 75.60
	14–3	–	5/8/0.67/82.50	–	–	–
15	15–1	1/10/0.17/ 86.00	–	–	6/1/0.10/ 72.60	–

(continued)

Table 6.5 (continued)

Job	Process	Processing time (min)/Unload power (kW)/Energy consumption (kWh)/ Unload noise (dB)				
		M1	M2	M3	M4	M5
	15–2	–	–	6/5/0.50/78.80	–	–
	15–3	–	–	–	6/1/0.10/ 72.60	–
16	16–1	–	–	4/5/0.50/78.80	–	–
	16–2	–	4/8/0.53/82.50	–	–	9/3/0.45/ 75.60
	16–3	2/10/0.33/ 86.00	–	–	10/1/0.17/ 72.60	–
17	17–1	–	–	–	–	10/3/0.50/ 75.60
	17–2	–	3/8/0.40/82.50	5/5/0.42/78.80	–	–
	17–3	3/10/0.50/ 86.00	–	–	–	8/3/0.40/ 75.60
18	18–1	4/10/0.67/ 86.00	–	–	–	–
	18–2	–	4/8/0.53/82.50	6/5/0.50/78.80	–	–
	18–3	–	–	–	6/1/0.10/ 72.60	–
19	19–1	5/10/0.83/ 86.00	6/8/0.80/82.50	–	–	–
	19–2	–	–	–	–	7/3/0.35/ 75.60
	19–3	–	5/8/0.67/82.50	–	–	–
20	20–1	–	–	3/5/0.25/78.80	10/1/0.17/ 72.60	–
	20–2	–	3/8/0.40/82.50	–	–	–
	20–3	3/10/0.50/ 86.00	–	–	–	7/3/0.35/ 75.60

$$\Delta = \frac{d_f + d_l + \sum_{i=1}^{n-1} |d_i - \overline{d}|}{d_f + d_l + (n-1) \times \overline{d}} \tag{6.24}$$

where d_i is the Euclidean distance between all two consecutive solutions in the Pareto front solution set, $\overline{d}$ is the average of d_i, d_f, and d_l are the Euclidean distances between the boundary point and the optimal boundary point of the Pareto front, respectively, and n is the number of Pareto fronts. For a uniformly distributed solution, the distribution factor is zero. Therefore, the smaller the distribution factor is, the better the distribution is.

Table 6.6 Scheduling schemes of instance 3

No	Weight of			T (min)	E (kWh)	N (dB)
	T	E	N			
1	0	0	1	171	25.10	78.42
2	0	0.2	0.8	165	25.03	78.57
3	0	0.4	0.6	155	25.01	78.75
4	0	0.6	0.4	157	24.95	78.78
5	0	0.8	0.2	144	24.83	79.17
6	0	1	0	146	24.66	79.25
7	0.2	0	0.8	141	25.29	78.75
8	0.2	0.2	0.6	125	25.16	78.94
9	0.2	0.4	0.4	138	24.71	79.06
10	0.2	0.6	0.2	132	24.81	79.14
11	0.2	0.8	0	126	24.96	79.58
12	0.4	0	0.6	98	25.80	79.48
13	0.4	0.2	0.4	92	25.78	79.60
14	0.4	0.4	0.2	96	25.58	79.67
15	0.4	0.6	0	102	25.45	80.01
16	0.6	0	0.4	80	26.55	80.16
17	0.6	0.2	0.2	81	26.32	80.27
18	0.6	0.4	0	84	25.94	80.33
19	0.8	0	0.2	67	27.15	81.25
20	0.8	0.2	0	68	26.94	81.24
21	1	0	0	63	27.65	81.64

For the above situations, the proposed algorithm is compared to a well-known MOEA like NSGA-II to see how well it works. The mean and standard deviation values for these methods across 30 separate runs are displayed in Tables 6.1, 6.2, 6.3 and 6.4. Tables 6.7, 6.8 and 6.9 show that for all criteria, the MOGA performs better than the NSGA-II. The MOGA is a very effective tool for handling this kind of scheduling issue.

6.5 Case Study

A vital part of an automobile engine is the cooling system (see Fig. 6.12 and Table 6.10). Engine problems account for around half of all car failures, and about half of these engine problems are brought on by cooling system problems. Cooling systems are crucial to the dependability of automobiles.

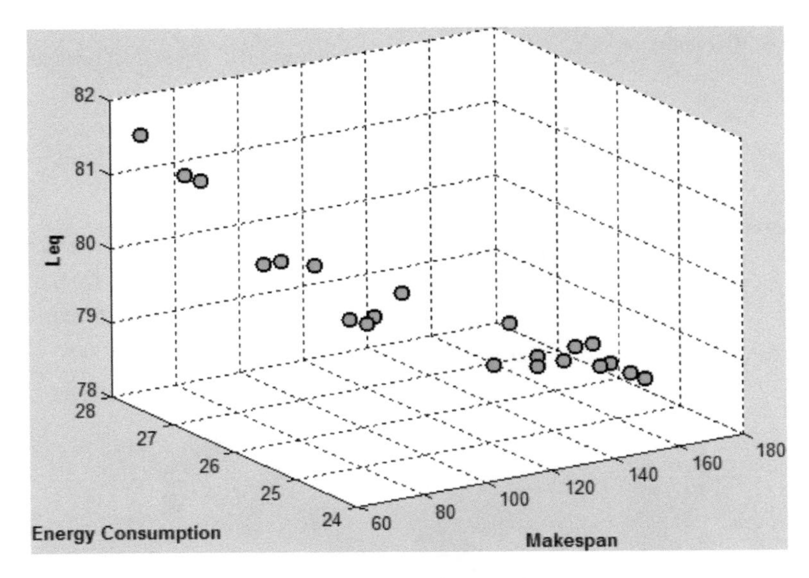

Fig. 6.10 Scatterplot of the scheduling schemes of instance 3

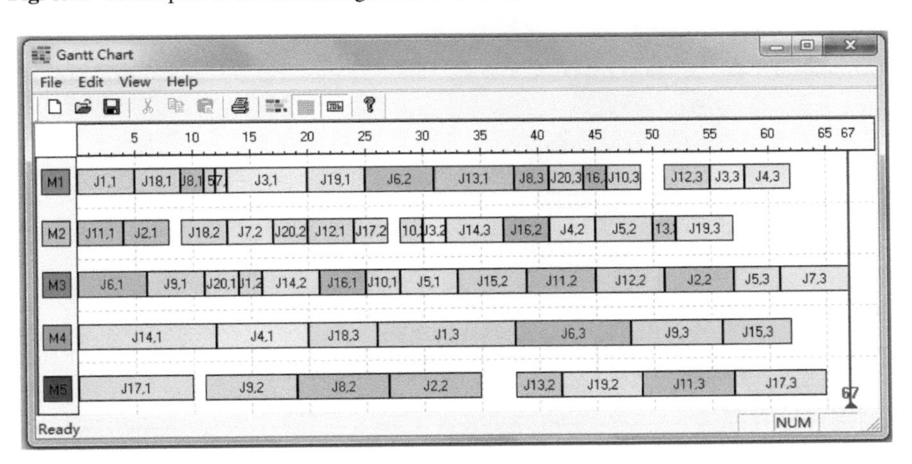

Fig. 6.11 Gantt chart of the 19th scheduling scheme of instance 3

Table 6.7 GD indicator

	MOGA (mean/std)	NSGA-II (mean/std)
Instance1	0.0038(0.0032)	0.0126(0.0210)
Instance2	0.0047(0.0013)	0.0081(0.0012)
Instance3	0.0025(0.0014)	0.0036(0.0013)

Table 6.8 IGD indicator

	MOGA (mean/std)	NSGA-II (mean/std)
Instance1	0.0221(0.0057)	0.0276(0.0089)
Instance2	0.0127(0.0054)	0.0196(0.0046)
Instance3	0.0146(0.0033)	0.0141(0.0027)

Table 6.9 SPREAD indicator

	MOGA (mean/std)	NSGA-II (mean/std)
Instance1	1.2560(0.1483)	1.3182(0.3744)
Instance2	0.8706(0.2031)	0.8096(0.3765)
Instance3	0.8507(0.3852)	0.9397(0.3008)

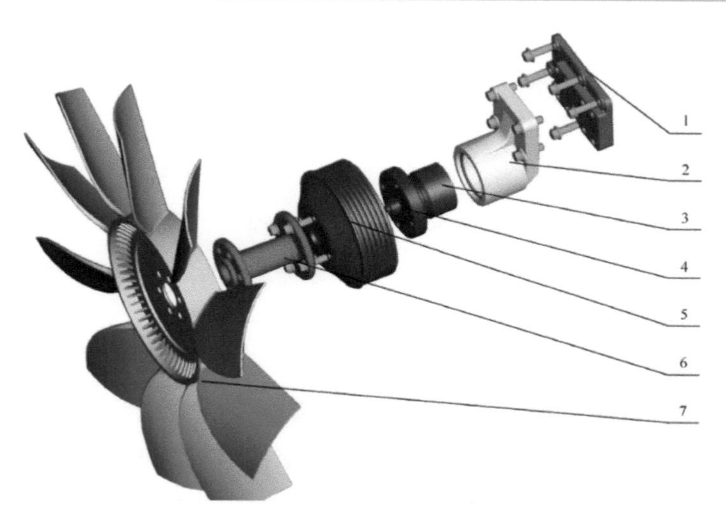

Fig. 6.12 Schematic diagram of an engine cooling system produced by a workshop

Table 6.10 Component parts of an automotive engine cooling system

Part No.	Part name	Process location
1	Adapter plate bracket	A workshop that fabricates engine cooling fan parts
2	Fan bracket	A workshop that fabricates engine cooling fan parts
3	Double-row deep-groove ball bearing	External collaboration
4	Fan hub	A workshop that fabricates engine cooling fan parts
5	Fan belt pulley	External collaboration
6	Adapter flange	A workshop that fabricates engine cooling fan parts
7	Cooling Fan	External collaboration

The engine factory of the DF Motor Company makes adapter flanges, fan hubs, adapter plate brackets, and fan brackets. Different machines can work on these four pieces, and the time it takes to work on them depends on the machine and how it works. The workshop is an example of a standard, flexible job shop. The priority of the jobs in the workshop determines the production schedule. On the basis of the monthly production plan, the workshop dispatcher or policymakers often organize the daily production operations. Factors like energy use and noise emissions are seldom taken into account and sometimes even sacrificed for the timely completion of the plan.

Eight different pieces of equipment are used in the workshop (see Table 6.11): two millers, two drilling machines, one machining center, one manual lathe, and two computer numerically controlled (CNC) lathes. Miller 1, Miller 2, Drilling 1, Drilling 2, Machining 1, Lathe 1, Numerical 1, and Numerical 2 are the serial numbers for these machines (M8). See Table 6.12 for the machine tools and accompanying machining process parameters.

Throughout the solution process, the maximum completion time, overall process energy consumption, and equivalent sound level A are employed as separate objectives and optimized. The outputs are as follows: $T_{\min} = 53.2$ and $T_{\max} = 82.8$, $E_{\min} = 28.91$ and $E_{\max} = 32.74$, and $N_{\min} = 84.0606$ and $N_{\max} = 85.1959$. As a result, the algorithm's fitness function is:

$$F(x) = w_1 \cdot \frac{T(x) - 53.2}{29.6} + w_2 \cdot \frac{E(x) - 28.91}{3.83} + w_3 \cdot \frac{N(x) - 84.0606}{1.1353} \tag{6.25}$$

During the algorithm's operation, the following pertinent parameters were established: the simplex lattice design order d ($d = 5$), number of individuals in the population (NIND = 300), maximum number of generations (MAXGEN = 200), the quality protection ratio G ($G = 10\%$), the crossover probability XOVR (XOVR = 0.9), and the mutation probability MUTR (MUTR = 0.8). The results of the various algorithm operations are displayed in Table 6.13 as solutions. Figure 6.13 shows the spatial distribution of the solution set, while Fig. 6.14 provides the Gantt chart of dispatch plan 17.

According to the calculated findings, the suggested low-carbon flexible job-shop scheduling problems (FJSP) model and associated algorithm may maximize the decision goal, thus resolving the process-based problem. The maximum process completion time, overall process energy consumption, and equivalent sound level A

Table 6.11 A workshop that fabricates component parts for engine cooling fans

Machine name	Serial numbered	Machine name	Serial numbered
Miller 1	M1	Machining center 1	M5
Miller 2	M2	Manual lathe 1	M6
Drilling machine 1	M3	CNC 1	M7
Drilling machine 2	M4	CNC 2	M8

Table 6.12 The machine tools and corresponding machining process parameters required for processing

Job	Process	Processing time (min)/Unload power (kW)/Energy consumption (kWh)/ Unload noise (dB)							
		Miller 1	Miller 2	Drilling machine 1	Drilling machine 2	Machining center 1	Manual lathe 1	CNC 1	CNC 2
Adapter plate bracket	1–1	8/10/ 1.20/ 85.20	8/10/ 1.20/ 85.20	–	–	6/7/1.20/ 81.50	–	–	–
	1–2	12/10/ 1.80/ 85.20	12/10/ 1.80/ 85.20	–	–	9/7/1.80/ 81.50	–	–	–
	1–3	6/10/ 0.9/ 85.20	6/10/ 0.9/ 85.20	–	–	7/7/1.40/ 81.50	–	–	–
	1–4	9/10/ 1.35/ 85.20	9/10/ 1.35/ 85.20	–	–	4/7/0.8/ 81.50	–	–	–
	1–5	3/10/ 0.45/ 85.20	3/10/ 0.45/ 85.20	–	–	2.7/7/0.54/ 81.50	–	–	–
	1–6	–	–	4/5/ 1.00/ 78.6	4/5/ 1.00/ 78.6	3/7/0.6/ 81.50	–	–	–
	1–7	–	–	3/5/ 0.75/ 78.6	3/5/ 0.75/ 78.6	2/7/0.4/ 81.50	–	–	–
	1–8	–	–	3/5/ 0.75/ 78.6	3/5/ 0.75/ 78.6	2/7/0.4/ 81.50	–	–	–
	1–9	–	–	5/5/ 1.25/ 78.6	5/5/ 1.25/ 78.6	3.8/7/0.76/ 81.50	–	–	–
	1–10	–	–	4/5/ 1.00/ 78.6	4/5/ 1.00/ 78.6	2.2/7/0.44/ 81.50	–	–	–
	1–11	–	–	6/5/ 1.50/ 78.6	6/5/ 1.50/ 78.6	4.3/7/0.86/ 81.50	–	–	–
	1–12	4.2/10/ 0.63/ 85.20	4.2/10/ 0.63/ 85.20	–	–	3.7/7/0.74/ 81.50	–	–	–
	1–13	2/10/ 0.3/ 85.20	2/10/ 0.3/ 85.20	–	–	1.8/7/0.36/ 81.50	–	–	–

(continued)

Table 6.12 (continued)

Job	Process	Processing time (min)/Unload power (kW)/Energy consumption (kWh)/ Unload noise (dB)							
		Miller 1	Miller 2	Drilling machine 1	Drilling machine 2	Machining center 1	Manual lathe 1	CNC 1	CNC 2
Fan bracket	2–1	3/10/ 0.45/ 85.20	3/10/ 0.45/ 85.20	–	–	2/7/0.4/ 81.50	–	–	–
	2–2	7/10/ 1.05/ 85.20	7/10/ 1.05/ 85.20	–		1.8/7/0.36/ 81.50	–	–	–
	2–3	5/10/ 0.75/ 85.20	5/10/ 0.75/ 85.20	–	–	2.5/7/0.5/ 81.50	–	–	–
	2–4	3/10/ 0.45/ 85.20	3/10/ 0.45/ 85.20	–	–	1.8/7/0.36/ 81.50	–	–	–
	2–5	3/10/ 0.45/ 85.20	3/10/ 0.45/ 85.20	–	–	1.8/7/0.36/ 81.50	–	–	–
	2–6	–	–	3/5/ 0.75/ 78.6	3/5/ 0.75/ 78.6	1.8/7/0.36/ 81.50	–	–	–
	2–7	–	–	2/5/0.5/ 78.6	2/5/0.5/ 78.6	1/7/0.2/ 81.50	–	–	–
	2–8	–	–	2/5/0.5/ 78.6	2/5/0.5/ 78.6	1/7/0.2/ 81.50	–	–	–
	2–9	–	–	–	–	5/7/1/ 81.50	–	2/8/ 0.67/ 86.70	2/8/ 0.67/ 86.70
	2–10	–	–	–	–	3.2/7/0.64/ 81.50	–	1/8/ 0.33/ 86.70	1/8/ 0.33/ 86.70
	2–11	–	–	–	–	2/7/0.4/ 81.50	–	1/8/ 0.33/ 86.70	1/8/ 0.33/ 86.70
	2–12	–	–	3/5/ 0.75/ 78.6	3/5/ 0.75/ 78.6	2.2/7/0.44/ 81.50	–	–	–
	2–13	–	–	3/5/ 0.75/ 78.6	3/5/ 0.75/ 78.6	2/7/0.4/ 81.50	–	–	–

(continued)

Table 6.12 (continued)

Job	Process	Processing time (min)/Unload power (kW)/Energy consumption (kWh)/ Unload noise (dB)							
		Miller 1	Miller 2	Drilling machine 1	Drilling machine 2	Machining center 1	Manual lathe 1	CNC 1	CNC 2
	2–14	5/10/ 0.75/ 85.20	5/10/ 0.75/ 85.20	–	–	1.8/7/0.36/ 81.50	–	–	–
	2–15	3/10/ 0.45/ 85.20	3/10/ 0.45/ 85.20	–	–	1.5/7/0.3/ 81.50	–	–	–
Fan hub	3–1	–	–	–	–	–	1/12/ 0.3/ 89.20	–	–
	3–2	–	–	–	–	–	0.5/12/ 0.15/ 89.20	–	–
	3–3	–	–	–	–	–	2.5/12/ 0.75/ 89.20	–	–
	3–4	–	–	–	–	–	0.5/12/ 0.15/ 89.20	–	–
	3–5	–	–	–	–	–	–	2/8/ 0.67/ 86.70	2/8/ 0.67/ 86.70
	3–6	–	–	–	–	–	–	1/8/ 0.33/ 86.70	1/8/ 0.33/ 86.70
	3–7	–	–	–	–	–	–	1/8/ 0.33/ 86.70	1/8/ 0.33/ 86.70
	3–8	–	–	–	–	–	–	1.5/8/ 0.50/ 86.70	1.5/8/ 0.50/ 86.70
	3–9	–	–	–	–	–	–	1/8/ 0.33/ 86.70	1/8/ 0.33/ 86.70
	3–10	–	–	–	–	–	–	3/8/1/ 86.70	3/8/1/ 86.70

(continued)

Table 6.12 (continued)

Job	Process	Processing time (min)/Unload power (kW)/Energy consumption (kWh)/ Unload noise (dB)							
		Miller 1	Miller 2	Drilling machine 1	Drilling machine 2	Machining center 1	Manual lathe 1	CNC 1	CNC 2
	3–11	–	–	–	–	–	–	2/8/ 0.67/ 86.70	2/8/ 0.67/ 86.70
	3–12	–	–	–	–	–	–	2/8/ 0.67/ 86.70	2/8/ 0.67/ 86.70
	3–13	–	–	–	–	–	–	2/8/ 0.67/ 86.70	2/8/ 0.67/ 86.70
	3–14	–	–	–	–	–	–	1.2/8/ 0.4/ 86.70	1.2/8/ 0.4/ 86.70
	3–15	–	–	1/5/ 0.25/ 78.6	1/5/ 0.25/ 78.6	1/7/0.2/ 81.50	–	–	–
Adapter flange	4–1		–	–	–	–	1.5/12/ 0.45/ 89.20	–	–
	4–2	–	–	–	–	–	3.5/12/ 1.05/ 89.20	–	–
	4–3	–	–	–	–	–	–	3/8/1/ 86.70	3/8/1/ 86.70
	4–4	–	–	–	–	–	–	2/8/ 0.67/ 86.70	2/8/ 0.67/ 86.70
	4–5	–	–	–	–	–	–	0.8/8/ 0.27/ 86.70	0.8/8/ 0.27/ 86.70
	4–6	–	–	–	–	–	–	0.5/8/ 0.17/ 86.70	0.5/8/ 0.17/ 86.70
	4–7	–	–	–	–	–	–	0.5/8/ 0.17/ 86.70	0.5/8/ 0.17/ 86.70

(continued)

Table 6.12 (continued)

Job	Process	Processing time (min)/Unload power (kW)/Energy consumption (kWh)/ Unload noise (dB)							
		Miller 1	Miller 2	Drilling machine 1	Drilling machine 2	Machining center 1	Manual lathe 1	CNC 1	CNC 2
	4–8	–	–	–	–	–	–	0.5/8/ 0.17/ 86.70	0.5/8/ 0.17/ 86.70
	4–9	–	–	–	–	–	–	0.5/8/ 0.17/ 86.70	0.5/8/ 0.17/ 86.70
	4–10	–	–	–	–	–	–	0.5/8/ 0.17/ 86.70	0.5/8/ 0.17/ 86.70
	4–11	–	–	–	–	–	–	2.2/8/ 0.73/ 86.70	2.2/8/ 0.73/ 86.70
	4–12	–	–	–	–	–	–	1.5/8/ 0.50/ 86.70	1.5/8/ 0.50/ 86.70
	4–13	–	–	2/5/0.5/ 78.6	2/5/0.5/ 78.6	1.5/7/0.3/ 81.50	–	–	–
	4–14	–	–	1/5/ 0.25/ 78.6	1/5/ 0.25/ 78.6	1/7/0.2/ 81.50	–	–	–

cannot all be optimized at the same time since doing so invariably makes the other two worse.

6.6 Conclusion

This research suggests a novel low-carbon mathematical scheduling model that takes into account productivity, energy efficiency, and noise reduction for the flexible work shop environment, given the urgent need for sustainable growth in the modern manufacturing industry. Evaluation techniques are offered for each of the production time, power, and noise aspects that are impacted by the processing machines in this model. To solve this mixed-integer programming model effectively, a multi-objective evolutionary algorithm based on a simplex lattice design is suggested. The fitness function, crossover and mutation operators, and encoding and decoding approach were created expressly to address the characteristics of this issue. Three issue examples with varying scales demonstrate the effectiveness of this strategy.

Table 6.13 Scheduling schemes of engineering case

No	Weight of			T (min)	E (kWh)	N (dB)
	T	E	N			
1	0	0	1	82.8	32.74	84.0606
2	0	0.2	0.8	73.3	31.26	84.2452
3	0	0.4	0.6	71.5	30.55	84.3856
4	0	0.6	0.4	68.2	29.44	84.6399
5	0	0.8	0.2	76.9	29.27	84.7796
6	0	1	0	69.4	28.91	84.5957
7	0.2	0	0.8	60.3	29.1	85.0893
8	0.2	0.2	0.6	60.8	30.62	84.5179
9	0.2	0.4	0.4	56.6	30.33	84.6217
10	0.2	0.6	0.2	63.1	28.95	84.9453
11	0.2	0.8	0	60.9	29.25	84.7320
12	0.4	0	0.6	55.3	29.68	84.9933
13	0.4	0.2	0.4	55.1	30.14	84.7812
14	0.4	0.4	0.2	55.6	31.23	84.6474
15	0.4	0.6	0	58.8	29.38	85.1959
16	0.6	0	0.4	53.6	32.24	84.7187
17	0.6	0.2	0.2	53.4	30.17	84.9967
18	0.6	0.4	0	53.5	30.81	84.7379
19	0.8	0	0.2	53.2	30.98	84.9308
20	0.8	0.2	0	53.5	31.79	84.9854
21	1	0	0	53.7	32.28	84.8572

The outcomes show how well the suggested model and approach for the low-carbon flexible job shop scheduling problem.

According to this study, there are three possible routes for further research on low-carbon scheduling issues. (1) There are several indications for low-carbon manufacturing. Future research might focus on other low-carbon metrics, including peak power and exhaust emissions. (2) In the future, further manufacturing plant varieties (such as open workshops) can be investigated. (3) By including dynamic elements in low-carbon scheduling, such as work delays, the priority of important tasks, and machine breakdowns, the model will become more accurate and reflect the processing conditions of real plants.

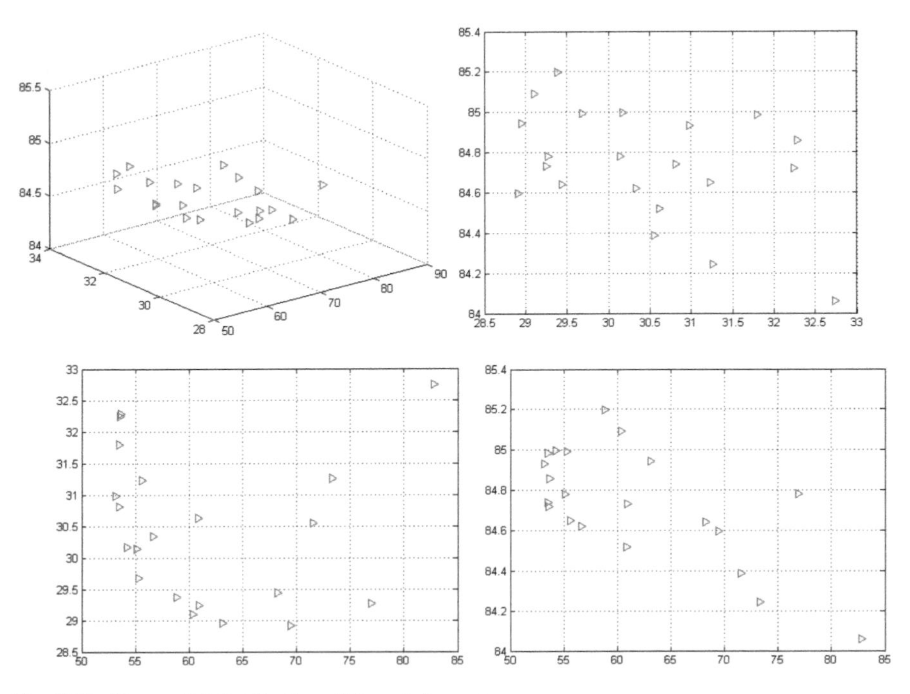

Fig. 6.13 The spatial distribution of the solution set

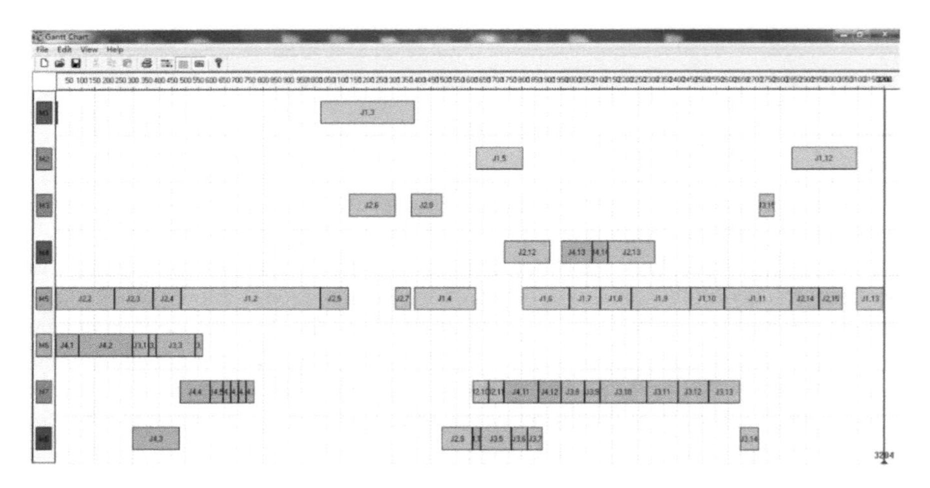

Fig. 6.14 Gantt chart of the 17th scheduling scheme of engineering case

References

1. Pinedo, M.: Scheduling Theory, Algorithms, and System, 2nd edn. Prentice Hall, Upper Saddle River, NJ (2002)
2. Low, C., Ji, M., Hsu, C., Su, C.: Minimizing the makespan in a single machine scheduling problems with flexible and periodic maintenance. Appl. Math. Model. **34**, 334–342 (2010)
3. Liu, F., Xu, Z.J.: Study on energy flow models of mechanical transmission systems. Chin. J. Mech. Eng. **6**, 215–219 (1993)
4. Cao, H.J.: Study on Process Planning Technologies for Green Manufacturing. Ph.D. Dissertation, Chongqing University, Chongqing (2004) (in Chinese)
5. Tandon, N.: Noise-reducing designs of machines and structures. Sadhana **25**(3), 331–339 (2000)
6. Olayinka, O.S., Abdullahi, S.A.: An overview of industrial employees' exposure to noise in sundry processing and manufacturing industries in Ilorin Metropolis, Nigeria. Ind. Health **47**, 123–133 (2009)
7. Li, N., Yang, Q.L., Zeng, L., Zhu, L.L., Tao, L.Y., Zhang, H., Zhao, Y.M.: Noise exposure assessment with task-based measurement in complex noise environment. Chin. Med. J. **124**, 1346–1351 (2011)
8. Garey, M.R., Johnson, D.S., Sethi, R.: The complexity of flow shop and job shop scheduling. Math. Oper. Res.Oper. Res. **1**, 117–129 (1978)
9. Azevedo, S., Cunha, L.M., Mahajan, P.V., Fonseca, S.C.: Application of simplex lattice design for development of moisture absorber for oyster mushrooms. Procedia Food Sci. **1**, 184–189 (2011)
10. Mandik, S.K., Adhikari, S., Deshpande, A.A.: Application of simplex lattice design in formulation and development of buoyant matrices of dipyridamole. J. Appl. Pharm. Sci. **2**, e7-111 (2012)
11. Li, X., Gao, L., Shao, X., Zhang, C., Wang, C.: Mathematical modeling and evolutionary algorithm-based approach for integrated process planning and scheduling. Comput. Oper. Res.. Oper. Res. **37**(20e), 656–667 (2010)
12. Deb, K., Pratap, A., Agarwal, S., Meyarivan, T.: A fast and elitist multiobjective genetic algorithm: NSGA-II. IEEE Trans. Evol. Computat. **6**, 182–197 (2002)
13. Zitzler, E., Thiele, L.: Multiobjective evolutionary algorithms: a comparative case study and the strength Pareto approach. IEEE Trans. Evol. Computat. **3**, 257–271 (1999)

Chapter 7
Green Scheduling in Welding Shop Environment

7.1 Brief Introduction

Researchers and businesses have paid close attention to sustainable scheduling. In sustainable scheduling problems, the objectives of economic and environmental impact should be considered concurrently [1]. The balance between economy (e.g., makespan) and environment (e.g., energy consumption or carbon emission) has been emphasized in most studies on sustainable scheduling problems [2]. However, another significant environmental issue, such as noise pollution caused by the manufacturing process, is frequently overlooked in the existing literature. As a result, this chapter investigated a welding shop scheduling problem (WSSP) that takes into account energy consumption and noise pollution in addition to productivity. The distinctive feature of WSSP is that multiple welders can work on the same task at the same time. As a result, WSSP is a novel scheduling problem that differs from traditional shop scheduling problems [3]. First, we present a new multi-objective WSSP mathematical model. The problem is then solved by using a novel hybrid multi-objective grey wolf algorithm (HMOGWO). This HMOGWO includes a new local search strategy to improve solutions' quality. Furthermore, a new energy-saving strategy is proposed to ensure welder longevity and energy efficiency. Finally, to validate the effectiveness of the proposed HMOGWO, we conduct comparison experiments with other well-known multi-objective evolutionary algorithms [4]. Experimental results mainfest that the proposed HMOGWO outperforms competitors on this problem. Furthermore, the case study demonstrates that this method can effectively solve the real-world welding shop scheduling problem.

C. Lu et al., *Intelligence Optimization for Green Scheduling in Manufacturing Systems*, Engineering Applications of Computational Methods 18,
https://doi.org/10.1007/978-981-99-6987-6_7

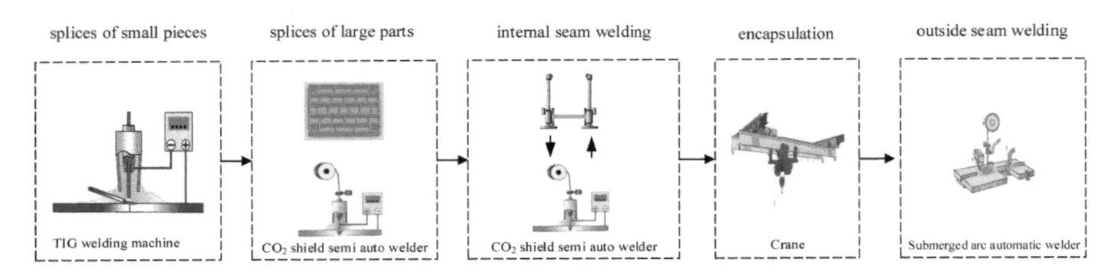

Fig. 7.1 The process flow chart for a welding plant

7.2 Problem Statement and Modeling

7.2.1 Problem Statement

Figure 7.1 presents a production process flow in a welding plant. The following is a description of this WSSP: A set of n tasks (or jobs) (i.e., $J = \{1, 2, \cdots, n\}$) should be completed in the same order through m stages. Each task $j \in J$ has m operations $O_{1j}, O_{2j}, \ldots, O_{\mathrm{ij}}, \ldots, O_{\mathrm{mj}}$, where O_{ij} is the operation of task j on the stage i. Multiple welders (or machines) can perform one operation O_{ij} at the same time [5]. When each operation is assigned to one and only one welder, each operation corresponds to a normal processing time. By allocating extra welders, normal operation processing times can be controlled (i.e., compressed). Once tasks begin to be processed, preemption and interruption are not permitted. Clearly, this problem is more complex than the traditional permutation flow shop scheduling problem (PFSP) [6]. Additionally, the features of the WSSP are presented as follows:

- This problem takes into account sequence dependent setup times. Specifically, the setup time is determined by the similarity of the two adjacent jobs.
- This problem takes into account transportation times. Transportation times are determined not only by the distance between successive stages (stations), but also by the task to be transported. Typically, the distance between successive stages is fixed. As a result, transportation times vary depending on the adjacent jobs [7].
- When considering setup and transportation times at the same time, overlapping is permitted. The starting time of one operation is determined by the completion times of its predecessor's setup and transportation. Figure 7.2 shows the overlapping of setup and transportation times.

7.2.2 Mathematical Modeling

The related notations and decision variables in this chapter are presented below.

(1) Notations

n : the total number of tasks (or jobs).

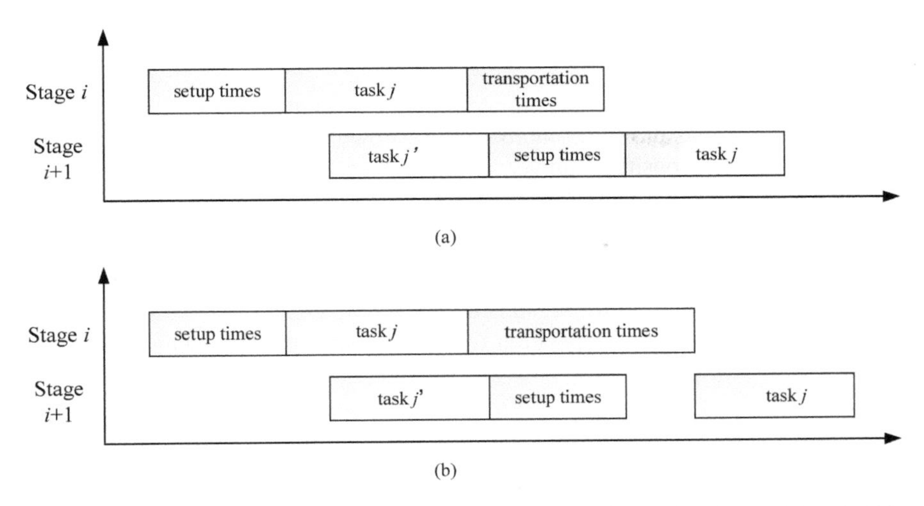

Fig. 7.2 Overlapping of setup and transportation times when **a** the starting time is determined by setup times and **b** the starting time is determined by transportation times

m : the total number of stages.
j : index of task, $j = 1, 2, \cdots, n$.
i : index of stage, $i = 1, 2, \cdots, m$.
$\pi(j)$: the job in the j-th position for a permutation of tasks.
$C_{\max}$: the makespan of the schedule, i.e., the completion time of the production process.
C_i : the completion time of all the tasks at stage i.
$\mathcal{M}_i$: a set of available welders (or machines) at stage i.
TM_i : the total number of available welders at stage i.
k : index of the welder, $k = 1, 2, \cdots, \text{TM}_i$.
O_{ij} : the operation of task j at stage i.
p_{ij} : the normal processing time of the operation O_{ij}. That is, the normal processing time of the operation should satisfy condition $N_{\text{ij}} = 1$.
p^a_{ij} : the actual processing time of the operation O_{ij}, namely, $p^a_{\text{ij}} = \frac{p_{\text{ij}}}{N_{\text{ij}}}$.
S_{ij} : the starting time of the operation O_{ij}.
ST_{i0j} : the setup time required to process task j first at stage i.
$\text{ST}_{\text{ijj}'}$: the setup time between task j and task j' at stage i.
$t_{\text{jii}'}$: the transportation time of task j from stage i to i'.
t_{j0i} : the transportation time of task j from warehouse to stage i.
t_{jiw} : the transportation time of task j from stage i to warehouse.
wt_{ij} : the idle time of operation $O_{i,j}$.
P_{basic} : the basic power during the whole production process (kW).
P^{trans}_j : the transportation power of the task j (kW).
P^{idle}_i : the idle power of a welder at stage i (kW).
P^{load}_i : the loading power of a welder at stage i (kW).

P_i^{setup} : the setup power of a welder at stage i (kW).
T_b : the breakeven duration.
L_{ij} : the equivalent continuous sound pressure level matrix (dB(A)).
T : the maximal allowable number to shut down a welder at each stage.
L : a very large positive number.

(2) Decision variables

π : a permutation of tasks.
N_{ij} : the actual quantity of the welders to process the O_{ij}.
S_{ij} : the starting time of the operation O_{ij}. It should meet the following conditions.

$$S_{ij} = \begin{cases} \max(t_{j0i}, ST_{i0j}) & \text{if } O_{ij} \text{ is the first operation of task } j \text{ and the first operation at stage } i \text{ (i.e., } i = 1\text{).} \\ \max\left(t_{j0i}, S_{ij'} + p^a_{ij'} + ST_{ij'j}\right) & \text{if } O_{ij} \text{ is the first operation of task } j \text{ and is started just after } O_{ij'} \text{ at stage } i \text{ (i.e., } i = 1\text{).} \\ \max\left(S_{i'j} + p^a_{i'j} + t_{ji'i}, ST_{i0j}\right) & \text{if } O_{i'j} \text{ is completed at stage } i' \text{ and then at stage } i \text{ and } O_{ij} \text{ is the first operation on stage } i. \\ \max\left(S_{i'j} + p^a_{i'j} + t_{ji'i}, S_{ij'} + p^a_{ij'} + ST_{ij'j}\right) & \text{if } O_{i'j} \text{ is completed at stage } i' \text{ and then at stage } i \text{ and } O_{ij} \text{ is processed just after } O_{ij'}. \end{cases}$$

$$x_{i'ij} = \begin{cases} 1, & \text{if } O_{i'j} \text{ precedes } O_{ij} \\ 0, & \text{if } O_{ij} \text{ precedes } O_{i'j} \end{cases}$$

$$y_{ij'j} = \begin{cases} 1, & \text{if } O_{ij'} \text{ precedes } O_{ij} \\ 0, & \text{if } O_{ij} \text{ precedes } O_{ij'} \end{cases}$$

$$z_{ijk} = \begin{cases} 1, & \text{if welder } k \text{ processes } O_{ij} \\ 0, & \text{otherwise} \end{cases}$$

The first objective of this problem is to minimize the makespan C_{max}. It is defined by the following formula:

$$\min f_1 = C_{max} = \max\{C_i | i = 1, \ldots, m\} \tag{7.1}$$

The second objective is to minimize noise pollution. It can be a logarithmic nonlinear summation of sound levels produced by at least one noise source. The noise pollution function is given as follow [8]:

$$L_{eq} = 10 \cdot \lg \frac{1}{T_e} \int_0^{T_e} 10^{0.1 \cdot L_i} dt = 10 \cdot \lg \frac{\sum_i t_i \cdot 10^{0.1 \cdot L_i}}{\sum_i t_i} \tag{7.2}$$

where L_{eq} is the equivalent continuous sound pressure level during the measured time interval (the unit is dB(A)), T_e is the noise exposure time, the i-th measured time interval is denoted by t_i, L_i denotes the equivalent continuous sound pressure level over a time interval t_i. Because multiple welders can process the same task at the same time, the overlapping of multiple noise sources is taken into account in this problem. In other words, this noise is linked to the machine assignment matrix. Equation (7.2) can thus be transformed into the following formula.

$$\min f_2 = L_{eq} = 10 \cdot \lg \frac{\sum_{i=1}^{m} \sum_{j=1}^{n} N_{ij} \cdot p_{ij}^{a} \cdot 10^{0.1 \cdot L_{ij}}}{\sum_{i=1}^{m} \sum_{j=1}^{n} p_{ij}^{a}} \quad (7.3)$$

The third objective is to minimize the total energy consumption (kWh). It is made up of the fundamental energy consumption E_{basic}, setup or preparation energy consumption E_{setup}, transportation energy consumption E_{trans}, and welding energy consumption E_{weld}. As a result, the third objective is as follows:

$$\min f_3 = E_{basic} + E_{setup} + E_{trans} + E_{weld} \quad (7.4)$$

The four types of energy consumption mentioned above are discussed in the following sections.

(1) **Energy consumption for basic stage (E_{basic})**

E_{basic} represents the basic energy consumption including control system, light, manpower, cutting fluid, and air conditioner. E_{basic} is determined by the basic power and makespan. It is written as follows:

$$E_{basic} = P_{basic} \times t_{basic} \quad (7.5)$$

(2) **Energy consumption for setup stage (E_{setup})**

E_{setup} represents the preparation energy consumption between the two consecutive tasks (or jobs). In general, the less energy the preparation work consumes, the smaller the physical property difference between immediate tasks. Thus, E_{setup} is associated with the schedule scenario. The formula of E_{setup} can be defined below:

$$E_{setup} = \sum_{i=1}^{m} \sum_{j=1}^{n} \sum_{j'=1}^{n} N_{ij} \cdot P_{i}^{setup} \cdot \mathrm{ST}_{i,j,j'} \quad (7.6)$$

(3) **Energy consumption for transportation stage (E_{trans})**

E_{trans} is affected by transportation times. The magnitude of transportation times is clearly determined by the distance between consecutive stations as well as the task type. The distance between consecutive stations, on the other hand, is usually fixed. Different transporters are required for different tasks [9]. As a result, transportation times vary depending on the task. E_{trans} can be written as follow:

$$E_{trans} = \sum_{j=1}^{n} \sum_{i=0}^{m} P_{j}^{trans} \cdot t_{jii'} \quad (7.7)$$

where $i' = i + 1$, when $i = 0$, it represents the task being moved from the warehouse to the first stage (i.e., t_{j01}); when $i = m$, it denotes that the task has been moved from the final stage to the warehouse (i.e., t_{jmw}).

(4) **Energy consumption for welding phase ($E_{\mathbf{weld}}$)**

Welders (or machines) typically have two states during the welding phase: loading and idle. According to the literature [10], energy consumption during the welding phase is as follows:

$$E_{\mathrm{weld}} = \sum_{i=1}^{m}\sum_{j=1}^{n} C_i \cdot N_{\mathrm{ij}} \cdot \left[P_i^{\mathrm{idle}} \cdot (1 - K_i) + P_i^{\mathrm{load}} \cdot K_i\right] \tag{7.8}$$

where K_i is the utilization factor that governs the loading and idle states, $K_i = \frac{\sum_{j=1}^{n} p_{\mathrm{ij}}^{a}}{C_i}$ in this chapter, C_i represents the total completion time of all tasks at stage i, P_i^{idle} represents the welder's idle power at stage i P_i^{load} is the welder's power at stage i during the loading state.

A turn off/on approach was typically used in previous research to determine whether a machine (or welder) is shut down or not in the idle state. This method can significantly reduce energy consumption. However, using this method frequently may shorten the life of a machine. To address this issue, the new energy-saving strategy is proposed:

Step 1: For each stage i ($\forall i = 1, 2, \cdots, m$), perform the following procedures on a task j ($\exists j = 1, 2, \ldots, n$).

Step 2: At stage i scan the current task j from left to right, and calculate the value $wt_{i,j} \times N_{\mathrm{ij}}$, where wt_{ij} is the idle time of operation. Then, put these values into a set and sort them in non-ascending order.

Step 3: Select the former elements from the as a new set. This set includes all idle times during which a machine is likely to be shut down.

Step 4: For all the welders at stage, perform the turn off/on strategy on a welder or welders at the current stage if the simultaneously satisfies the following conditions: (namely,). Otherwise, keep the machine idle.

As a result, the welding phase's energy consumption can be expressed as follows:

$$E_{\mathrm{weld}} = \sum_{i=1}^{m}\sum_{j=1}^{n} C_i \cdot N_{\mathrm{ij}} \cdot \left[P_i^{\mathrm{idle}} \cdot (1 - K_i) \cdot Z_{\mathrm{ij}} + P_i^{\mathrm{load}} \cdot K_i\right] \tag{7.9}$$

$$Z_{\mathrm{ij}} = \begin{cases} 0, \text{ if } wt_i > T_b \wedge \mathrm{wt}_{\mathrm{ij}} \in \mathrm{WT}_i \\ 1, \text{ otherwise} \end{cases}$$

To summarize, the following is a mathematical model for the WSSP aimed at minimizing the makespan, noise pollution, and energy consumption:

$$\begin{cases} \min f_1 = C_{\max} \\ \min f_2 = 10 \cdot \lg \frac{\sum_{i=1}^{m} \sum_{j=1}^{n} p_{ij}^{a} \cdot 10^{0.1 \cdot L_{ij}}}{\sum_{i=1}^{m} \sum_{j=1}^{n} p_{ij}^{a}} \\ \min f_3 = E_{\text{basic}} + E_{\text{setup}} + E_{\text{trans}} + E_{\text{weld}} \end{cases} \tag{7.10}$$

Subject to

$$\sum_{\mathcal{M}_i} z_{ijk} = N_{ij} \geq 1, \forall i = 1, 2, \cdots, m;\ j = 1, 2, \cdots, n;\ k = 1, 2, \cdots, \text{TM}_i \tag{7.11}$$

$$N_{ij} \leq \text{TM}_i, \forall i = 1, 2, \cdots, m;\ j = 1, 2, \cdots, n \tag{7.12}$$

$$S_{ij} \geq t_{j0i}, \forall i = 1, 2, \cdots, m;\ j = 1, 2, \cdots, n \tag{7.13}$$

$$S_{ij} \geq ST_{i0j}, \forall i = 1, 2, \cdots, m;\ j = 1, 2, \cdots, n \tag{7.14}$$

$$L\left(1 - x_{i'ij}\right) + S_{ij} \geq S_{i'j} + \frac{p_{i'j}}{N_{i'j}} + t_{ji'i}, \forall i, i' = 1, 2, \cdots, m;\ j = 1, 2, \cdots, n \tag{7.15}$$

$$L \cdot x_{i'ij} + S_{i'j} \geq S_{ij} + \frac{p_{ij}}{N_{ij}} + t_{jii'}, \forall i, i' = 1, 2, \cdots, m;\ j = 1, 2, \cdots, n \tag{7.16}$$

$$L\left(1 - y_{ij'j}\right) + S_{ij} \geq S_{ij'} + \frac{p_{ij'}}{N_{ij'}} + ST_{ij'j}, \forall i = 1, 2, \cdots, m;\ j, j' = 1, 2, \cdots, n \tag{7.17}$$

$$L \cdot y_{ij'j} + S_{ij'} \geq S_{ij} + \frac{p_{ij}}{N_{ij}} + ST_{ijj'}, \forall i = 1, 2, \cdots, m;\ j, j' = 1, 2, \cdots, n \tag{7.18}$$

$$C_{\max} \geq S_{ij} + \frac{p_{ij}}{N_{ij}} + t_{jiw}, \forall i = 1, 2, \cdots, m;\ j = 1, 2, \cdots, n \tag{7.19}$$

$$x_{i'ij} \in \{0, 1\}, y_{ij'j} \in \{0, 1\}, N_{ij} \in N^{+}, \forall i = 1, 2, \cdots, m;\ j = 1, 2, \cdots, n \tag{7.20}$$

The objectives of Eq. (7.10) are to minimize makespan, noise pollution, and total energy consumption. Constraint (7.11) ensures that at least one welder can process each operation. Constraint (7.12) ensures that the number of machines used on each operation is limited to its maximum value. Constraint (7.13) ensures that a task's current operation cannot begin until it is transported from the warehouse to the stage performing its first operation. Constraint (7.14) states that the task's first operation cannot begin until its setup time has been completed. Constraints (7.15)–(7.16) ensure the priority relationship among tasks' operations. The priority relationship among operations on the same stage is implied by constraints (7.17)–(7.18). Constraint (7.19) specifies the time required to complete each task. The range of decision variables is imposed by constraint (7.20).

7.3 Proposed Algorithm

To solve this WSSP, a hybrid multi-objective grey wolf optimizer with genetic algorithm (HMOGWO) is proposed in this section. The following are the primary reasons for using this HMOGWO. First, GA and GWO are capable of resolving such a problem [11]. Second, the hybrid algorithm is capable of balancing global and local searches [12]. Global search operators, such as the genetic operator and the grey wolf search operator, can be used to explore unknown areas of the search space. Local search based on the problem property, on the other hand, can be used to exploit the potential promising solution in the neighborhood of the high-quality solutions discovered. For these reasons, we propose a hybrid multi-objective optimization algorithm for this scheduling problem. The flow chart of the proposed HMOGWO is shown in Fig. 7.3. The main improvement strategies include: encoding and decoding, initialization, the update operator, and the local search. The sections that follow provide a detailed description of these procedures.

7.3.1 Encoding and Decoding

When using the proposed algorithm to solve this WSSP, the encoding scheme is critical. This WSSP must address both the task (or job) processing sequence and the machine assignment matrix at the same time. As a result, a two-layer encoding scheme is used in this chapter. The task processing sequence is represented by the first part, denoted by π. The machine assignment matrix is represented by the second part, denoted by N.To explain this encoding scheme, one example with 4 tasks and 5 stages is presented in Fig. 7.4.

In Fig. 7.4, the first part π is [2-4-3-1]. The second part N is a matrix with 5 stages and 4 tasks. The element N_{ij} represents the quantity of machines assigned to the O_{ij}. Figure 7.5 shows a Gantt chart of this solution. N_{23} represents that the quantity of machines assigned to task 3 at stage 2 (i.e., O_{23}) is 2. The O_{23} is represented by the red box with one white bar, which denotes the additional one welder to process this operation. The box without white bars (e.g., O_{12}) represents the operation with normal processing times. For example, the operation O_{12} is processed by one and only one welder.

7.3.2 Initialization

In the scheduling problem, an efficient initialization procedure is critical. We proposed an improved initialization strategy to ensure population quality and diversity. This initialization strategy's main procedure is as follows: To begin, generate an initial population at random and then randomly select two solutions from the

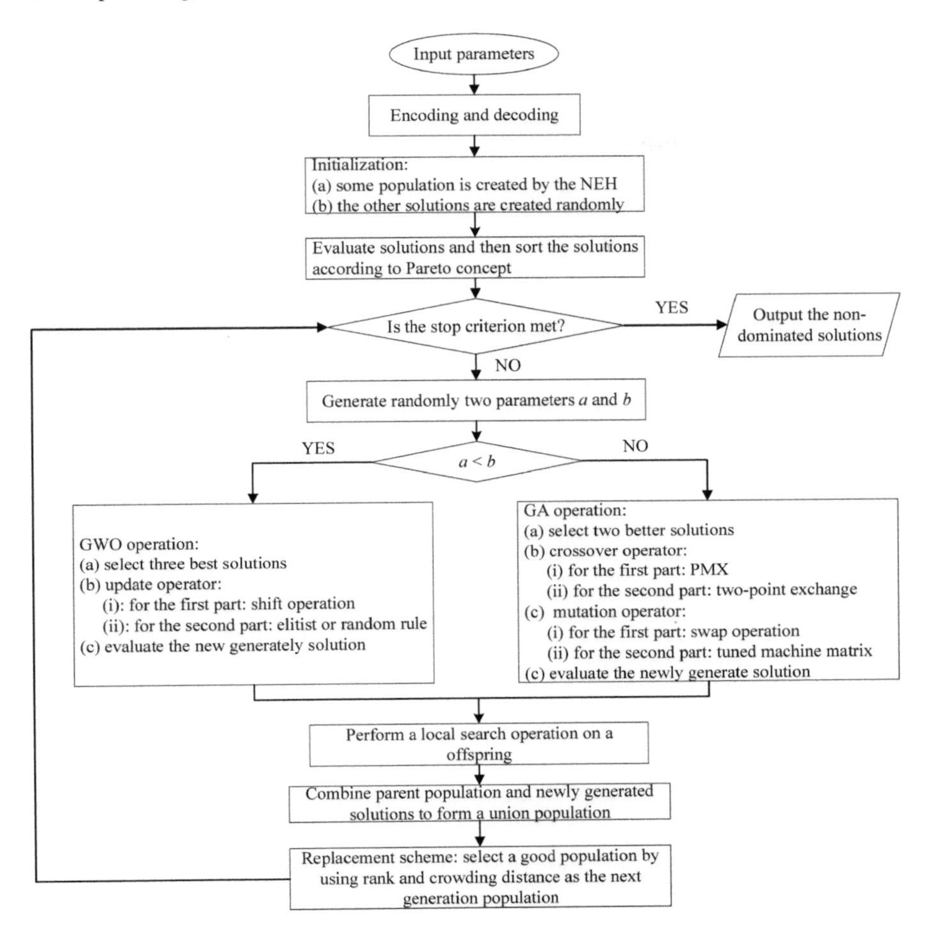

Fig. 7.3 Flow chart of the proposed HMOGWO

population. The quantity of machine matrix is all set to 1 for the second part of one solution. The quantity of machine matrix is all set to the maximum allowable value in the other solution. Then, on one-third of the initial population, apply the NEH method to different machine assign matrices. Finally, on a new initial population, use a fast non-dominated sorting method.

7.3.3 Update Operation

The proposed algorithm's update operator combines the genetic operator and the grey wolf search operator. This combination of two search operators has the potential to increase search diversity. They are thoroughly described below.

Fig. 7.4 Encoding of a solution

$$\begin{matrix} J_2 & J_4 & J_3 & J_1 \end{matrix}$$
$$\pi = [\,2 \;\; 4 \;\; 3 \;\; 1\,]$$

$$\begin{matrix} & J_1 & J_2 & J_3 & J_4 \end{matrix}$$
$$N = \begin{matrix} stage1 \\ stage2 \\ stage3 \\ stage4 \\ stage5 \end{matrix}\begin{bmatrix} 1 & 1 & 3 & 2 \\ 1 & 1 & 2 & 1 \\ 2 & 2 & 3 & 1 \\ 1 & 1 & 1 & 1 \\ 2 & 2 & 3 & 1 \end{bmatrix}$$

Before performing an update operation on the genetic operator, use a binary tournament selection method to select two good solutions. To ensure solution feasibility, we use partially matched crossover (PMX) to update the first part of the two solutions [13]. Meanwhile, the second part of the two solutions is updated using two-point crossover. This unique genetic operator is as follows:

For the first part:

Step 1: Step 1: Choose two positions on two parents at random and define the area between the two positions as a matching substring.

Step 2: Step 2: Produce temporary offspring by swapping the matching substrings of two parents.

Step 3: Step 3: Create a mapped relationship between the tasks that are in conflict. When the same task index is assigned more than once, the sequence's mapped relationship is defined.

Step 4: Step 4: Using the mapped relationship, make the task sequence feasible without changing the substring.

For the second part:

Step 1: Step 1: Choose two positions at random and define the substring between them as an exchange area.

Step 2: On the second part, swap the areas between the two positions.

Jobs	Normal processing times				
	Stage 1	*Stage 2*	*Stage 3*	*Stage 4*	*Stage 5*
1	10	15	30	20	25
2	30	10	20	25	15
3	35	10	25	5	30
4	10	30	20	10	15

Jobs	Transportation times					
	w to s1	*s1 to s2*	*s2 to s3*	*s3 to s4*	*s4 to s5*	*s5 to w*
1	6	10	15	20	10	14
2	5	10	20	15	15	14
3	4	15	10	25	10	14
4	3	10	10	15	5	14

Jobs	Setup times			
	1	2	3	4
1	-	10	20	15
2	12	-	10	20
3	5	9	-	16
4	10	5	9	-

Jobs	Setup times				
	Stage 1	*Stage 2*	*Stage 3*	*Stage 4*	*Stage 5*
1	10	5	14	13	6
2	4	8	7	10	18
3	8	10	5	12	7
4	9	18	7	16	12

(a) instance

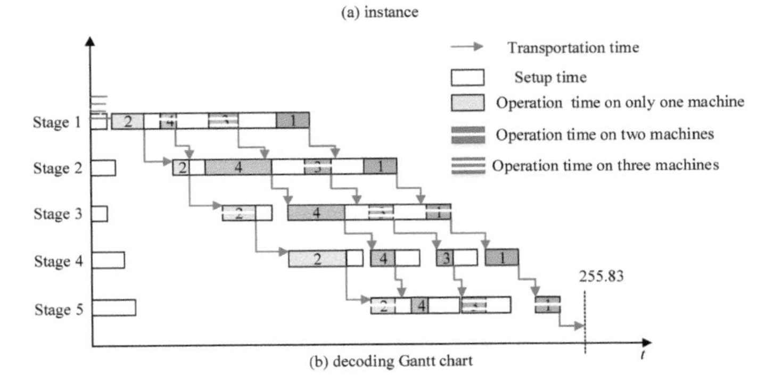

(b) decoding Gantt chart

Fig. 7.5 Gantt chart of a solution in an instance

An example of this crossover operation is shown in Fig. 7.6.

The mutation operation can be divided into two parts. The following are the detailed procedures:

Step 1: Step 1: Select two positions at random for the first part of a solution, then swap the positions of two corresponding tasks.

Step 2: Step 2: The machine quantity in this machine assignment matrix will be fine-tuned within its range for the second part of a solution.

For the search operator of GWO, the main idea of this GWO is that the search process is guided by the best three wolves (or solutions) denoted by symbols α, β, and δ [14]. This search operator is given as follows:

The following formula yields a new task sequence for the first part:

$$\pi_i^{t+1} = \begin{cases} \text{shift}\left(\pi_i^t, \left(\pi_\alpha^t - \pi_i^t\right)\right) & \text{if rand} < \frac{1}{3} \\ \text{shift}\left(\pi_i^t, \left(\pi_\beta^t - \pi_i^t\right)\right) & \text{else if rand} < \frac{2}{3} \\ \text{shift}\left(\pi_i^t, \left(\pi_\delta^t - \pi_i^t\right)\right) & \text{otherwise} \end{cases} \tag{7.21}$$

where shift(x, d) represents that the element x can be shifted to the right or left with $|d|$ units, in detail, the element x can be shifted to the right with $|d|$ units if d is

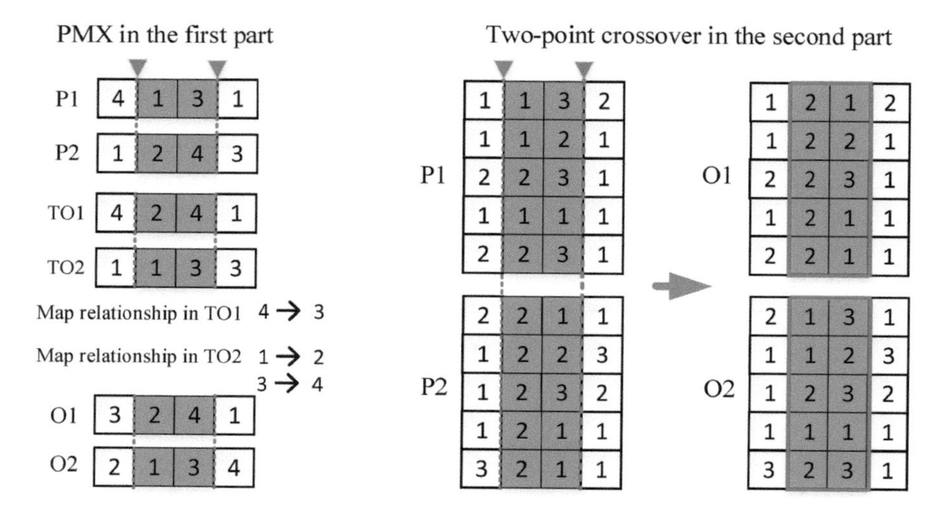

Fig. 7.6. Crossover operation in the two parts of a solution

a negative integer number. Otherwise, the element x can be shifted to the left with $|d|$ units. It should be noted that if the shift units exceed the boundary range, the boundary element is connected in the opposite direction to that, rand is a uniform random value between [0,1], π_i^t is a solution corresponding to a permutation π in the t iteration.

The following equation generates a new machine assignment matrix for the second part:

$$N_{i,j}^{t+1} = \begin{cases} N_{\text{ij}}^t \in N(\alpha)_{\text{ij}}^t \text{ or } N(\beta)_{\text{ij}}^t \text{ or } N(\delta)_{\text{ij}}^t & \text{if rand} \leq a \\ N_{\text{ij}}^t \in N^t & \text{otherwise} \end{cases} \tag{7.22}$$

where $N_{i,j}^t$ is the quantity of machines assigned on the operation O_{ij} in the t iteration, the newly generated $N_{i,j}^{t+1}$ is obtained from two rules: (a) elitist selection and (b) random selection.

7.3.4 *The New Local Search*

According to the property of this problem, a local search based on a critical path or a non-critical path is designed. The neighbor of a schedule is obtained in this local search by adjusting the quantity of machine (welder) on a critical path or non-critical path. The critical path is the longest route from the beginning to the end of the entire manufacturing process. There are one or more critical paths in a schedule. In single-objective scheduling optimization problems, the local search approach is effective.

In MOPs, however, any improvement in one objective would result in a deterioration in at least another. As a result, the previous local algorithm is not directly applied to this problem. To improve the quality of solutions, we proposed a new local search algorithm based on the problem properties. This WSSP has the following properties:

Property 7.1 Obviously, increasing one welder (if it has one) on a critical path without affecting the task sequence (i.e., the first part) can improve makespan.

Proof 7.1 For a solution $\mathbf{x} = (\pi, N)$, where $\pi = (\pi(1), \pi(2), \cdots, \pi(n))$ represents a permutation on a set of tasks (i.e., the first part of a solution), N denotes a machine assignment matrix (i.e., the second part of a solution). The completion time $C_{\max}(\pi, N)$ (i.e., makespan) of a given schedule can be defined by the following expression.

$$\begin{aligned} C_{\max}(\pi, N) = \max_{1 \le t_1 \le t_2 \cdots \le t_m} \Bigg(& S_{1,\pi(j)} + \sum_{j=1}^{t_1} p^a_{1,\pi(j)} + \sum_{j=1}^{t_1-1} ST_{1,\pi(j),\pi(j+1)} \\ & + t_{\pi(j),1,2} + \sum_{j=t_1}^{t_2} p^a_{2,\pi(j)} + \sum_{j=t_1}^{t_2-1} ST_{2,\pi(j),\pi(j+1)} + t_{\pi(j),2,3} +, \cdots, \\ & + \sum_{j=t_{m-1}}^{t_m} p^a_{m,\pi(j)} + \sum_{j=t_{m-1}}^{t_m-1} ST_{m,\pi(j),\pi(j+1)} + t_{\pi(j),m,w} \Bigg) \end{aligned} \tag{7.23}$$

A sequence of $t = (t_1, t_2, \cdots, t_{m-1})$, such that $1 \le t_1 \le t_2 \le \cdots \le t_{m-1} \le t_m = n$, is called a path in a permutation π. The length of $C(t, N)$ is defined as follow:

$$\begin{aligned} C(t, N) = & S_{1,\pi(j)} + \sum_{j=1}^{t_1} p^a_{1,\pi(j)} + \sum_{j=1}^{t_1-1} ST_{1,\pi(j),\pi(j+1)} + t_{\pi(j),1,2} \\ & + \sum_{j=t_1}^{t_2} p^a_{2,\pi(j)} + \sum_{j=t_1}^{t_2-1} ST_{2,\pi(j),\pi(j+1)} + t_{\pi(j),2,3} +, \cdots, \\ & + \sum_{j=t_{m-1}}^{t_m} p^a_{m,\pi(j)} + \sum_{j=t_{m-1}}^{t_m-1} ST_{m,\pi(j),\pi(j+1)} + t_{\pi(j),m,w} \end{aligned} \tag{7.24}$$

For any π, it has $C_{max}(\pi, N) \ge C(t, N)$, where $t = (t_1, t_2, \cdots, t_{m-1})$ is any path in π. A path $u = (u_1, u_2, \cdots, u_{m-1})$ is called a critical path if it satisfies the following formula.

$$C_{\max}(\pi, N) = C(u, N) = S_{1,\pi(j)} + \sum_{j=1}^{u_1} p^a_{1,\pi(j)} + \sum_{j=1}^{u_1-1} ST_{1,\pi(j),\pi(j+1)}$$

$$+ t_{\pi(j),1,2} + \sum_{j=u_1}^{u_2} p^a_{2,\pi(j)} + \sum_{j=u_1}^{u_2-1} ST_{2,\pi(j),\pi(j+1)} + t_{\pi(j),2,3} +, \cdots,$$

$$+ \sum_{j=u_{m-1}}^{n} p^a_{m,\pi(j)} + \sum_{j=u_{m-1}}^{n-1} ST_{m,\pi(j),\pi(j+1)} + t_{\pi(j),m,w} \tag{7.25}$$

Let $N_{i,\pi(j)}$ represent the number of machines on the critical path. When the $N_{i,\pi(j)}$ is increased to $N'_{i,\pi(j)}$ by one (if it can), the new actual processing time $p'^a_{1,\pi(j)} = \frac{p_{i,\pi(j)}}{N_{i,\pi(j)}'} < p^a_{1,\pi(j)} = \frac{p_{i,\pi(j)}}{N_{i,\pi(j)}}$ and it will decreases. That is, $C_{\max}(\pi, N') \leq C_{\max}(\pi, N)$. Because a schedule may contain multiple critical paths, repeat the preceding procedure until $C_{\max}(\pi, N') < C_{\max}(\pi, N)$ (if it can).

Property 7.2 When reducing additional machines on a non-critical path with a feasible idle time interval (if it has one) under the same makespan, total energy consumption and noise pollution can be reduced simultaneously.

Proof 7.2 Given an idle time interval $\left[t^E_{i,\pi(j)}, t^S_{i,\pi(j+1)}\right]$, starting at $t^S_{i,\pi(j+1)}$ and finishing at $t^E_{i,\pi(j)}$ for two consecutive tasks at stage *i*. If the time interval meets the following conditions, the number of machines on the non-critical path can be reduced by one.

$$\begin{cases} t^E_{i,\pi(j)} + \mathrm{ST}_{i,\pi(j),\pi(j+1)} + \frac{p_{i,\pi(j)}}{N^{\max}_{i,\pi(j)}} \leq t^S_{i,\pi(j+1)} \\ t^E_{i,\pi(j)} + t_{\pi(j),i,i+1} + \frac{p_{i,\pi(j)}}{N^{\max}_{i,\pi(j)}} \leq t^S_{i,\pi(j+1)} \end{cases}$$
$$i = 1, 2, \ldots, m-1;\ j = 1, 2, \ldots, n-1 \tag{7.26}$$

where $N^{\max}_{i,\pi(j)}$ is the maximal allowable quantity of machines for the operation $O_{i,\pi(j)}$. When a given time interval satisfies the above constraint, the number of machines on the non-critical path can be reduced by one (if it can). The energy consumption and noise criteria are obviously proportional to the number of machines. The number of machines is reduced by one, and as a result, energy consumption and noise pollution are reduced further. Because the previous critical path remains unchanged, and the makespan criterion is unaffected by the decrease in the number of machines on the non-critical path. As a result, after adjusting the number of machines, the new solution outperforms the previous solution.

The pseudocode of the proposed local search is provided in Algorithm 7.1.

7.4 Experiments

To demonstrate the HMOGWO's effectiveness on this WSSP, it is compared to two other well-known MOEAs: NSGA-II and SPEA2. In addition, we investigate the impact of the local search and NEH initialization strategies: MOGWO1 and

MOGWO2. Java is used to code all algorithms. The experiments are carried out on a computer equipped with an Intel Core i5, 2.39 GHz, 4 GB RAM, and the Windows 8 operating system.

7.4.1 Parameter Settings

The instances are randomly generated as shown in Table 7.1. Each instance is defined by the symbol "WSSP_*n*_*m*". For instance, "WSSP_20_5" denotes that the instance is characterized by 20 tasks and 5 stages. These instances include different problem scales. To make a fair comparison, all MOEAs use the same stop condition (i.e., the maximal number of function evaluations (NFEs)). The maximal NFEs is set to 500,000. All the MOEAs adopt the proposed encoding and decoding scheme. The other parameter settings for MOEAs are as follows. The population size is 100, and external archive size is 100. The PMX and two-point crossover probability of SPEA2 and NSGA-II is 0.9. The swap and fine-tuned mutation probability is 0.2.

Algorithm 1: Local Search

Table 7.1 Parameter settings for instances

Input variables	Distribution
Number of tasks (n)	20, 40, 60,80, 100
Number of stage (m)	5, 10
Number of available machines at each stage ($\mathcal{M}$)	Discrete uniform [1, 4]
Normal processing time (p)	Discrete uniform [30,50] min
Setup time (sT)	Discrete uniform [5,20] min
Transportation time (t)	Discrete uniform [10,20] min
Noise (L_{ij})	Discrete uniform [70,80] dB
The transportation power of task (P_j^{trans})	Continuous uniform [5, 10] kW
The idle power of a welder (P_i^{idle})	Continuous uniform [3, 10] kW
The loading power of a welder (P_i^{load})	Continuous uniform [20,30] kW
The basic power (P_{basic})	5 kW
The breakeven duration (T_b)	5 min
The maximal allowable number to shut down a welder (T) at each stage	4

Input: a existent schedule **x**
Output: a new schedule **x**$'$
1. Select a denoted by $\mathbf{x} = (\pi, N)$ solution-and count the corresponding critical machine assign matrix.
2. **When** $NFEs \geq \frac{3 \cdot NFEs_{max}}{4}$, execute the procedures following . //$NFEs$ present number of function evaluations; $NFEs_{max}$ is the maximal $NFEs$.
3. **If** random number < 0.9
   3.1 **If** rand < 0.5 // rand is a random number from 0 to 1.
   According to problem property 1 generate a new solution $\mathbf{x}' = (\pi, N')$. Its procedure is as follows:
   For each stage i=1 to m // m is the total number of stages.
   For each job j=1 to n// n is the total number of jobs.
   If $O_{i,\pi(j)}$ is on the critical path
   Find the welders with corresponding quantity, namely, $N_{i,\pi(j)}$. // N_i^{max} is the maximal value.
   Then, $N'_{i,\pi(j)} = N_{i,\pi(j)} + 1$. If $N'_{i,\pi(j)} > N_i^{max}$, then $N'_{i,\pi(j)} = N_i^{max}$.
   Break.
   End if
   End for
   End for
   3.2 **Else**
   according to problem property 2 (if it has a suitable time interval) generate a new solution $\mathbf{x}' = (\pi, N')$-. Its procedure is as follows:
   For each stage i=1 to m -1// m is the total number of stages.
   For each job j=1 to n-1// n is the total number of jobs.
   Compute interval of idle time $wt_{i,\pi(j)} = \left[t^E_{i,\pi(j)}, t^S_{i,\pi(j+1)}\right]$.
   If $O_{i,\pi(j)}$ is on the non-critical path and the $wt_{i,\pi(j)}$ meets the Eq. (26).
   Find the welders with corresponding quantity, namely, $N_{i,\pi(j)}$. // N_i^{max} is the minimal value.
   Then, $N'_{i,\pi(j)} = N_{i,\pi(j)} - 1$. If $N'_{i,\pi(j)} < N_i^{min}$, then $N'_{i,\pi(j)} = N_i^{min}$.
   Break.
   End if
   End for
   End for
   3.3 **End If**
4. **Else** // avoid local convergence
   in the first part of a solution select two different elements randomly, and insert the latter element before the former element.
5. **End If**
6. **End When**

7.4.2 Comparison of HMOGWO with Other Algorithms

In this section, some performance metrics should be used to evaluate the algorithm's performance. Because the inverted generational distance (IGD) metric is a comprehensive performance indicator, it can be considered the primary performance metric [14]. The IGD with a low value is preferable. Each algorithm was run 30 times on each problem independently.

The statistical results obtained by various MOEAs are shown in Table 7.2, where "mean" and "std" represent the average and standard deviation values, respectively. The best results are highlighted in bold. Table 7.2 shows that the proposed HMOGWO algorithm outperforms five other MOEAs in the majority of cases. Furthermore, in order to demonstrate the effectiveness of the improvement strategies, we assess the contribution of each improvement strategy separately. MOGWO is HMOGWO without the NEH and local search strategies. MOGWO1 is the HMOGWO without a NEH initialization strategy. The HMOGWO without a local search strategy is represented by MOGWO2. The results of Table 7.2 clearly show that MOGWO2 outperforms MOGWO1. This means that, when compared to the local search strategy, the improved NEH initialization strategy has a greater positive effect on the overall performance of the HMOGWO. In most cases, HMOGWO outperforms MOGWO1 and MOGWO2. It means that combining the two strategies at the same time yields the best results. Furthermore, MOGWO1 and MOGWO2 outperform MOGWO. This implies that each improvement strategy has a positive effect on the algorithm's behavior and that combining the two strategies can help to improve the algorithm's performance even further.

Because MOEAs are stochastic, a Wilcoxon sign rank test is used to determine whether there is a significant difference between them. The level of confidence is set to 95% (corresponding to $a = 0.05$). The Wilcoxon sign rank test results, including R^+, R^-, and p-values, are shown in Table 7.3. The symbol " + " ("−") indicates whether the proposed HMOGWO outperforms or underperforms its counterpart. The " = " symbol indicates that there is no discernible difference between the proposal and the compared algorithm. R^+ represents the sum of ranks for instances where the

Table 7.2 Mean and standard deviation value of IGD metric on different MOEAs

Problems	MOEAs (mean/std)					
	NSGA-II	SPEA2	MOGWO	MOGWO1	MOGWO2	HMOGWO
WSSP_20_5	2.98e-02/ 1.9e-03	2.91e-02/ 2.4e-03	1.08e-02/ 9.4e-04	1.21e-02/ 3.0e-03	8.51e-03/ 1.3e-03	**7.75e-03/ 1.1e-03**
WSSP_20_10	1.99e-02/ 1.5e-03	1.95e-02/ 1.3e-03	1.13e-02/ 1.9e-03	1.34e-02/ 2.5e-03	8.55e-03/ 1.3e-03	**7.48e-03/ 1.4e-03**
WSSP_40_5	2.01e-02/ 2.0e-03	1.85e-02/ 1.0e-03	1.03e-02/ 1.3e-03	1.60e-02/ 2.3e-03	**5.34e-03/ 7.9e-04**	5.58e-03/ 8.9e-04
WSSP_40_10	3.06e-02/ 4.8e-03	2.68e-02/ 1.8e-03	1.44e-02/ 2.2e-03	1.58e-02/ 1.7e-03	1.01e-02/ 1.7e-03	**9.75e-03/ 1.5e-03**
WSSP_60_5	1.26e-02/ 1.4e-03	1.12e-02/ 7.6e-04	6.45e-03/ 9.2e-04	1.55e-02/ 2.7e-03	5.79e-03/ 1.0e-03	**5.51e-03/ 1.1e-03**
WSSP_60_10	2.30e-02/ 2.0e-03	2.02e-02/ 7.5e-04	1.61e-02/ 9.9e-04	2.32e-02/ 2.2e-03	1.24e-02/ 2.0e-03	**1.20e-02/ 1.7e-03**
WSSP_80_5	4.17e-02/ 1.7e-03	4.00e-02/ 1.3e-03	2.60e-02/ 2.1e-03	1.20e-02/ 1.6e-03	**5.97e-03/ 1.9e-03**	6.26e-03/ 1.6e-03
WSSP_80_10	3.25e-02/ 2.1e-03	2.69e-02/ 1.2e-03	1.45e-02/ 3.2e-03	1.63e-02/ 2.1e-03	1.04e-02/ 1.3e-03	**9.90e-03/ 1.8e-03**
WSSP_100_5	3.45e-02/ 8.7e-04	3.53e-02/ 4.5e-03	1.09e-02/ 1.8e-03	1.29e-02/ 7.8e-04	**9.61e-03/ 1.4e-03**	9.65e-03/ 1.6e-03
WSSP_100_10	1.95e + 00/ 7.7e-02	1.96e + 00/ 8.6e-02	1.07e + 00/ 1.4e-01	7.34e-01/ 1.0e-01	**4.34e-01/ 2.8e-01**	5.83e-01/ 7.4e-01

HMOGWO outperforms its competitor, while R- represents the opposite. The count of ratios for significant difference is recorded in the bottom row of this table. This table shows that the HMOGWO has more " + " counts than other MOEAs, particularly NSGA-II, SPEA2, MOGWO, and MOGWO1. Furthermore, in most cases, HMOGWO is superior to MOGWO2. This means that the proposal's superiority is overwhelming. The following are the reasons for the proposed HMOGWO's good performance. First, an improved NEH strategy improves convergence and diversity at the outset. Second, at a later stage, a local search scheme can effectively exploit promising solutions. As a result, the proposed HMOGWO can effectively address this WSSP.

7.4.3 *Case Study*

The HMOGWO algorithm is used in this section to solve a WSSP from a real-world welding workshop in China.

(1) **Introduction to case**

This real-world case was inspired by the need for a practical welding production with 10 tasks and 5 stages. Q235 low-carbon steel plate is commonly used for welding. Figure 7.7 depicts the crane structure's real-world welding process in a workshop. In this case, T_b is set to 5 min, and T is set to 4. $P_j^{\text{trans}} = [28, 30, 35, 30, 28, 30, 30, 35, 35, 35]$ kW, where $j = 1, \ldots, 10$. Tables 7.4, 7.5, 7.6, 7.7 and 7.8 show the other related data.

(2) **Results**

HMOGWO is tested against NSGA-II and SPEA2 in 30 independent runs. Table 7.9 summarizes the statistical findings. The Wilcoxon sign rank test results, including R+ , R−, and p-values, are shown in Table 7.10. The best IGD metric result is highlighted in bold font. On this case, we can see that the HMOGWO significantly outperforms other MOEAs.

The Pareto fronts (PFs) obtained by various MOEAs are presented to visualize the performance of these MOEAs. In terms of solution's quality and coverage, the proposed HMOGWO can outperform its competitors. Figure 7.8a depicts a three-dimensional plot with three objectives. It is difficult to determine which algorithm is the best. As a result, we compare the PFs discovered by various MOEAs under a 2-dimensional view angle. Figure 7.8b shows a plot with makespan and noise pollution criteria. The makespan and noise pollution are clearly at odds with one another. In terms of distribution and solution coverage, the HMOGWO outperforms its competitors. Figure 7.8c presents a graphical view only considering makespan and energy consumption. These two goals are in direct opposition to one another. Furthermore, when compared to other MOEAs, the HMOGWO can approach the true PF more closely. Figure 7.8d gives a 2-dimensional view involving the noise pollution and energy consumption. It demonstrates that the HMOGWO outperforms

Table 7.3 Wilcoxon signed rank test results for each instance (a level of significant $\alpha = 0.05$)

Problems	HMOGWO vs. NSGA-II			HMOGWO vs. SPEA2			HMOGWO vs. MOGWO			HMOGWO vs. MOGWO1			HMOGWO vs. MOGWO2		
	R^+	R^-	*p*-value/ win	R^+	R^-	*p*-value/ win	R^+	R^-	*p*-value/ win	R^+	R^-	*p*-value/ win	R^+	R^-	*p*-value/ win
WSSP_ 20_5	465	0	1.73e-06/ +	465	0	1.73e-06/ +	465	0	1.73e-06/ +	452	13	6.34e-06/ +	330	135	4.49e-02/ +
WSSP_ 20_10	465	0	1.73e-06/ +	465	0	1.73e-06/ +	465	0	1.73e-06/ +	460	5	2.88e-06/ +	377	88	3.00e-03/ +
WSSP_ 40_5	465	0	1.73e-06/ +	465	0	1.73e-06/ +	465	0	1.73e-06/ +	465	0	1.73e-06/ +	191	274	3.93e-01/ =
WSSP_ 40_10	465	0	1.73e-06/ +	465	0	1.73e-06/ +	463	2	2.13e-06/ +	465	0	1.73e-06/ +	296	169	1.91e-01/ =
WSSP_ 60_5	465	0	1.73e-06/ +	465	0	1.73e-06/ +	396	69	7.71e-04/ +	465	0	1.73e-06/ +	281	184	3.19e-01/ =
WSSP_ 60_10	465	0	1.73e-06/ +	465	0	1.73e-06/ +	463	2	2.13e-06/ +	465	0	1.73e-06/ +	274	191	3.93e-01/ =
WSSP_ 80_5	465	0	1.73e-06/ +	465	0	1.73e-06/ +	465	0	1.73e-06/ +	465	0	1.73e-06/ +	192	273	4.05e-01/ =
WSSP_ 80_10	465	0	1.73e-06/ +	465	0	1.73e-06/ +	455	10	4.73e-06/ +	465	0	1.73e-06/ +	279	186	3.39e-01/ =
WSSP_ 100_5	465	0	1.73e-06/ +	465	0	1.73e-06/ +	356	109	1.11e-02/ +	455	10	4.73e-06/ +	239	226	8.94e-01/ =
WSSP_ 100_10	435	30	2.56e-06/ +	435	30	2.56e-06/ +	434	31	2.85e-06/ +	357	108	2.60e-03/ +	191	274	5.66e-01/ =
+ / = /−	10/0/0			10/0/0			10/0/0			10/0/0			2/8/0		

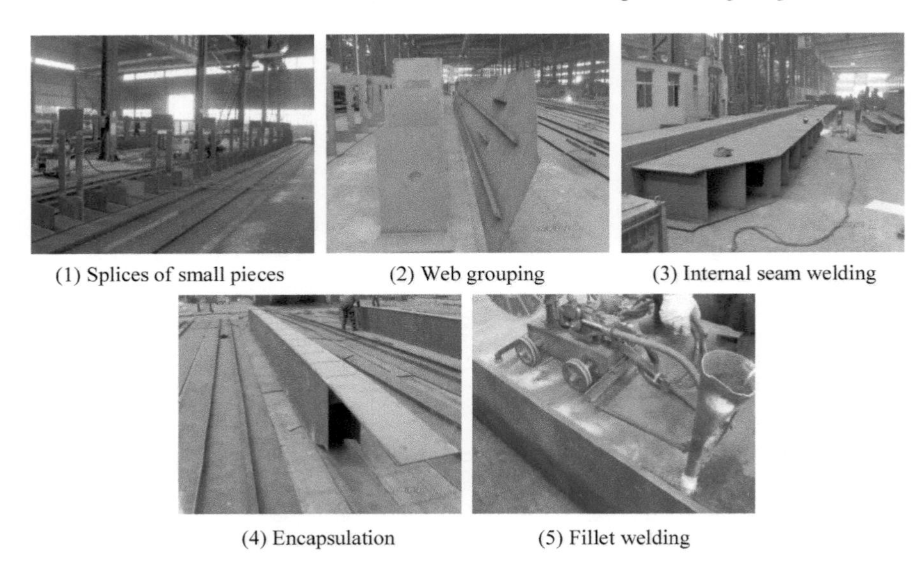

(1) Splices of small pieces (2) Web grouping (3) Internal seam welding

(4) Encapsulation (5) Fillet welding

Fig. 7.7 The welding process of the crane structure in a rea-world workshop

Table 7.4 Available welders in the welding workshop

Machine type	Available machine quantity	Stage	load power (kW)	Idle power (kW)	Setup power (kW)
TIG welding machine	3	Slicing	28	3.6	10
CO_2 shield semi-auto welder	4	Web grouping	30	4	10
CO_2 shield semi-auto welder	4	Internal seam weld	35	6	15
Crane	1	Encapsulation	30	10	25
Submerged arc automatic welder	3	Filet welding	28	5	10

its competitors in terms of solution distribution. These observations imply that the three objectives of the studied problem are frequently at odds with one another. Table 7.11 also includes the three PFs and corresponding solutions at points A, B, and C. The Gantt charts of the solutions of point A, B, and C are plotted in Fig. 7.9. The corresponding result, according to the practical production schedule (First In First Out rule), is (518.0 min, 77.74 dB, 1077.99 kWh). In terms of makespan, the proposed HMOGWO can increase productivity by 32.82% when compared to

Table 7.5 The normal processing times of tasks on each stage

Task type	Span (m)	Task NO	Processing time (min)/ noise (dB)				
			Slicing (Stage 1)	Web grouping (Stage 2)	Internal seam weld (Stage 3)	Encapsulation (Stage 4)	Filet welding (Stage 5)
5tA5A6	22.5	1	20/80	25/58	30/66	20/73	24/80
5tA5A6	25.5	2	31/72.4	40/68	49/72	25/80	35/80
5tA5A6	28.5	3	35/63	40/65	45/72	36/80	39/82
10tA5A6	22.5	4	10/55	12/69	14/70	11/72	14/71
10tA5A6	25.5	5	29/42	35/78	41/74	27/64	33/72.4
10tA5A6	28.5	6	20/50	25/66.8	30/75	20/72	24/78
20t16tA5	22.5	7	31/67	40/70	49/71	25/81	35/72
20t16tA5	25.5	8	35/63	40/56	45/79	36/81	39/80
20t16tA5	28.5	9	10/62	12/62	14/77	11/72	14/78
20t16tA6	25.5	e	29/60	35/55	41/78	27/73	33/75

Table 7.6 The setup times of the first task to processed on each stage

Setup time(min)	task 1	task 2	task 3	task 4	task 5	task 6	task 7	task 8	task 9	task e
Slicing	10	4	8	9	14	11	12	6	4	2
Web grouping	5	8	10	18	7	5	6	15	10	5
Internal seam weld	14	7	5	7	4	15	7	17	6	7
Encapsulation	13	10	12	16	9	3	10	2	14	8
Filet welding	6	18	7	12	15	9	12	7	4	11

practical production. When compared to the practical schedule, noise pollution can be reduced by 2.7%. When compared to the practical schedule, the proposed HMOGWO can save nearly 14.93% of energy consumption.

In this section, we also examine various energy-saving scenarios. Scenarios 1 and 2 represent, respectively, the proposed energy-saving strategy and the turn on/off strategy. Both energy-saving strategies can significantly reduce energy consumption. However, the contributions of these two scenarios to energy savings differ. Two energy-saving scenarios have the same task sequence and machine matrix to allow for a fair comparison. Figure 7.10 presents the results generated by the HMOGWO with two energy-saving scenarios. In terms of energy consumption, it clearly shows that scenario 1 outperforms scenario 2 in the same environment. The proposed energy-saving strategy can save 2.54% of total energy consumption, indicating its superiority. Furthermore, the proposed energy-saving strategy not only reduces energy consumption but also increases welder's life.

Table 7.7 The setup times between two consecutive tasks

Setup time (min)	Task 1	Task 2	Task 3	Task 4	Task 5	Task 6	Task 7	Task 8	Task 9	Task 10
Task 1	–	10	20	15	10	5	4	12	11	8
Task 2	12	–	10	20	6	9	10	18	2	13
Task 3	5	9	–	16	8	10	5	8	15	20
Task 4	10	5	9	–	14	17	6	10	12	15
Task 5	7	5	3	15	–	10	8	5	10	15
Task 6	15	10	8	5	8	–	8	12	10	12
Task 7	7	5	3	6	9	10	–	11	10	5
Task 8	11	15	10	6	8	5	12	–	12	7
Task 9	15	10	13	8	9	5	6	12	–	10
Task 10	12	8	10	6	9	5	10	15	12	–

Table 7.8 The transportation times of task between two consecutive stages or stage and warehouse

Task No.	Transportation time(min)					
	Warehouse to stage 1	Stage 1 to stage 2	Stage 2 to stage 3	Stage 3 to stage 4	Stage 4 to stage 5	Stage 5 to warehouse
Task 1	4	10	15	20	10	14
Task 2	5	21	20	19	15	14
Task 3	4	15	10	25	16	14
Task 4	3	10	12	14	5	14
Task 5	6	9	5	12	17	14
Task 6	4	14	15	20	10	14
Task 7	5	21	15	24	15	14
Task 8	7	20	20	15	26	14
Task 9	10	5	8	24	10	14
Task 10	8	9	15	21	17	14

Table 7.9 Mean and standard deviation values of IGD metric for SPEA2, NSGA-II, and HMOGWO

MOEAs	IGD
	Mean (std)
SPEA2	8.67e − 03(7.5e − 04)
NSGA-II	8.42e − 03(7.1e − 04)
HMOGWO	**2.55e − 03(1.3e − 04)**

Table 7.10 Wilcoxon signed rank test results for this case (a level of significant $\alpha = 0.05$)

Problem	HMOGWO versus NSGA-II			HMOGWO versus SPEA2		
	R^+	R^-	*p*-value/win	R^+	R^-	*p*-value/win
Case	465	0	1.73e-06/ +	465	0	1.73e-06/ +

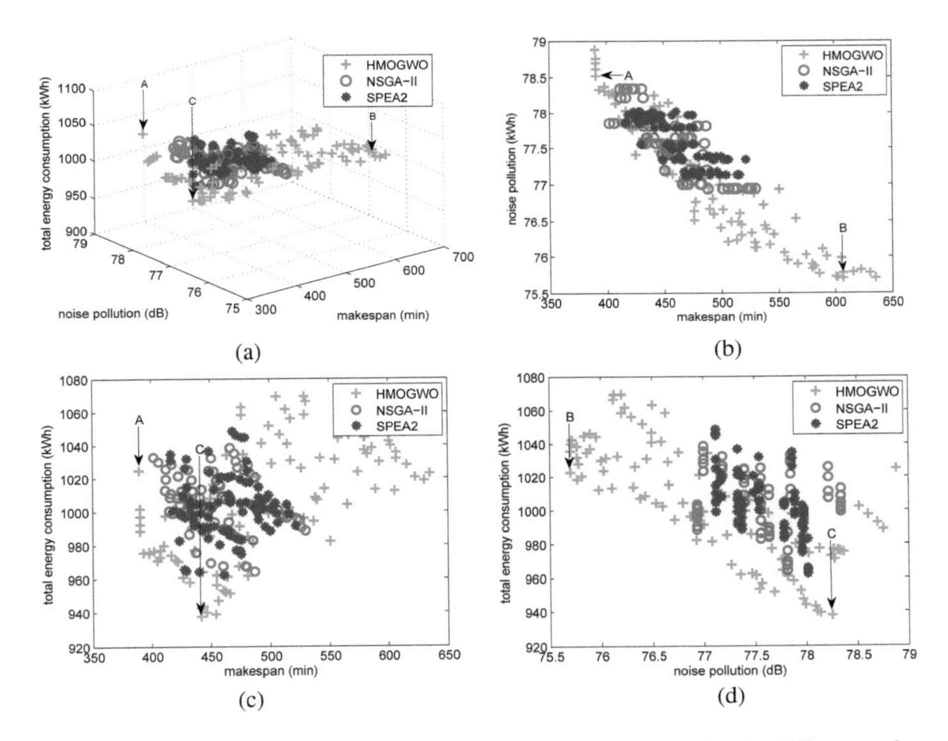

Fig. 7.8 Pareto front (PF) obtained by HMOGWO, NSGA-II, and SPEA2 under different angles, **a** PF by different MOEAs with three criteria, **b** PF with makespan and noise, **c** PF with makespan and energy consumption, **d** PF with noise and energy consumption

Table 7.11 The corresponding extreme results by HMOGWO on the scheduling problem

No.	The first part:π	The second part: N	f_1 (min)	f_2 (dB)	f_3 (kWh)
A	[1–10]	$\begin{bmatrix} 1,1,1,3,3,3,3,3,3,3 \\ 1,1,4,4,4,4,4,4,1,4 \\ 3,4,4,4,4,4,2,4,1,4 \\ 1,1,1,1,1,1,1,1,1,1 \\ 3,3,3,1,3,1,3,3,1,1 \end{bmatrix}$	390.00	78.41	1038.74
B	[1–10]	$\begin{bmatrix} 1,1,1,1,1,1,1,1,1,1 \\ 1,1,1,1,1,1,1,1,1,1 \\ 1,1,1,1,1,1,1,1,1,1 \\ 1,1,1,1,1,1,1,1,1,1 \\ 1,1,1,1,1,1,1,1,1,1 \end{bmatrix}$	607.00	75.70	1035.66
C	[1–10]	$\begin{bmatrix} 1,1,1,2,1,1,1,1,1,1 \\ 1,4,4,1,4,4,4,3,4,4 \\ 3,4,4,2,4,4,4,4,4,4 \\ 1,1,1,1,1,1,1,1,1,1 \\ 3,3,3,1,3,3,3,3,1,3 \end{bmatrix}$	441.25	78.25	937.98

7.5 Chapter Conclusion

In this chapter, we investigate a real-world welding shop scheduling problem (WSSP) from a novel perspective on noise pollution. To begin, a new mathematical model for the WSSP is developed. The goals of this model are to minimize makespan, total energy consumption, and noise pollution all at the same time. Then, to solve this scheduling problem, a hybrid multi-objective grey wolf optimization algorithm (HMOGWO) is developed. Furthermore, the experiment is carried out by comparing the results obtained by various MOEAs on this problem. According to the empirical results, HMOGWO outperforms the other MOEAs on this problem. Finally, the HMOGWO is applied successfully to a real-world case.

Future research will look at more complex scheduling problems, such as dynamic scheduling. Furthermore, a more high-performance algorithm or strategy is being developed to address this issue.

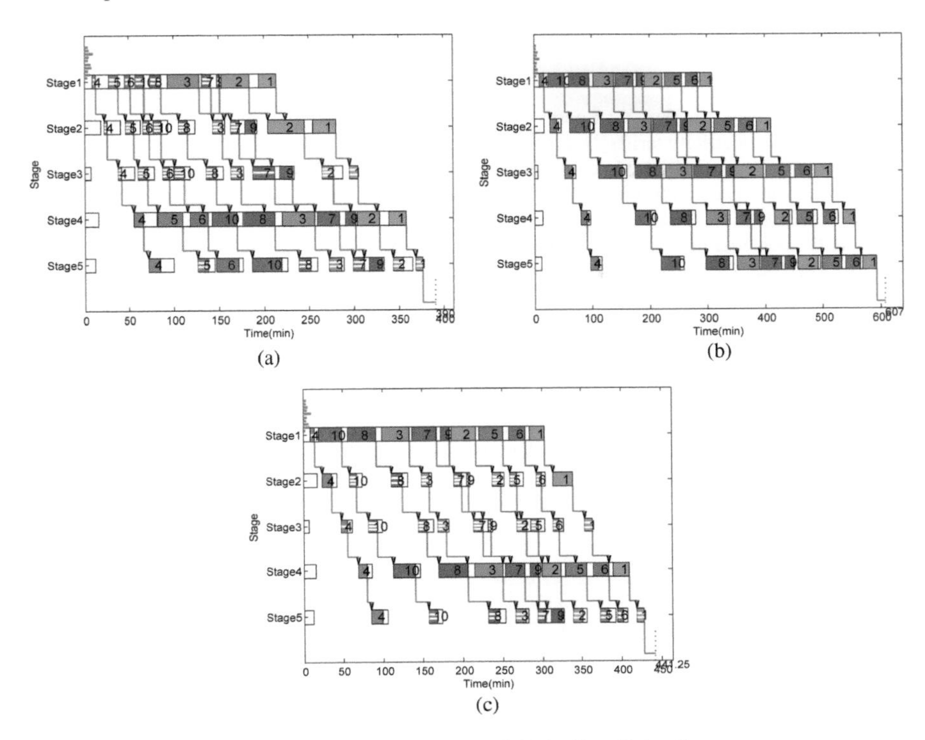

Fig. 7.9 Gantt charts for three solutions, **a** Point A, **b** Point B, **c** Point C

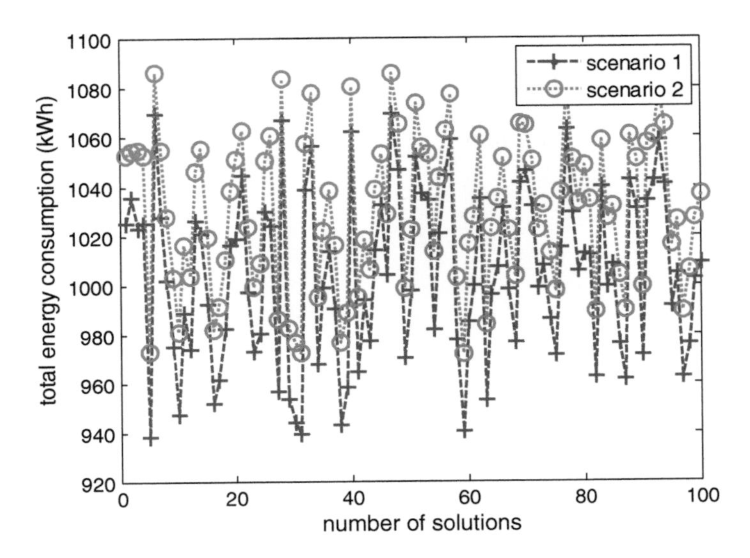

Fig. 7.10 Energy consumption comparison between scenario 1 and scenario 2

References

1. Che, A., Zhang, S., Wu, X.: Energy-conscious unrelated parallel machine scheduling under time-of-use electricity tariffs[J]. J. Clean. Prod. **156**, 688–697 (2017)
2. Gahm, C., Denz, F., Dirr, M., et al.: Energy-efficient scheduling in manufacturing companies: a review and research framework[J]. Eur. J. Oper. Res. **248**(3), 744–757 (2016)
3. Lei, D., Zheng, Y., Guo, X.: A shuffled frog-leaping algorithm for flexible job shop scheduling with the consideration of energy consumption[J]. Int. J. Prod. Res. **55**(11), 3126–3140 (2017)
4. Liu, G., Zhou, Y., Yang, H.: Minimizing energy consumption and tardiness penalty for fuzzy flow shop scheduling with state-dependent setup time[J]. J. Clean. Prod. **147**, 470–484 (2017)
5. Lu, C., Gao, L., Li, X., et al.: Energy-efficient multi-pass turning operation using multi-objective backtracking search algorithm[J]. J. Clean. Prod. **137**, 1516–1531 (2016)
6. Lu, C., Gao, L., Li, X., et al.: A hybrid multi-objective grey wolf optimizer for dynamic scheduling in a real-world welding industry[J]. Eng. Appl. Artif. Intell. **57**, 61–79 (2017)
7. Lu, C., Li, X., Gao, L., et al.: An effective multi-objective discrete virus optimization algorithm for flexible job-shop scheduling problem with controllable processing times[J]. Comput. Ind. Eng. **104**, 156–174 (2017)
8. Yin, L., Li, X., Gao, L., et al.: A novel mathematical model and multi-objective method for the low-carbon flexible job shop scheduling problem[J]. Sustain. Comput.: Inf. Syst. **13**, 15–30 (2017)
9. Lu, C., Xiao, S., Li, X., et al.: An effective multi-objective discrete grey wolf optimizer for a real-world scheduling problem in welding production[J]. Adv. Eng. Softw. **99**, 161–176 (2016)
10. Yan, W., Zhang, H., Jiang, Z., et al.: Multi-objective optimization of arc welding parameters: the trade-offs between energy and thermal efficiency[J]. J. Clean. Prod. **140**, 1842–1849 (2017)
11. Lu, C., Gao, L., Li, X., et al.: Energy-efficient permutation flow shop scheduling problem using a hybrid multi-objective backtracking search algorithm[J]. J. Clean. Prod. **144**, 228–238 (2017)
12. Tang, L., Wang, X.: A hybrid multiobjective evolutionary algorithm for multiobjective optimization problems[J]. IEEE Trans. Evol. Comput. **17**(1), 20–45 (2013)
13. Yin, L., Li, X., Gao, L., et al.: Energy-efficient job shop scheduling problem with variable spindle speed using a novel multi-objective algorithm[J]. Adv. Mech. Eng. **9**(4), 2071938683 (2017)
14. Cao, B., Zhao, J., Yang, P., et al.: Distributed parallel cooperative coevolutionary multi-objective large-scale immune algorithm for deployment of wireless sensor networks[J]. Futur. Gener. Comput. Syst. **82**, 256–267 (2018)

Chapter 8
Green Scheduling in Distributed Permutation Flow Shop with Non-identical Factories

8.1 Brief Introduction

With the requirements of economic globalization and sustainable development, sustainable scheduling of distributed manufacturing has attracted attention from researchers [1, 2]. Most existing research has focused on distributed manufacturing with the identical factory. However, the non-identical factory has not yet been well studied. This part is attempt to present a novel mathematical model of a sustainable distributed permutation flow shop scheduling problem with a non-identical factory (DPFSP-NF). The objective of this problem is to minimize makespan, negative social impact (NSI), and total energy consumption (TEC). Then, we designed a new energy conservation strategy and embedded in the model to reduce TEC. A knowledge-based multi-objective memetic optimization algorithm (KMMOA) is developed to address the DPFSP-NF. In the KMMOA, first, a cooperative initialization mechanism is designed to yield initial solutions with good diversity and convergence. Then, the knowledge-based local search operator based on several properties of DPFSP-NF is proposed. In the end, we compare KMMOA to its variants and other well-known multi-objective optimization algorithms. The experiment results demonstrate that each improvement of the KMMOA is effective, and the KMMOA can obtain a better solution for the DPFSP-NF.

C. Lu et al., *Intelligence Optimization for Green Scheduling in Manufacturing Systems*, Engineering Applications of Computational Methods 18,
https://doi.org/10.1007/978-981-99-6987-6_8

8.2 Problem Statement and Modeling

8.2.1 Problem Statement

The DPFSP-NF can be described as follows. There are F non-identical factories, each of which has a set of m different machines. A set of n jobs have to be distributed on F non-identical factories, with m operations per job. Each factory can handle any job. However, if a job is assigned to any one of F non-identical factories, all operations of the job must be processed in the assigned factory and cannot be transferred to another factory. All jobs should be continuously processed and not preempted or interrupted. It should be noted that non-identical factories may have different machines. In other words, the processing time of job on the machine varies from factory to factory. The setup times and transport times are negligible.

To plainly describe the problem, a instance is given. Suppose the instance contains 4 jobs, 2 factories, and 2 machines in each factory. The processing time and penalty coefficient of jobs are recorded in Table 8.1. Assume that $P_{i,f}^{\text{run}} = 2$, $P_{i,f}^{\text{ilde}} = 1$, and 1. Other parameter settings are the same as in Sect. 8.4.1. Let a job permutation $\pi = [1\text{–}4]$ represents a solution. According to the decoding scheme in Sect. 8.3.1, we can get the Gantt chart of π as shown in Fig. 8.1. The calculation of the three criteria of the above solution is as follows: The first objective makespan: $C_{\max} = \max\{C_1, C_2\} = 12$. The second objective NSI:

$$f_2 = \sum\nolimits_{f=1}^{F} \sum\nolimits_{i=1}^{m} n \sum\nolimits_{j=1}^{n} \sum\nolimits_{k=1}^{n} p_{i,j,f} \cdot w_{i,j,f} \cdot x_{j,k,f} = 4 \times 0.5 + 5 \times 0.7 + 5 \times 0.7 + 2 \times 0.2 + 4 \times 0.5 + 3 \times 0.4 + 3 \times 0.4 + 5 \times 0.7 = 17.8$$

$$\text{The third objective TEC} : f_3 = \text{TEC} = \text{EC}_w + \text{EC}_s = \sum\nolimits_{f=1}^{F} \sum\nolimits_{i=1}^{m} \sum\nolimits_{j=1}^{n_f} P_{i,f}^{\text{run}} \cdot p_{i,j,f} + \sum\nolimits_{f=1}^{F} \sum\nolimits_{i=1}^{m} \sum\nolimits_{j=1}^{n_f} P_{i,f}^{\text{ilde}} \cdot t_{i,j,f}(1 - Z_{i,j,f}) + E_{i,f}^{\text{off\_on}} \cdot Z_{i,j,f} = 2 \times 4 + 2 \times 5 + 2 \times 5 + 2 \times 2 + 2 \times 4 + 2 \times 3 + 2 \times 3 + 2 \times 5 + 0 = 62$$

8.2.2 Mathematical Modeling

The related notation of DPFSP-NF is given as follows:

n : the number of jobs.
m : the number of machines in each factory.

Table 8.1 Processing time and penalty coefficient of jobs

Processing time/penalty coefficient	Factory 1		Factory 2	
Job	Machine 1	Machine 2	Machine 1	Machine 2
1	4/0.5	5/0.7	3/0.4	6/0.9
2	3/0.4	4/0.5	4/0.5	3/0.4
3	5/0.7	2/0.2	6/0.9	2/0.2
4	2/0.2	6/0.8	3/0.4	5/0.7

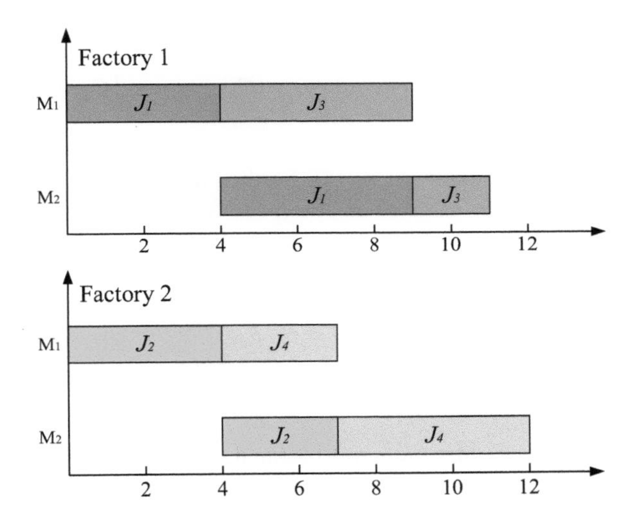

Fig. 8.1 Gantt chart of a solution

F : the number of factories.
j : index of jobs, $j = 1, 2, \cdots, n$.
i : index of machines, $i = 1, 2, \cdots, m$.
f : index of factories, $f = 1, 2, \cdots, F$.
k : index for the position of the job in a schedule, $k = 1, 2, \cdots, n$.
n_f : the number of jobs assigned to factory f.
C_f : the maximal completion time of jobs in the factory f.
$C_{i,f}$: the completion time of jobs on machine i in the factory f.
$C_{k,i,f}$: the completion time of the job in position k on machine i in factory f.
$C_{\max}$: the maximal completion time of all the factories.
$O_{i,j,f}$: the operation of job j on machine i in factory f.
$p_{i,j,f}$: the processing time of the operation $O_{i,j,f}$.
$t_{i,j,f}$: the idle time of the operation $O_{i,j,f}$.
$P_{i,f}^{\text{idle}}$: the idle energy consumption of a machine i per unit time in factory f.
$P_{i,f}^{\text{run}}$: the run energy consumption of a machine i per unit time in factory f.
$w_{i,j,f}$: the penalty coefficient of operation $O_{i,j,f}$.
$E_{i,f}^{\text{off\_on}}$: the energy consumption required to turn off/on the machine i in factory f.

EC_w : energy consumption during the working phase.
EC_s : energy consumption during the idle phase.
T : the maximal allowable number of switching off a machine.
T_b : the breakeven duration.
π : a permutation of jobs.
$S_{k,i,f}$: the starting time of the job in position k on machine i at factory f.

The sustainability of DPFSP-NF should comply with the TBL [3]. Therefore, DPFSP-NF involves three optimization objectives: makespan, NSI, and TEC.

(1) Economic criterion

$$\min f_1 = C_{\max} = \max\{C_f | f = 1, \ldots, F\} \tag{8.1}$$

In most cases, the optimization objective of the shop scheduling problem is to maximize economic benefits [4]. Further, the makespan can map the production benefit or economic value of an enterprise to a certain extent. Therefore, we can improve economic value by minimizing makespan.

(2) NSI criterion

$$\min f_2 = \sum_{f=1}^{F} \sum_{i=1}^{m} \sum_{j=1}^{n} \sum_{k=1}^{n} p_{i,j,f} \cdot w_{i,j,f} \cdot x_{j,k,f} \tag{8.2}$$

The social criterion in this part mainly refer to the working environment of workers [5]. If the working environment includes noise pollution, high-risk operation, overwork and other factors, the working efficiency of workers will decline, and will also have a negative impact on the company's image. The social criterion is usually related to the operation and processing times of jobs [6]. Thus, the penalty coefficient (i.e.,$w_{i,j,f}$) is used to express the *NSI*. The worse the working environment, the greater the penalty coefficient.

(3) TEC criterion

$$\min f_3 = \mathrm{TEC} = \sum_{f=1}^{F} \sum_{i=1}^{m} \int_{t=0}^{C_{i,f}} \theta_i(t)\mathrm{d}t = \mathrm{EC}_w + \mathrm{EC}_s \tag{8.3}$$

$$z_{i,j,f} = \begin{cases} 1, & \text{if machine } i \text{ to process job } j \text{ is switched off in factory } f \\ 0, & \text{otherwise} \end{cases} \tag{8.4}$$

In the process of production, huge energy consumption will promote global warming and cause a series of environmental problems. In order to comply with the strategy of sustainable development, we cannot ignore the impact of production on the environment while seeking the maximization of economic benefits. Thus, we set environmental guideline, which is to minimize the *TEC*. *TEC* objective is calculated through Eq. (8.3), where $\theta_i(t)$ represents the instantaneous power of machine i at time t, $\mathrm{EC}_w = \sum_{f=1}^{F}\sum_{i=1}^{m}\sum_{j=1}^{n_f} P_{i,f}^{\mathrm{run}} \cdot p_{i,j,f}$, and $\mathrm{EC}_s = \sum_{f=1}^{F}\sum_{i=1}^{m}\sum_{j=1}^{n_f} P_{i,f}^{\mathrm{ilde}} \times t_{i,j,f}(1 - Z_{i,j,f}) + E_{i,f}^{\mathrm{off\_on}} \times Z_{i,j,f}$.

The above process reduces energy consumption by shutting down the machine when it is idle [7]. However, frequent switching on and off of the machine will damage the machine tool [8]. To address this issue, a new energy conservation rule is designed below:

Step 1: Scan each machine assigned on each factory, and then calculate the idle time $t_{i,j,f}$ of each machine to process $O_{i,j,f}$.

Step 2: Sort these idle time values based on a non-ascending order for each factory, and then select the former or biggest T members from these idle times. Note that T is the maximal allowable number of switching off each machine.

Step 3: Switch off the machine assigned on the current factory, if the idle time $t_{i,j,f}$ meets the following two conditions: (1) $t_{i,j,f} \geq T_b$ and (2) $t_{i,j,f}$ is a member of a set of the biggest T idle time members.

DPFSP-NF aims to minimize the makespan, NSI, and TEC at the same time. As shown in Eqs. (8.2), (8.3), and (8.5), it is the objective function of the problem. In addition, the MILP model of the DPFSP-NF also includes the following constraints:

$$\sum_{k=1}^{n}\sum_{f=1}^{F} x_{j,k,f} = 1, \forall j \tag{8.5}$$

$$\sum_{j=1}^{n}\sum_{f=1}^{F} x_{j,k,f} = 1, \forall k \tag{8.6}$$

$$C_{k,i,f} = S_{k,i,f} + \sum_{j=1}^{n} x_{j,k,f} \cdot p_{i,j,f}, \forall i, k, f \tag{8.7}$$

$$C_{k,i,f} \geq C_{k,i-1,f} + \sum_{j=1}^{n} x_{j,k,f} \cdot p_{i,j,f}, \forall i > 1, k, f \tag{8.8}$$

$$C_{k,i,f} \geq C_{k-1,i,f} + \sum_{j=1}^{n} x_{j,k,f} \cdot p_{i,j,f}, \forall k > 1, i, f \tag{8.9}$$

$$C_{\max} \geq C_{k,m,f}, \forall k, f \tag{8.10}$$

$$C_{k,i,f} \geq 0, \forall k, i, f \tag{8.11}$$

$$x_{j,k,f} \in \{0, 1\}, \forall j, k, f \tag{8.12}$$

Constraint (8.5) guarantees that each job is required to occupy exactly one position in the processing sequence in the assigned factory. Constraint (8.6) ensures that n

positions among all the $n \cdot F$ possible positions should be occupied. Constraint (8.7) defines that the completion time of each operation equal to the sum of starting times and processing times of its operation. Constraint (8.8) ensures that the current operation of one job in the position k cannot start until its previous operation is completed in the same factory. The operation of the job cannot begin before its predecessor on the same machine in the same factory has finished, according to constraint (8.9). Constraint (8.10) defines the boundary of makespan. Constraint (8.11) ensures the completion time of each operation is a non-negative number. Constraint (8.12) expresses the binary decision variables.

8.3 Proposed Algorithm

In this part, a knowledge-based multi-objective optimization algorithm (KMMOA) is proposed to solve this DPFSP-NF problem. The global and local search can be balanced by the memetic algorithm. While a knowledge-based local search is utilized to mine the potential promising solutions [9].

Algorithm 8.1 shows a framework of the KMMOA. The following subsections provide a specific description of these components.

Algorithm 8.1: Framework of KMMOA Input: PopSize, Maximum NFEs, crossover and mutation probability

Output: a set of non-dominated solutions

1. $P_0 \leftarrow$ Initialize population (PopSize)
2. $t \leftarrow 0$
3. **While** $t \leq MaxNFEs$ **do**
4. $S_t \leftarrow$ Selection (P_t)
5. $S_{t'} \leftarrow$ Crossover (S_t)
6. $S_t^{''} \leftarrow$ Mutation $(S_{t'})$
7. $Q_t \leftarrow$ Local Search $(S_t^{''})$
8. $C_t \leftarrow$ Combination (P_t, Q_t)
9. $\{F_1, F_2, \cdots\} \leftarrow$ Fast Non-Dominated Sort (C_t)
10. $P_{t+1} \leftarrow \emptyset$
11. $i \leftarrow 1$
12. **While** $|P_{t+1}| + |F_i| \leq PopSize$ **do**
13. Crowding Distance Assignment (F_i)
14. $P_{t+1} \leftarrow$ Combination (P_{t+1}, F_i)
15. $i++$
16. **End while**
17. $P_{t+1} \leftarrow$ Elitist Strategy (P_{t+1})
18. Fast Non-Dominated Sort (F_i)
19. $t++$
20. **End while**

8.3.1 Encoding and Decoding

For the DPFSP-NF, we adopt a permutation-based representation as the solution encoding method; that is, element in the solution is the number of the job, and the sequence of the number is also the sequence of the operation processing. The decoding of DPFSP-NF requires to address two issues: (1) Job is assigned to the factory; (2) determine the processing sequence of operations in each factory. In order to balance the load between all factories, we adopt the earliest completion factory (*ECF*) rule to decode. More specifically, *ECF* rule stipulates that each job in the queue is assigned to the factory that can handle the job at the earliest completion time [10].

8.3.2 Initialization

For the algorithm, the quality of initial solution will affect the performance of the algorithm. The DPFSP-NF problem has three objectives (i.e., makespan, *TEC*, and *NSI*). Therefore, adopt the strategy of cooperative initialization, that is, the initial solution is as close as possible to the three targets. In practical terms, the cooperative initialization strategy is as follows. For the goal of minimizing the maximum completion time, we adopt the usual the NEH rule to obtain an initial solution. Testing each job in all possible positions of a job permutation, then, inserting it the position with the minimum total penalty coefficient, helps to reduce the *NSI*. In the above way, we can get the initial solution based on the second objective. Similarly, we can get the initial solution to minimize the TEC by inserting the jobs one by one until the total energy consumption reaches the minimum. In order to ensure the diversity of the population, the rest of the solutions are randomly generated. See Algorithm 8.2 for pseudocode of algorithm.

Algorithm 8.2. The Cooperative Initialization **Input**: one initial population $\mathbf{X} = [\pi_1, \pi_2, \cdots, \pi_{ps}]$ //π is a job permutation, $\boldsymbol{ps}$ is the population size

Output: one improved initial population $\mathbf{X}'$

1. Generate one initial population $\mathbf{X}$ randomly
2. Obtain the first solution π_1 based on a non-ascending order of the total processing times of jobs
3. **for** $j = 1$ to n

   Check job j in all possible positions of a job permutation π_1

   Insert job j into the position with minimum the makespan to generate a new permutation π'_1
4. **end for**
5. Obtain the second solution π_2 based on a non-ascending order of the total penalty coefficient of jobs
6. **for** $j = 1$ to n

   Check job j in all possible positions of π_2

Insert job j into the position with the minimum *NSI* to generate a new permutation π'_2

7. **end for**
8. Obtain the third solution π_3 based on a non-ascending order of the working energy consumption of jobs
9. **for** $j = 1$ to n

Check job j in all possible positions of π_3

Insert job j into the position with the minimum *TEC* to generate a new permutation π'_3

10. **end for**

8.3.3 Update Operation

Similar to genetic algorithm, the algorithm also adopts the operation of selection, crossover, and mutation when updating the population [11]. First, we need to select two parents from the initial population by binary tournament selection. Then, we adopt the partially mapped crossover (PMX) [12] for crossover operation, and swap mutation operators for mutation operation.

The procedure of PMX is as follows:

Step 1: Arbitrarily select two positions on parents, define elements between two positions as a matching substring.
Step 2: By switching the matching substrings from two parents, we can get a temporary solution.
Step 3: The relationships of jobs in conflict are determined and mapped.
Step 4: According to the mapped relationship, the job permutation is possible without altering the substring.

The swap mutation operator is to randomly select two positions in a test plan and exchange two positions [13].

8.3.4 Local Search

Search heuristic is often used to solve single-objective optimization problems, but when sampling local search in multi-objective problems (SOP) [14], the optimization of a single target will have a negative impact on the other target value. Therefore, we improve the SOP local search heuristic based on the characteristics of DPFSP-NF. Several features of DPFSP-NF are as follows:

Property 1 This multi-objective function of DPFSP-NF is briefly defined as: min $f(\pi) = [f_1(\pi), f_2(\pi), f_3(\pi)]$, where f_1, f_2, f_3 represents makespan, *NSI*, and *TEC*, respectively. One solution $\boldsymbol{a}$ will dominate another solution $\boldsymbol{b}$ (denoted as $\boldsymbol{a} \prec \boldsymbol{b}$), if and only if $\forall l \in \{1, 2, 3\}$, $f_l(\boldsymbol{a}) \leq f_l(\boldsymbol{b})$ and $\exists g \in \{1, 2, 3\}$, $f_g(\boldsymbol{a}) < f_g(\boldsymbol{b})$.

Property 2 Makespan criterion can only be reduced by moving jobs in the critical factory, if there is only one critical factory (the factory with the maximum completion time is called a critical factory).

Proof For each factory f $(f \in \{1, \ldots, F\})$, there is a corresponding maximum completion time $C_{\max,f}$. Thus, the makespan of a schedule $\boldsymbol{\pi}$ can be represented by $C_{\max}(\pi) = \max\{C_{\max,f}(\pi) | f = 1, \ldots, F\}$. Suppose that the factory $f^c \in \{1, \ldots, F\}$ is only one critical factory and others are non-critical factories. Obviously, we can obtain $C_{\max}(\boldsymbol{\pi}) = C_{\max,f^c}(\boldsymbol{\pi})$ according to the previous definition. By the reverse proving, if job j does not belong to the critical factory f^c, moving the job j will generate a new schedule $\boldsymbol{\pi}'$ and have the following two conditions: (1) if $C_{\max}(\boldsymbol{\pi}') = \max\{C_{\max,f}(\boldsymbol{\pi}') | f = 1, \ldots, F \text{ and } f \neq f^c\} \leq C_{\max,f^c}(\boldsymbol{\pi})$, then $C_{\max}(\boldsymbol{\pi}') = C_{\max,f^c}(\boldsymbol{\pi})$; (2) if $C_{\max}(\boldsymbol{\pi}') = \max\{C_{\max,f}(\boldsymbol{\pi}') | f = 1, \ldots, F \text{ and } f \neq f^c\} > C_{\max,f^c}(\boldsymbol{\pi})$, then $C_{\max}(\boldsymbol{\pi}') > C_{\max}(\boldsymbol{\pi}) = C_{\max,f^c}(\boldsymbol{\pi})$. On the whole, $C_{\max}(\boldsymbol{\pi}) \leq C_{\max}(\boldsymbol{\pi}')$. Therefore, we can conclude that the makespan can only be reduced by moving jobs in the critical factory.

Based on the above characteristics, we improve the local search. *TEC* or *NSI* may be reduced by swapping the jobs with the maximum and the minimum *TEC* or *NSI* in the non-critical factory without affecting the makespan. For the critical factory, we can move critical jobs to reduce makespan. Therefore, the improved local search consists of three neighborhoods:

In *insertion on the critical factory*, extract one job in the critical factory, and insert this job into other all possible positions in this critical factory.

In *swap for NSI on the non-critical factory*, swap one job with the maximum *NSI* and the other job with minimum *NSI* in the non-critical factory.

In *swap for TEC on the non-critical factory*, swap one job with the maximum *TEC* and the other job with minimum *TEC* in the non-critical factory.

A trial solution π can get three groups of solutions after the above three local searches, and then we will compare the trial solution π with all the solutions in the three groups (set_1, set_2, set_3), if a new solution dominates the original solution, it will substitute the original solution and add a temporary archive Q. The pseudocode of the proposed local search is presented in Algorithm 8.3.

Algorithm 8.3: Knowledge-based local search **Input**: a trial solution $\boldsymbol{\pi}$

Output: a set of solutions in Q

1. $set_1 \leftarrow$ *insertion on the critical factory* $(\boldsymbol{\pi})$
2. $set_2 \leftarrow$ *swap for NSI on the non-critical factory* $(\boldsymbol{\pi})$
3. $set_3 \leftarrow$ *swap for TEC on the non-critical factory* $(\boldsymbol{\pi})$
4. **for** $\boldsymbol{\pi}' \in \{set_1, set_2, set_3, \boldsymbol{\pi}\}$
5.   **if** $\mathbf{f}(\boldsymbol{\pi}') \prec \mathbf{f}(\boldsymbol{\pi})$ **then**
6. $\boldsymbol{\pi} \leftarrow \boldsymbol{\pi}'$
7.   Insert $\boldsymbol{\pi}'$ into Q

8. **end if**
9. **end for**

8.4 Experiments

8.4.1 Instances and Performance Metrics

To verify the performance of the improved algorithm, we generated 600 test instances in total. We define the number of jobs as n, the number of machines as m and the number of stage as factories F, where $n = \{40, 60, 80, 100, 200\}$, $m = \{4, 8, 16\}$, and $F = \{2, 3, 4, 5\}$. The normal processing time $p_{i,j}$ follows discrete uniform distribution [10,100]. The actual processing time $p_{i,j,f} = p_{i,j}/r$, where r is a random number and generated by a discrete uniform distribution [1, 10]. The penalty coefficient $w_{i,j,f} = |\sin(\pi \cdot r/n \cdot m \cdot F)|$. The energy power during work mode is set as $P_{i,f}^{\text{run}} = 2 \times r^2$. The idle power is set as $P_{i,f}^{idle} = 3$. The energy consumption required to turn off/on the machine $E_{i,f}^{\text{off\_on}} = 12$. The breakeven duration $T_b = 4$ and the maximum allowable number to close a machine $T = 3$. The performance evaluation indicators of the algorithm include Spread [15], Generation Distance (GD) [16], and Inverted Generation Distance (IGD) [17].

8.4.2 Parameter Calibration

We adopted a Taguchi approach of design-of-experiments (DOE) experimental design the best parameter configuration of the algorithm. The main parameters involved in the algorithm include population size (PS), probabilities of crossover and mutation (p_c and p_m). The level of each parameter is as follows: PS $=$ $\{40, 60, 80, 100\}$, $p_c = \{0.7, 0.8, 0.9, 1.0\}$, and $p_m = \{0.1, 0.2, 0.3, 0.4\}$. One orthogonal array $L_{16}(4^3)$ is adopted in this parameter calibration experiment, the maximum number of function evaluations (NFEs) is equal to 50,000, and the measurement adopted is the comprehensive performance metric IGD. Table 8.2 gives the response variable value and the significance rank of these parameters. The larger the data value corresponding to the parameter, the more significant the influence of the parameter is. Figure 8.2 also shows the main effect plot of three parameters.

According to the experimental results, PS has the most significant impact on the algorithm, and p_m has the least impact. We can get the best configuration from the main effect plot, the best configuration of parametric values is set as PS $= 100$, p_c $= 1.0$, and $p_m = 0.2$.

Table 8.2 Average response variable value and rank of each parameter

Level	PS	p_c	p_m
1	0.011321	0.010749	0.010407
2	0.011321	0.010467	0.010357
3	0.009904	0.010208	0.010247
4	0.009722	0.009826	0.010240
Delta	0.001599	0.000923	0.000167
Rank	1	2	3

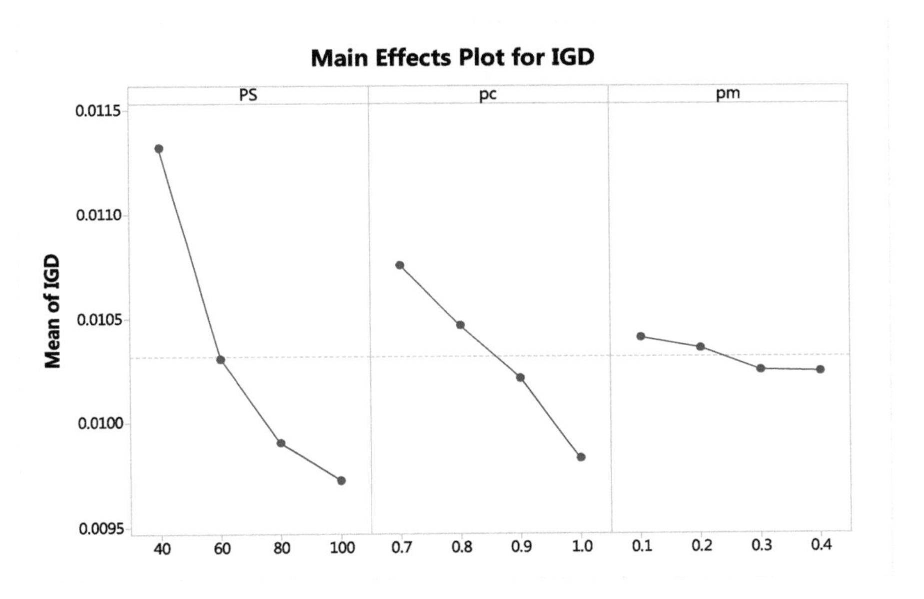

Fig. 8.2 Main effects plot for IGD of each parameter

8.4.3 *Effectiveness of Initialization Strategy*

Figures 8.3, 8.4 and 8.5 show the statistical result distribution boxplot over 20 independent replicates between two initialization components on instances grouped by different factories ($n = 40, 60, 80, 100$). The experiment compares the cooperative initialization with random initialization. Initialization_1 represents the cooperative initialization strategy, while Initialization_2 denotes the random initialization strategy. Therefore, we can conclude that the collaborative initialization generates high-quality initial solutions at the initialization stage, and the way to improve the quality is to spend a little time improving the quality of initial solutions at the late stage of the search progress. It is worthy to design the cooperative initialization strategy to enhance the behavior of KMMOA

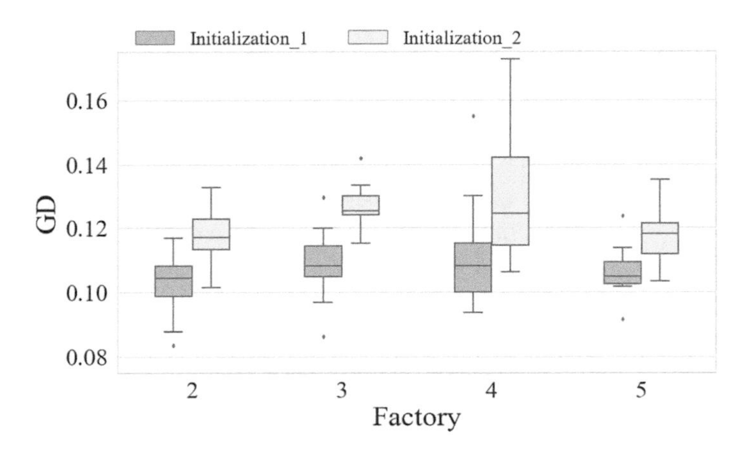

Fig. 8.3 Distribution boxplot for GD metric

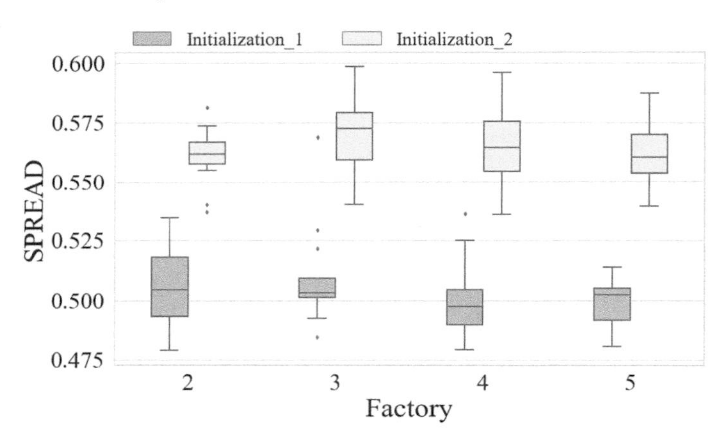

Fig. 8.4 Distribution boxplot for spread metric

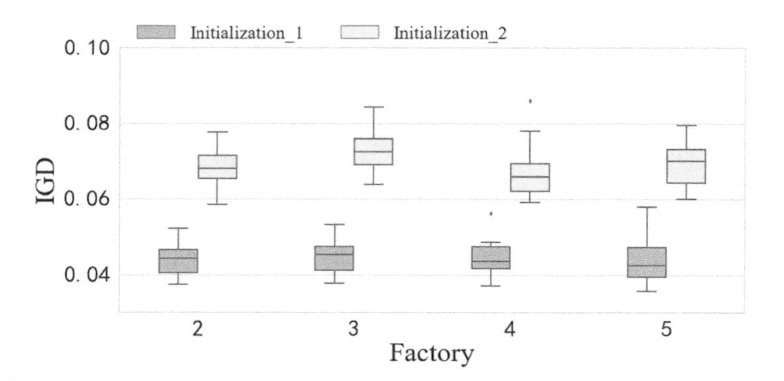

Fig. 8.5 Distribution boxplot for IGD metric

8.4.4 *Effectiveness of Local Search*

We validate the effectiveness of the proposed local search by comparing KMMOA with KMMOA without local searches (denoted as KMMOA1). They adopt the same termination criterion (i.e., maximum NFEs = 50,000); then the mean results of three metrics are listed in Tables 8.3, 8.4 and 8.5. A Wilcoxon signed-rank test is conducted on these results obtained by KMMOA and KMMOA1. Table 8.6 provides the related p values on instances grouped by different factories.

As observed in Tables 8.3 and 8.6, KMMOA achieves significantly smaller values than KMMOA1 in all instances for the GD metric. This demonstrates that our proposed local search helps to improve the convergence to the true Pareto set. Regarding the Spread, as observed in Tables 8.4 and 8.6, KMMOA is capable of obtaining significantly smaller results than KMMOA1 in all instances. It can be observed from Tables 8.5 and 8.6 that KMMOA significantly outperforms KMMOA2 in all instances, we can infer that this local search can maintain the diversity of solutions. Experimental result confirms the effectiveness of the proposed local search, which can contribute to the performance improvement of the algorithm.

8.4.5 *Effectiveness of Energy-Saving Strategy*

In this section, we conduct a comparison experiment between the proposed energy-saving strategy (denoted as strategy_2) and the previous energy-saving strategy(denoted as strategy_1), using the results of TEC as a measure. These instances are divided into four groups according to the number of factories. *TEC* values are processed by the normalization method. Figure 8.6 presents the curve of average results of *TEC* obtained by two kinds of strategies on four groups of instances.

As shown in Fig. 8.6, TEC values of strategy_ 2 are lower than strategy_ 1. In other words, the proposed rules for energy conservation can contribute to the sustainability of manufacturing systems. Therefore, our energy-saving strategy reduces TEC without affecting the lifetime of the machine.

8.4.6 *Comparison of Algorithms*

In this section, we compare KMMOA with several common algorithms to measure their performance. The algorithms we used for comparison are NSGA-II, MOEA/D, SPEA2, CMPSO, and MO-LR. They all adopt the same termination criterion (i.e., maximal NFEs is set to 50,000) among all algorithms. Meanwhile, they also use the above-mentioned encoding and decoding mechanism, crossover, and mutation (if they exist), and energy-saving strategy. Tables 8.7, 8.8 and 8.9 record the mean

Table 8.3 Mean values of GD metric between KMMOA and KMMOA1

(n, m)	KMMOA	KMMOA1	KMMOA	KMMOA1	KMMOA	KMMOA1	KMMOA	KMMOA1
	$F = 2$		$F = 3$		$F = 4$		$F = 5$	
(40,4)	**5.50e-3**	5.86e-3	**7.42e-3**	9.16e-3	**8.37e-3**	1.07e-2	**1.01e-2**	1.15e-2
(40,8)	**7.36e-3**	8.58e-3	**9.26e-3**	1.15e-2	**9.29e-3**	1.12e-2	**9.50e-3**	1.06e-2
(40,16)	**9.66e-3**	1.17e-2	**1.20e-2**	1.49e-2	**1.14e-2**	1.41e-2	**1.21e-2**	1.43e-2
(60,4)	**8.45e-3**	1.04e02	**1.30e-2**	1.77e-2	**1.28e-2**	1.85e-2	**1.52e-2**	1.97e-2
(60,8)	**1.03e-2**	1.52e-2	**1.28e-2**	1.70e-2	**1.47e-2**	2.01e-2	**1.44e-2**	2.16e-2
(60,16)	**1.22e-2**	1.70e-2	**1.36e-2**	1.90e-2	**1.49e-2**	2.20e-2	**1.64e-2**	2.35e-2
(80,4)	**1.01e-2**	1.45e-2	**1.41e-2**	2.06e-2	**1.48e-2**	2.30e-2	**1.71e-2**	2.58e-2
(80,8)	**1.14e-2**	1.85e-2	**1.35e-2**	2.23e-2	**1.85e-2**	2.79e-2	**1.93e-2**	2.79e-2
(80,16)	**1.54e-2**	2.26e-2	**1.97e-2**	3.06e-2	**1.77e-2**	2.63e-2	**1.88e-2**	3.09e-2
(100,4)	**1.19e-2**	1.74e-2	**1.81e-2**	2.81e-2	**2.12e-2**	3.36e-2	**1.99e-2**	3.11e-2
(100,8)	**1.49e-2**	2.54e-2	**1.95e-2**	3.45e-2	**2.12e-2**	3.45e-2	**2.11e-2**	3.29e-2
(100,16)	**1.56e-2**	2.58e-2	**1.92e-2**	3.29e-2	**2.38e-2**	4.00e-2	**2.06e-2**	3.08e-2

Table 8.4 Mean values of spread metric between KMMOA and KMMOA2

(n, m)	KMMOA	KMMOA1	KMMOA	KMMOA1	KMMOA	KMMOA1	KMMOA	KMMOA1
	$F = 2$		$F = 3$		$F = 4$		$F = 5$	
(40,4)	**6.75e-1**	7.41e-1	**7.39e-1**	8.44e-1	**8.13e-1**	9.20e-1	**8.79e-1**	9.63e-1
(40,8)	**6.59e-1**	7.37e-1	**7.15e-1**	8.07e-1	**7.35e-1**	8.39e-1	**7.75e-1**	8.75e-1
(40,16)	**6.85e-1**	7.63e-1	**7.36e-1**	8.25e-1	**7.81e-1**	8.73e-1	**8.09e-1**	9.00e-1
(60,4)	**6.97e-1**	7.92e-1	**8.03e-1**	9.41e-1	**8.76e-1**	1.00e + 0	**9.48e-1**	1.04e + 0
(60,8)	**7.13e-1**	8.06e-1	**7.69e-1**	9.12e01	**8.17e-1**	9.32e-1	**8.57e-1**	9.67e-1
(60,16)	**6.85e-1**	7.69e-1	**7.45e-1**	8.49e01	**7.82e-1**	8.93e-1	**8.32e-1**	9.31e-1
(80,4)	**7.11e-1**	8.00e-1	**8.32e-1**	9.48e-1	**9.04e-1**	1.02e + 0	**9.66e-1**	1.07e + 0
(80,8)	**6.99e-1**	8.19e-1	**7.82e-1**	9.04e-1	**8.62e-1**	9.89e-1	**8.61e-1**	9.91e-1
(80,16)	**7.30e-1**	8.22e-1	**7.63e-1**	8.82e-1	**8.08e-1**	9.10e-1	**8.70e-1**	9.62e-1
(100,4)	**7.07e-1**	8.09e-1	**8.02e-1**	9.41e-1	**9.12e-1**	1.03e + 0	**9.49e-1**	1.05e + 0
(100,8)	**7.39e-1**	8.74e-1	**8.34e-1**	9.44e-1	**9.09e-1**	1.03e + 0	**9.40e-1**	1.02e + 0
(100,16)	**6.99e-1**	8.13e-1	**7.96e-1**	9.14e-1	**8.37e-1**	9.29e-1	**8.62e-1**	9.53e-1

The better results are presented in bold format

Table 8.5 Mean values of IGD metric between KMMOA and KMMOA1

(n, m)	KMMOA	KMMOA1	KMMOA	KMMOA1	KMMOA	KMMOA1	KMMOA	KMMOA1
	$F = 2$		$F = 3$		$F = 4$		$F = 5$	
(40,4)	**4.00e-3**	4.59e-3	**5.81e-3**	6.95e-3	**6.66e-3**	7.90e-3	**7.81e-3**	9.06e-3
(40,8)	**5.21e-3**	6.14e-3	**7.09e-3**	8.67e-3	**6.88e-3**	8.36e-3	**6.60e-3**	7.93e-3
(40,16)	**6.73e-3**	8.17e-3	**8.67e-3**	1.06e-2	**8.63e-3**	1.04e-2	**9.34e-3**	1.10e-2
(60,4)	**6.49e-3**	8.05e-3	**1.00e-2**	1.28e-2	**1.10e-2**	1.44e-2	**1.27e-2**	1.56e-2
(60,8)	**8.13e-3**	1.08e-2	**1.01e-2**	1.26e-2	**1.17e-2**	1.50e-2	**1.17e-2**	1.57e-2
(60,16)	**9.16e-3**	1.20e-2	**1.05e-2**	1.36e-2	**1.19e-2**	1.59e-2	**1.21e-2**	1.57e-2
(80,4)	**8.04e-3**	1.08e-2	**1.18e-2**	1.52e-2	**1.23e-2**	1.70e-2	**1.36e-2**	1.87e-2
(80,8)	**9.39e-3**	1.34e-2	**1.11e-2**	1.55e-2	**1.40e-2**	2.01e-2	**1.52e-2**	2.08e-2
(80,16)	**1.16e-2**	1.60e-2	**1.62e-2**	2.41e-2	**1.44e-2**	1.97e-2	**1.53e-2**	2.19e-2
(100,4)	**9.02e-3**	1.23e-2	**1.38e-2**	1.94e-2	**1.74e-2**	2.56e-2	**1.55e-2**	2.18e-2
(100,8)	**1.20e-2**	1.77e-2	**1.54e-2**	2.31e-2	**1.75e-2**	2.52e-2	**1.59e-2**	2.34e-2
(100,16)	**1.28e-2**	1.83e-2	**1.55e-2**	2.25e-2	**1.84e-2**	2.76e-2	**1.58e-2**	2.16e-2

The better results are presented in bold format

Table 8.6 P values of all metrics between KMMOA and KMMOA1 on different factories

Number of factories	GD	Spread	IGD
F = 2	4.88e-4	4.88e-4	4.88e-4
F = 3	4.88e-4	4.88e-4	4.88e-4
F = 4	4.88e-4	4.88e-4	4.88e-4
F = 5	4.88e-4	4.88e-4	4.88e-4

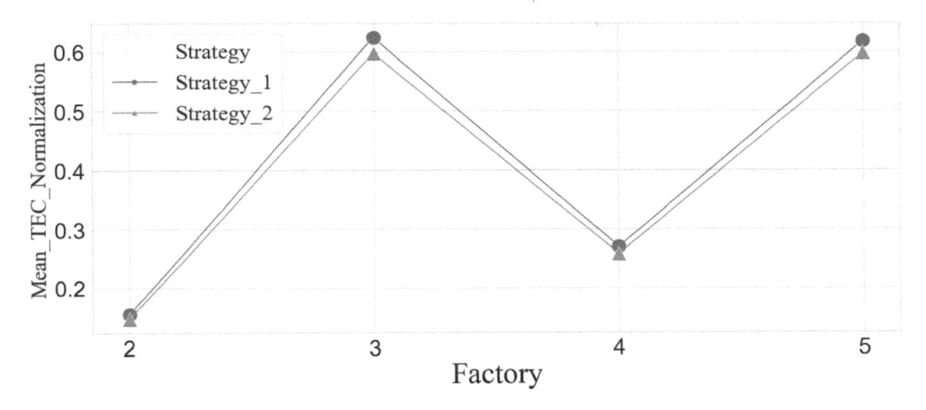

Fig. 8.6 Curve of average results of TEC by two kinds of strategies

values of three metrics in instances grouped by different factories. KMMOA performs best in Spread and IGD metrics and is similar to MOEA/D results in GD metrics. To check whether the difference between statistical results is significant or not, a Wilcoxon signed-rank test at 0.05 significance level is conducted on these results. Table 8.10 summarizes the related *p* values among the compared algorithms on instances grouped by different factories.

As shown in Table 8.10, the superiority of the proposal is overwhelming from a statistical standpoint. Furthermore, Fig. 8.7 presents the Pareto approximation obtained by different algorithms on one randomly selected instance from different perspectives.

Figure 8.7 shows Pareto approximation in three-dimensional space, and other subgraphs present Pareto approximation among two objectives in two-dimensional space. The experimental results show that the Pareto frontier generated by KMMOA is closer to the lower left corner of the graph, that is, KMMOA can find a better solution. Although MOEA/D can find several better non-dominated solutions than KMMOA concerning *TEC* objective in Fig. 8.7b, d, it is not good for other two objectives. Taking these considerations into account, we can formally conclude that KMMOA has the strong search ability in terms of exploration and exploitation

Table 8.7 Mean value of GD metric among all compared MOEAs

(n, m)	KMMOA	NSGA-II	MOEA/D	SPEA2	CMPSO	MO-LR	KMMOA	NSGA-II	MOEA/D	SPEA2	CMPSO	MO-LR
	$F = 2$						$F = 3$					
(40,4)	5.30e-3	6.35e-3	**2.97e-3**	4.89e-3	4.80e-3	6.20e-3	7.70e-3	9.95e-3	**6.12e-3**	8.18e-3	7.74e-3	9.74e-3
(40,8)	6.63e-3	9.44e-3	**4.46e-3**	7.09e-3	7.32e-3	9.54e-3	9.03e-3	1.21e-2	**7.51e-3**	1.02e-2	1.05e-2	1.25e-2
(40,16)	9.58e-3	1.21e-2	**5.84e-3**	9.26e-3	9.31e-3	1.19e-2	1.11e-2	1.46e-2	**1.03e-2**	1.24e-2	1.28e-2	1.43e-2
(60,4)	7.74e-3	9.90e-3	**5.08e-3**	8.25e-3	8.31e-3	1.04e-2	1.10e-2	1.64e-2	**9.79e-3**	1.32e-2	1.39e-2	1.63e-2
(60,8)	9.84e-3	1.43e-2	**8.86e-3**	1.08e-2	1.11e-2	1.48e-2	1.16e-2	1.66e-2	**9.22e-3**	1.35e-2	1.40e-2	1.60e-2
(60,16)	1.15e-2	1.71e-2	**1.04e-2**	1.38e-2	1.39e-2	1.69e-2	**1.33e-2**	1.84e-2	1.55e-2	1.61e-2	1.63e-2	1.79e-2
(80,4)	1.03e-2	1.42e-2	**6.94e-3**	1.16e-2	1.27e-2	1.43e-2	**1.21e-2**	1.99e-2	1.22e-2	1.72e-2	1.75e-2	1.91e-2
(80,8)	1.06e-2	1.74e-2	**9.57e-3**	1.37e-2	1.38e-2	1.78e-2	1.30e-2	2.00e-2	**1.06e-2**	1.72e-2	1.76e-2	1.99e-2
(80,16)	**1.27e-2**	2.10e-2	1.33e-2	1.71e-2	1.81e-2	1.99e-2	**1.58e-2**	2.43e-2	1.63e-2	2.31e-2	2.28e-2	2.44e-2
(100,4)	9.84e-3	1.62e-2	**7.25e-3**	1.30e-2	1.39e-2	1.63e-2	1.54e-2	2.38e-2	1.29e-2	2.28e-2	2.11e-2	2.38e-2
(100,8)	1.35e-2	2.17e-2	**1.26e-2**	1.94e-2	2.00e-2	2.21e-2	**1.76e-2**	3.04e-2	2.26e-2	2.87e-2	2.85e-2	2.98e-2
(100,16)	1.26e-2	2.20e-2	**1.21e-2**	1.88e-2	1.82e-2	2.12e-2	**1.56e-2**	2.51e-2	1.99e-2	2.36e-2	2.35e-2	2.57e-2
(200,4)	1.80e-2	2.92e-2	**1.63e-2**	2.89e-2	2.87e-2	3.00e-2	**2.13e-2**	2.62e-2	2.53e-2	2.39e-2	2.28e-2	2.26e-2
(200,8)	**2.03e-2**	3.41e-2	2.45e-2	3.44e-2	3.26e-2	3.60e-2	**2.92e-2**	4.53e-2	3.70e-2	4.67e-2	4.53e-2	4.59e-2
(200,16)	**1.82e-2**	3.11e-2	2.69e-2	2.96e-2	3.21e-2	3.10e-2	**2.35e-2**	3.81e-2	2.92e-2	3.82e-2	3.76e-2	3.83e-2
(500,4)	**3.15e-2**	5.16e-2	5.56e-2	5.87e-2	4.55e-2	5.32e-2	**2.53e-2**	3.09e-2	3.04e-2	4.02e-2	3.42e-2	3.82e-2
(500,8)	**2.81e-2**	4.29e-2	3.72e-2	4.77e-2	4.74e-2	5.27e-2	**2.39e-2**	4.45e-2	4.08e-2	4.37e-2	4.14e-2	4.73e-2
(500,16)	**2.76e-2**	4.21e-2	2.87e-2	4.07e-2	4.55e-2	4.53e-2	**3.10e-2**	3.97e-2	4.18e-2	4.53e-2	4.63e-2	4.43e-2
(n, m)	KMMOA	NSGA-II	MOEA/D	SPEA2	CMPSO	MO-LR	KMMOA	NSGA-II	MOEA/D	SPEA2	CMPSO	MO-LR
	$F = 4$						$F = 5$					
(40,4)	8.80e-3	1.14e-2	**7.15e-3**	9.45e-3	9.59e-3	1.13e-2	9.99e-3	1.21e-2	**8.87e-3**	1.09e-2	1.12e-2	1.24e-2

(continued)

Table 8.7 (continued)

(*n*, *m*)	KMMOA	NSGA-II	MOEA/D	SPEA2	CMPSO	MO-LR	KMMOA	NSGA-II	MOEA/D	SPEA2	CMPSO	MO-LR
	$F = 2$						$F = 3$					
(40,8)	8.58e-3	1.17e-2	**6.61e-3**	9.62e-3	9.32e-3	1.17e-2	9.18e-3	1.11e-2	**6.07e-3**	9.78e-3	9.63e-3	1.14e-2
(40,16)	1.09e-2	1.54e-2	**8.75e-3**	1.31e-2	1.31e-2	1.52e-2	1.11e-2	1.42e-2	**1.03e-2**	1.24e-2	1.27e-2	1.45e-2
(60,4)	1.21e-2	1.72e-2	**1.19e-2**	1.53e-2	1.57e-2	1.66e-2	1.51e-2	2.17e-2	**1.25e-2**	1.83e-2	1.92e-2	2.17e-2
(60,8)	**1.30e-2**	1.74e-2	1.35e-2	1.60e-2	1.67e-2	1.82e-2	**1.34e-2**	2.04e-2	1.35e-2	1.75e-2	1.79e-2	1.87e-2
(60,16)	1.41e-2	2.10e-2	1.39e-2	1.89e-2	**1.32e-2**	2.07e-2	1.52e-2	2.25e-2	**1.39e-2**	1.89e-2	2.00e-2	2.25e-2
(80,4)	1.33e-2	2.07e-2	1.32e-2	**1.21e-2**	1.94e-2	2.03e-2	1.49e-2	2.26e-2	**1.38e-2**	2.15e-2	2.03e-2	2.28e-2
(80,8)	1.67e-2	2.64e-2	**1.51e-2**	2.37e-2	2.48e-2	2.58e-2	**1.50e-2**	2.35e-2	1.54e-2	2.28e-2	2.20e-2	2.34e-2
(80,16)	**1.78e-2**	2.68e-2	1.85e-2	2.65e-2	2.54e-2	2.81e-2	**1.61e-2**	2.62e-2	1.62e-2	2.36e-2	2.41e-2	2.64e-2
(100,4)	1.65e-2	2.45e-2	**1.46e-2**	2.42e-2	2.27e-2	2.39e-2	**1.75e-2**	2.70e-2	1.85e-2	2.41e-2	2.54e-2	2.50e-2
(100,8)	1.80e-2	2.78e-2	**1.49e-2**	2.70e-2	2.68e-2	2.85e-2	1.74e-2	2.70e-2	**1.56e-2**	2.59e-2	2.53e-2	2.63e-2
(100,16)	**1.91e-2**	3.00e-2	2.26e-2	2.81e-2	2.86e-2	2.97e-2	**1.89e-2**	3.03e-2	2.43e-2	2.90e-2	2.92e-2	2.92e-2
(200,4)	**2.33e-2**	3.63e-2	3.00e-2	3.77e-2	3.71e-2	3.71e-2	**2.40e-2**	3.61e-2	3.31e-2	3.91e-2	4.04e-2	3.68e-2
(200,8)	**2.54e-2**	3.85e-2	2.99e-2	4.00e-2	3.99e-2	3.87e-2	**2.92e-2**	4.46e-2	4.24e-2	4.55e-2	4.72e-2	4.48e-2
(200,16)	**2.41e-2**	3.95e-2	3.31e-2	3.82e-2	3.98e-2	3.71e-2	**2.40e-2**	4.04e-2	3.29e-2	3.99e-2	3.95e-2	3.74e-2
(500,4)	**3.94e-2**	4.22e-2	3.96e-2	5.17e-2	4.75e-2	4.01e-2	**2.44e-2**	4.75e-2	3.20e-2	4.31e-2	4.77e-2	4.20e-2
(500,8)	**2.98e-2**	4.41e-2	3.86e-2	5.00e-2	4.85e-2	5.00e-2	**4.45e-2**	6.10e-2	5.56e-2	5.42e-2	5.45e-2	5.86e-2
(500,16)	**2.96e-2**	5.50e-2	5.17e-2	5.40e-2	5.74e-2	4.75e-2	**2.60e-2**	4.40e-2	4.90e-2	4.42e-2	3.91e-2	4.12e-2

The better results are presented in bold format

Table 8.8 Mean value of Spread metric among all compared MOEAs

(n, m)	KMMOA	NSGA-II	MOEA/D	SPEA2	CMPSO	MO-LR	KMMOA	NSGA-II	MOEA/D	SPEA2	CMPSO	MO-LR
	$F = 2$						$F = 3$					
(40,4)	**6.50e-1**	7.49e-1	1.00e + 0	6.46e-1	6.45e-1	7.46e-1	**7.06e-1**	8.34e-1	1.00e + 0	7.19e-1	7.12e-01	8.30e-01
(40,8)	**6.39e-1**	7.44e-1	1.00e + 0	6.46e-1	6.41e-1	7.35e-1	**6.91e-1**	8.08e-1	1.00e + 0	7.05e-1	6.96e-01	8.17e-01
(40,16)	6.69e-1	7.75e-1	1.00e + 0	6.77e-1	**6.68e-1**	7.73e-1	**7.12e-1**	8.25e-1	1.00e + 0	7.42e-1	7.24e-01	8.31e-01
(60,4)	6.70e-1	7.85e-1	1.00e + 0	6.67e-1	**6.62e-1**	7.78e-1	7.77e-1	8.91e-1	1.00e + 0	**7.75e-1**	7.87e-01	8.91e-01
(60,8)	**6.86e-1**	8.14e-1	1.00e + 0	6.91e-1	7.07e-1	8.11e-1	**7.43e-1**	8.77e-1	1.00e + 0	7.82e-1	7.84e-01	8.73e-01
(60,16)	6.64e-1	7.93e-1	1.00e + 0	6.65e-1	**6.59e-1**	7.88e-1	**7.32e-1**	8.64e-1	1.00e + 0	7.43e-1	7.41e-01	8.57e-01
(80,4)	6.95e-1	8.23e-1	1.00e + 0	6.88e-1	**6.75e-1**	8.12e-1	**8.07e-1**	9.27e-1	1.00e + 0	8.39e-1	8.23e-01	9.25e-01
(80,8)	**6.88e-1**	8.12e-1	1.00e + 0	6.97e-1	6.98e-1	8.14e-1	**7.51e-1**	8.83e-1	1.00e + 0	7.96e-1	7.98e-01	8.89e-01
(80,16)	7.09e-1	8.58e-1	1.00e + 0	7.36e-1	**7.07e-1**	8.50e-1	**7.49e-1**	8.85e-1	1.00e + 0	7.66e-1	7.73e-01	8.97e-01
(100,4)	6.84e-1	8.21e-1	1.00e + 0	**6.68e-1**	6.78e-1	8.21e-1	**8.02e-1**	9.40e-1	1.00e + 0	8.57e-1	8.38e-01	9.24e-01
(100,8)	**7.20e-1**	8.54e-1	1.00e + 0	7.41e-1	7.32e-1	8.58e-1	**8.13e-1**	9.43e-1	1.00e + 0	8.56e-1	8.67e-01	9.30e-01
(100,16)	**7.05e-1**	8.47e-1	1.00e + 0	7.14e-1	7.18e-1	8.47e-1	**7.82e-1**	9.23e-1	1.00e + 0	8.09e-1	8.26e-01	9.04e-01
(200,4)	**7.69e-1**	9.25e-1	1.00e + 0	8.37e-1	8.59e-1	9.33e-1	**8.32e-1**	9.72e-1	1.00e + 0	8.89e-1	8.66e-01	9.26e-01
(200,8)	**7.76e-1**	9.43e-1	1.00e + 0	8.39e-1	8.37e-1	9.30e-1	**8.94e-1**	9.91e-1	1.00e + 0	9.60e-1	9.25e-01	1.00e + 00
(200,16)	**7.64e-1**	9.08e-1	1.00e + 0	7.99e-1	7.82e-1	9.05e-1	**8.64e-1**	9.82e-1	1.00e + 0	8.97e-1	8.91e-01	9.86e-01
(500,4)	1.0e + 0	1.0e + 0	1.00e + 0	**9.57e-1**	1.0e + 0	1.0e + 0	**9.10e-1**	1.00e + 0	1.00e + 0	9.61e-1	9.25e-01	9.61e-01
(500,8)	9.10e-1	1.0e + 0	1.00e + 0	9.61e-1	**9.05e-1**	9.61e-1	1.00e + 0	1.00e + 0	1.00e + 0	1.00e + 0	1.00e + 0	1.00e + 0
(500,16)	**8.54e-1**	1.0e + 0	1.00e + 0	9.10e-1	8.57e-1	9.96e-1	9.87e-1	1.00e + 0	1.00e + 0	**9.15e-1**	9.83e-1	1.00e + 0
(n, m)	KMMOA	NSGA-II	MOEA/D	SPEA2	CMPSO	MO-LR	KMMOA	NSGA-II	MOEA/D	SPEA2	CMPSO	MO-LR
	$F = 4$						$F = 5$					
(40,4)	**7.87e-1**	8.99e-1	1.00e + 0	8.01e-1	8.17e-1	9.07e-1	**8.39e-1**	9.35e-1	1.00e + 0	8.59e-1	8.56e-1	9.31e-1

(continued)

Table 8.8 (continued)

(n, m)	KMMOA	NSGA-II	MOEA/D	SPEA2	CMPSO	MO-LR	KMMOA	NSGA-II	MOEA/D	SPEA2	CMPSO	MO-LR
	$F = 2$						$F = 3$					
(40,8)	**7.15e-1**	8.34e-1	1.00e + 0	7.36e-1	7.38e-1	8.31e-1	**7.46e-1**	8.66e-1	1.00e + 0	7.82e-1	7.74e-1	8.64e-1
(40,16)	**7.57e-1**	8.72e-1	1.00e + 0	7.87e-1	7.95e-1	8.72e-1	**7.72e-1**	8.80e-1	1.00e + 0	8.15e-1	8.16e-1	8.83e-1
(60,4)	**8.74e-1**	9.54e-1	1.00e + 0	9.12e-1	9.16e-1	9.71e-1	**9.39e-1**	1.00e + 0	1.00e + 0	9.88e-1	9.83e-1	1.00e + 0
(60,8)	**7.96e-1**	9.22e-1	1.00e + 0	8.29e-1	8.41e-1	9.12e-1	**8.48e-1**	9.34e-1	1.00e + 0	8.88e-1	8.70e-1	9.39e-1
(60,16)	**7.65e-1**	9.00e-1	1.00e + 0	8.01e-1	7.80e-1	8.88e-1	**8.08e-1**	9.11e-1	1.00e + 0	8.49e-1	8.39e-1	9.05e-1
(80,4)	**8.80e-1**	9.75e-1	1.00e + 0	9.23e-1	9.28e-1	9.72e-1	**9.58e-1**	1.00e + 0	1.00e + 0	1.00e + 0	9.82e-1	1.00e + 0
(80,8)	**8.43e-1**	9.59e-1	1.00e + 0	8.98e-1	8.87e-1	9.64e-1	**8.46e-1**	9.63e-1	1.00e + 0	9.03e-1	9.14e-1	9.70e-1
(80,16)	**8.08e-1**	9.27e-1	1.00e + 0	8.25e-1	8.26e-1	9.20e-1	**8.41e-1**	9.45e-1	1.00e + 0	8.93e-1	8.81e-1	9.53e-1
(100,4)	**8.98e-1**	9.85e-1	1.00e + 0	9.59e-1	9.50e-1	9.86e-1	**9.51e-1**	1.00e + 0	1.00e + 0	9.97e-1	9.88e-1	1.00e + 0
(100,8)	**8.85e-1**	9.87e-1	1.00e + 0	9.40e-1	9.42e-1	9.91e-1	**9.28e-1**	1.00e + 0	1.00e + 0	9.79e-1	9.66e-1	1.00e + 0
(100,16)	**8.34e-1**	9.35e-1	1.00e + 0	8.44e-1	8.47e-1	9.39e-1	8.61e-1	9.56e-1	1.00e + 0	**8.52e-1**	8.72e-1	9.49e-1
(200,4)	1.0e + 0	1.0e + 0	1.00e + 0	1.0e + 0	1.0e + 0	1.00e + 0	1.00e + 0	1.00e + 0	1.00e + 0	1.00e + 0	1.00e + 0	1.00e + 0
(200,8)	**9.25e-1**	1.0e + 0	1.00e + 0	9.98e-1	1.0e + 0	1.00e + 0	**9.86e-1**	1.00e + 0	1.00e + 0	1.00e + 0	1.00e + 0	1.00e + 0
(200,16)	**9.29e-1**	1.0e + 0	1.00e + 0	9.43e-1	9.44e-1	1.00e + 0	**9.58e-1**	1.00e + 0	1.00e + 0	9.62e-1	9.78e-1	1.00e + 0
(500,4)	1.0e + 0	1.0e + 0	1.0e + 0	1.0e + 0	1.0e + 0	1.00e + 0	1.00e + 0	1.00e + 0	1.00e + 0	1.00e + 0	1.00e + 0	1.00e + 0
(500,8)	1.0e + 0	1.0e + 0	1.0e + 0	1.0e + 0	1.0e + 0	1.00e + 0	**9.95e-1**	1.00e + 0	1.00e + 0	1.00e + 0	1.00e + 0	1.00e + 0
(500,16)	1.0e + 0	1.0e + 0	1.0e + 0	1.0e + 0	1.0e + 0	1.00e + 0	1.00e + 0	**9.88e-1**	1.00e + 0	1.00e + 0	1.00e + 0	1.00e + 0

The better results are presented in bold format

Table 8.9 Mean value of IGD metric among all compared MOEAs

(*n*, *m*)	KMMOA	NSGA-II	MOEA/D DD	SPEA2	CMPSO	MO-LR	KMMOA	NSGA-II	MOEA/D	SPEA2	CMPSO	MO-LR
	$F = 2$						$F = 3$					
(40,4)	**3.30e-3**	3.48e-3	1.08e-2	3.98e-3	4.07e-3	3.59e-3	6.55e-3	**6.36e-3**	1.58e-2	7.14e-3	7.37e-3	6.41e-3
(40,8)	**5.02e-3**	5.52e-3	1.41e-2	6.05e-3	5.96e-3	5.30e-3	**7.31e-3**	7.91e-3	1.89e-2	8.33e-3	8.33e-3	7.77e-3
(40,16)	**5.48e-3**	6.14e-3	1.53e-2	6.79e-3	6.73e-3	6.22e-3	**8.41e-3**	9.25e-3	2.30e-2	9.89e-3	9.84e-3	9.39e-3
(60,4)	6.58e-3	6.47e-3	1.49e-2	7.62e-3	7.29e-3	**6.54e-3**	**1.06e-2**	1.12e-2	2.39e-2	1.17e-2	1.19e-2	1.08e-2
(60,8)	**8.78e-3**	9.64e-3	2.22e-2	1.00e-2	1.03e-2	9.80e-3	**1.04e-2**	1.06e-2	2.04e-2	1.13e-2	1.14e-2	1.06e-2
(60,16)	**9.57e-3**	1.07e-2	2.44e-2	1.06e-2	1.11e-2	1.05e-2	**1.06e-2**	1.21e-2	3.65e-2	1.24e-2	1.23e-2	1.20e-2
(80,4)	**9.31e-3**	9.59e-3	2.13e-2	1.04e-2	1.09e-2	9.57e-3	**1.10e-2**	1.21e-2	2.85e-2	1.28e-2	1.28e-2	1.22e-2
(80,8)	**1.03e-2**	1.08e-2	2.54e-2	1.14e-2	1.13e-2	1.10e-2	**1.07e-2**	1.25e-2	3.11e-2	1.29e-2	1.30e-2	1.26e-2
(80,16)	**1.23e-2**	1.33e-2	3.18e-2	1.35e-2	1.40e-2	1.35e-2	**1.33e-2**	1.58e-2	3.78e-2	1.65e-2	1.63e-2	1.55e-2
(e0,4)	1.16e-2	**1.12e-2**	1.91e-2	1.29e-2	1.25e-2	1.14e-2	**1.55e-2**	1.68e-2	3.07e-2	1.75e-2	1.73e-2	1.70e-2
(e0,8)	**1.33e-2**	1.51e-2	3.43e-2	1.54e-2	1.57e-2	1.52e-2	**1.46e-2**	1.92e-2	4.89e-2	1.94e-2	1.91e-2	1.90e-2
(e0,16)	**1.24e-2**	1.38e-2	3.21e-2	1.46e-2	1.43e-2	1.38e-2	**1.34e-2**	1.56e-2	3.80e-2	1.61e-2	1.64e-2	1.59e-2
(200,4)	**1.71e-2**	1.96e-2	3.77e-2	2.15e-2	2.18e-2	2.09e-2	**1.78e-2**	2.06e-2	2.87e-2	2.01e-2	1.99e-2	2.25e-2
(200,8)	**1.80e-2**	2.35e-2	5.48e-2	2.60e-2	2.44e-2	2.42e-2	**2.44e-2**	3.23e-2	7.31e-2	3.46e-2	3.35e-2	3.25e-2
(200,16)	**1.69e-2**	2.08e-2	5.24e-2	2.21e-2	2.28e-2	2.15e-2	**1.91e-2**	2.41e-2	5.64e-2	2.54e-2	2.49e-2	2.42e-2
(500,4)	**2.84e-2**	3.75e-2	9.11e-2	4.36e-2	3.66e-2	4.09e-2	**2.13e-2**	2.40e-2	6.06e-2	3.03e-2	2.77e-2	2.63e-2
(500,8)	**2.75e-2**	3.63e-2	8.70e-2	3.89e-2	3.95e-2	4.27e-2	**1.91e-2**	2.81e-2	6.52e-2	2.80e-2	3.02e-2	2.68e-2
(500,16)	**3.23e-2**	3.99e-2	8.26e-2	4.05e-2	4.39e-2	4.24e-2	**3.03e-2**	4.17e-2	1.02e-2	4.73e-2	4.60e-2	4.69e-2
(n, m)	KMMOA	NSGA-II	MOEA/D	SPEA2	CMPSO	MO-LR	KMMOA	NSGA-II	MOEA/D	SPEA2	CMPSO	MO-LR
	$F = 4$						$F = 5$					
(40,4)	**6.41e-3**	6.84e-3	1.85e-2	7.50e-3	7.34e-3	6.79e-3	**7.44e-3**	8.01e-3	2.08e-2	8.52e-3	8.64e-3	8.03e-3

(continued)

Table 8.9 (continued)

(n, m)	KMMOA	NSGA-II	MOEA/D DD	SPEA2	CMPSO	MO-LR	KMMOA	NSGA-II	MOEA/D	SPEA2	CMPSO	MO-LR
	$F = 2$						$F = 3$					
(40,8)	**6.55e-3**	6.90e-3	2.04e-2	7.52e-3	7.33e-3	6.94e-3	**6.98e-3**	7.30e-3	1.69e-2	7.91e-3	7.86e-3	7.28e-3
(40,16)	**8.25e-3**	8.88e-3	1.87e-2	9.05e-3	9.69e-3	8.57e-3	**8.46e-3**	9.13e-3	2.21e-2	9.67e-3	9.88e-3	9.07e-3
(60,4)	**1.12e-2**	1.20e-2	2.78e-2	1.30e-2	1.28e-2	1.21e-2	**1.31e-2**	1.47e-2	3.25e-2	1.55e-2	1.50e-2	1.41e-2
(60,8)	**1.01e-2**	1.10e-2	2.66e-2	1.15e-2	1.17e-2	1.07e-2	**1.13e-2**	1.30e-2	3.27e-2	1.34e-2	1.34e-2	1.23e-2
(60,16)	**1.03e-2**	1.19e-2	3.27e-2	1.26e-2	1.24e-2	1.20e-2	**1.29e-2**	1.43e-2	3.16e-2	1.48e-2	1.46e-2	1.45e-2
(80,4)	**1.15e-2**	1.28e-2	3.00e-2	1.30e-2	1.34e-2	1.26e-2	**1.27e-2**	1.36e-2	2.66e-2	1.41e-2	1.42e-2	1.34e-2
(80,8)	**1.48e-2**	1.65e-2	3.61e-2	1.72e-2	1.74e-2	1.68e-2	**1.25e-2**	1.45e-2	3.30e-2	1.50e-2	1.50e-2	1.50e-2
(80,16)	**1.44e-2**	1.77e-2	4.66e-2	1.88e-2	1.84e-2	1.88e-2	**1.39e-2**	1.69e-2	4.10e-2	1.70e-2	1.68e-2	1.67e-2
(e0,4)	**1.46e-2**	1.64e-2	3.65e-2	1.72e-2	1.69e-2	1.67e-2	**1.55e-2**	1.76e-2	4.11e-2	1.85e-2	1.86e-2	1.75e-2
(e0,8)	**1.65e-2**	1.99e-2	4.30e-2	2.03e-2	2.05e-2	2.00e-2	**1.57e-2**	1.84e-2	4.35e-2	1.93e-2	1.92e-2	1.82e-2
(e0,16)	**1.53e-2**	1.93e-2	4.72e-2	1.93e-2	1.96e-2	1.86e-2	**1.58e-2**	2.06e-2	5.25e-2	2.07e-2	2.09e-2	1.98e-2
(200,4)	**1.66e-2**	2.29e-2	5.66e-2	2.43e-2	2.40e-2	2.33e-2	**1.80e-2**	2.35e-2	5.76e-2	2.57e-2	2.59e-2	2.43e-2
(200,8)	**2.33e-2**	2.98e-2	6.74e-2	3.19e-2	3.21e-2	3.00e-2	**2.39e-2**	3.15e-2	7.49e-2	3.34e-2	3.42e-2	3.21e-2
(200,16)	**1.88e-2**	2.41e-2	5.93e-2	2.50e-2	2.57e-2	2.34e-2	**1.91e-2**	2.58e-2	6.07e-2	2.65e-2	2.61e-2	2.44e-2
(500,4)	**3.07e-2**	3.52e-2	8.12e-2	4.16e-2	3.95e-2	3.34e-2	**2.35e-2**	3.01e-2	6.55e-2	3.11e-2	3.21e-2	2.95e-2
(500,8)	**2.04e-2**	2.48e-2	6.46e-2	2.98e-2	2.86e-2	2.74e-2	**3.87e-2**	4.97e-2	1.07e-1	4.57e-2	5.18e-2	4.62e-2
(500,16)	**2.06e-2**	2.99e-2	7.77e-2	2.96e-2	3.48e-2	2.64e-2	**1.92e-2**	2.52e-2	6.50e-2	2.78e-2	2.67e-2	2.66e-2

The better results are presented in bold format

Table 8.10 *P* values of all metrics among all MOEAs on different factories

No. of factories	GD					Spread					IGD				
	KMMOA vs. NSGAII	KMMOA vs. MOEA/D	KMMOA vs. SPEA2	KMMOA vs. CMPSO	KMMOA vs. MO-LR	KMMOA vs. NSGAII	KMMOA vs. MOEA/D	KMMOA vs. SPEA2	KMMOA vs. CMPSO	KMMOA vs. MO-LR	KMMOA vs. NSGAII	KMMOA vs. MOEA/D	KMMOA vs. SPEA2	KMMOA vs. CMPSO	KMMOA vs. MO-LR
F = 2	1.96e-4	6.31e-1	3.27e-4	3.26e-4	3.26e-5	2.92e-4	2.93e-4	2.63e-2	2.55e-1	2.91e-4	4.54e-4	1.96e-4	1.96e-4	1.96e-4	3.27e-4
F = 3	1.96e-4	7.78e-2	1.96e-4	1.96e-4	1.96e-4	2.92e-4	2.93e-4	5.60e-3	3.51e-4	2.91e-4	2.32e-4	8.63e-4	1.96e-4	1.96e-4	2.32e-4
F = 4	1.96e-4	3.48e-1	3.26e-4	3.26e-4	1.96e-4	1.22e-4	1.22e-4	1.22e-4	1.22e-4	1.22e-4	1.96e-4	1.96e-4	3.26e-4	3.26e-4	1.96e-4
F = 5	1.96e-4	1.84e-1	1.96e-4	1.96e-4	1.96e-4	1.22e-4	1.22e-4	1.22e-4	1.22e-4	1.22e-4	1.96e-4	1.96e-4	1.96e-4	1.96e-4	1.96e-4

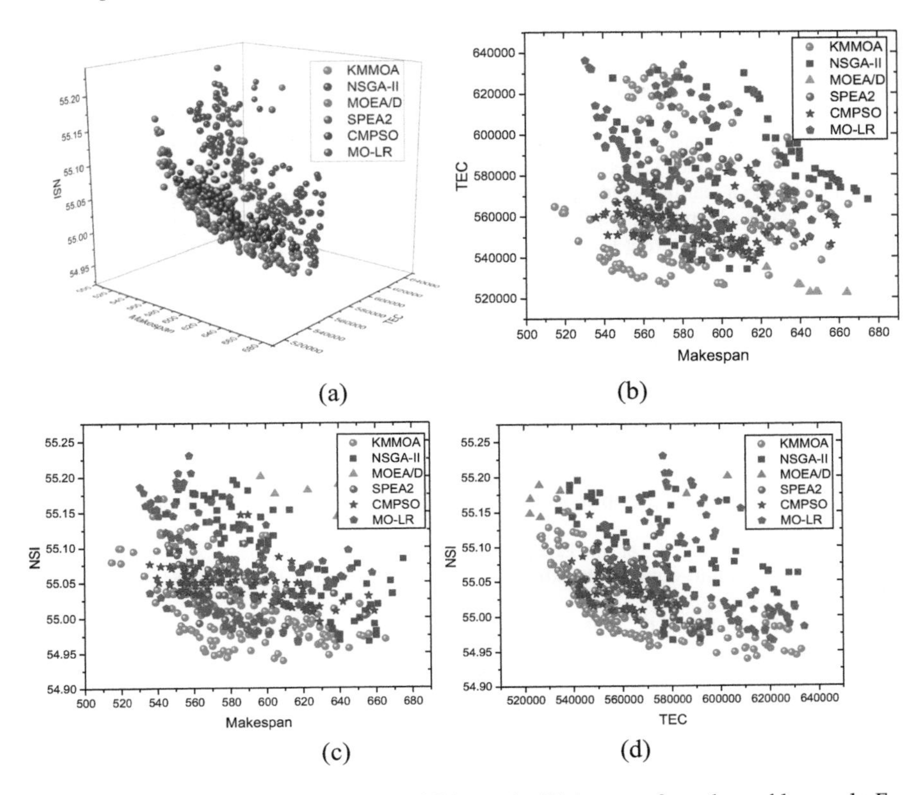

Fig. 8.7 Pareto front found by different MOEAs on the 5th instance from the problem scale $F = 3, m = 8, n = 80$

8.5 Chapter Conclusion

In this chapter, we explain DPFSP-NF, which has three objectives (include makespan, *TEC*, and *NSI*), involves multiple factories, and each has a different machine. We also designed KMMOA to solve DPFSP-NF. This algorithm uses a cooperative initialization heuristic to generate high-quality initial solutions and also presents an effective energy-saving strategy. To improve the exploitation ability, a knowledge-based local search is presented and conducted on the solutions in the offspring. From a series of experiments, we can prove the high efficiency of KMMOA in solving DPFSP-NF. The distributed difference factory problem of hybrid-flow shop, job-shop and dynamic shop scheduling is more similar to the actual production process, which is worth studying. However, KMMOA sacrifices a lot of computing time in achieving good performance. In order to further optimize the algorithm, extremely high computational efficiency is a problem worth studying. Additionally, big data technology and deep learning techniques may be used to address dynamic dispersed shop scheduling issues, which is a crucial direction.

References

1. Yw, A., Kang, L.A., Sg, A., et al.: Analysis of energy saving potentials in intelligent manufacturing: a case study of bakery plants[J]. Energy **172**, 477–486 (2019)
2. Li, X., Lu, C., Gao, L., et al.: An effective multiobjective algorithm for energy-efficient scheduling in a real-life welding shop[J]. IEEE Trans. Industr. Inf. **17**(10), 6687–6696 (2018)
3. Stoycheva, S., Marchese, D., Paul, C., et al.: Multi-criteria decision analysis framework for sustainable manufacturing in automotive industry[J]. J. Clean. Prod. **187**(JUN.20), 257–272 (2018)
4. Wen, X., Li, X., Gao, L., et al.: Modified honey bees mating optimization algorithm for multi-objective uncertain integrated process planning and scheduling problem[J]. Int. J. Adv. Rob. Syst. **17**(3), 255688469 (2020)
5. Elkington, J.: Cannibals with forks: the triple bottom line of 21st century business. Gabriola Island, BC[J]. Environ. Qual. Manage., **8**(1), 37–51 (1998)
6. Amrina, E., Vilsi, A.L.: Key Performance indicators for sustainable manufacturing evaluation in cement industry[J]. Procedia Cirp. **26**, 19–23 (2015)
7. Liu, Y., Dong, H., Lohse, N., et al.: A multi-objective genetic algorithm for optimisation of energy consumption and shop floor production performance[J]. Int. J. Prod. Econ. **179**, 259–272 (2016)
8. Lu, C., Gao, L., Li, X., et al.: Energy-efficient permutation flow shop scheduling problem using a hybrid multi-objective backtracking search algorithm[J]. J. Clean. Prod., **144**(FEB.15), 228–238 (2017)
9. Chao, L., Li, X., Liang, G., et al.: An effective multi-objective discrete virus optimization algorithm for flexible job-shop scheduling problem with controllable processing times[J]. Comput. Indus. Eng., **104**(FEB), 156–174 (2017)
10. Xu, Y., Wang, L., Wang, S., et al.: An effective hybrid immune algorithm for solving the distributed permutation flow-shop scheduling problem[J]. Eng. Optim. **46**(9), 1269–1283 (2014)
11. Luo, G., Wen, X., Li, H., et al.: An effective multi-objective genetic algorithm based on immune principle and external archive for multi-objective integrated process planning and scheduling[J]. Int. J. Adv. Manuf. Technol. **91**(9–12), 3145–3158 (2017)
12. Goldberg, D.E., Alleles, R.L.: Loci and the traveling salesman problem[J]. Inventiones Mathematicae (1985)
13. Chao, L., Liang, G., Quanke, P., Li, X., Zheng, J.: A multi-objective cellular grey wolf optimizer for hybrid flowshop scheduling problem considering noise pollution[J]. Appl. Soft Comput. **75**, 728–749 (2019)
14. Peng, K.Q.L.X.: A multi-start variable neighbourhood descent algorithm for hybrid flowshop rescheduling[J]. Swarm Evol. Comput. **45**, 92–112 (2019)
15. Deb, K., Pratap, A., Agarwal, S., et al.: A fast and elitist multiobjective genetic algorithm: NSGA-II[J]. IEEE Trans. Evol. Comput. **6**(2), 182–197 (2002)
16. Zitzler, E., Deb, K., Thiele, L.: Comparison of multiobjective evolutionary algorithms: empirical results[J]. Evol. Comput. **8**(2), 173–195 (2000)
17. Zitzler, E., Thiele, L.: Multiobjective evolutionary algorithms: a comparative case study and the strength Pareto approach[J]. IEEE Trans. Evol. Comput. **3**(4), 257–271 (1999)

Chapter 9
Green Scheduling in Distributed Permutation Flow Shop with Limited Buffers

9.1 Brief Introduction

With the advancement of economic globalization and green manufacturing, energy-efficient scheduling of distributed production systems has become a common practice among large corporations [1–6]. However, the energy-efficient scheduling of the distributed permutation flow shop problem with limited buffers (DPFSP_LB) receives insufficient attention in the relevant literature. This chapter is thus the first attempt to investigate this DPFSP_LB with the goal of minimizing makespan and total energy consumption (TEC). A Pareto-based collaborative multi-objective optimization algorithm (CMOA) is proposed to solve this energy-efficient DPFSP_LB.

At first, the proposed CMOA reduces TEC by using a speed scaling strategy based on problem property. Second, a collaborative initialization strategy for generating a high-quality initial population is presented. Third, three DPFSP_LB properties are used to create a collaborative search operator and a knowledge-based local search operator. Finally, we test the effectiveness of each CMOA improvement component on instances and compare it to other well-known multi-objective optimization algorithms [7]. The results of the experiments show that CMOA is effective in solving this energy-efficient DPFSP_LB. The CMOA, in particular, is capable of obtaining excellent results on all problems concerning the comprehensive metric, and it is also competitive with its rivals in terms of the convergence metric.

C. Lu et al., *Intelligence Optimization for Green Scheduling in Manufacturing Systems*, Engineering Applications of Computational Methods 18,
https://doi.org/10.1007/978-981-99-6987-6_9

9.2 Problem Statement and Modeling

9.2.1 Problem Statement

The following are some notations for the DPFSP_LB problem:

n : number of jobs.
m : number of machines or stages in each factory.
f : number of factories.
j : index of jobs, $j = 1, 2, \ldots, n$.
i : index of machines, $i = 1, 2, \ldots, m$.
k : index of factories, $k = 1, 2, \ldots, f$.
J_j : the job in the j-th position.
M_i : the i-th machine.
F_k : the k-the factory.
N : a set of jobs, $J_j \in N$.
M : a set of machines, $M_i \in M$.
F : a set of factories, $F_k \in F$.
V : a set of machine processing speeds of operations,$v_{i,j} \in V$.
B_i :the limited buffer between the machine M_i and M_{i+1}.
$O_{i,j}$: the operation of the job J_j on the machine M_i.
$t_{i,j}$: the standard processing time of $O_{i,j}$.
$v_{i,j}$: the machine processing speed of $O_{i,j}$.
$p_{i,j}$: the actual processing time of $O_{i,j}$.
n_k : the number of jobs assigned to the factory F_k.
π : a schedule of jobs in each factory, $\pi = \{\pi_k | k = 1, 2, \ldots, f\}$.
$J_{\pi_k(h)}$: the job in the h-th position in π_k, where $h = 1, 2, \ldots, n_k$.
π_k : the sequence of jobs in factory F_k, i.e., $\pi_k = \left[J_{\pi_k(1)}, \ldots, J_{\pi_k(h)}, \ldots, J_{\pi_k(nk)}\right]$.
$d_{i,\pi_k(h)}$: the departure time of operation $O_{i,\pi_k(h)}$ from the machine M_i.
$S_{i,\pi_k(h)}$: the starting time of $O_{i,\pi_k(h)}$.
$C_{i,\pi_k(h)}$: the completion time of $O_{i,\pi_k(h)}$.
$C^k_{\max}$: the maximum completion time of the factory F_k.
$C_{\max}$: the maximum completion time among all factories, i.e., makespan.
P_i^{run} :the energy consumption of the machine M_i per unit time during the run or processing phase.
P_i^{idle} : the energy consumption of the machine M_i per unit time during the idle phase.
P_s : the energy consumption of the storage machine per unit time in each buffer.

Energy-efficient DPFSP-LB is stated as follows: A set $N = \{J_j | j = 1, \ldots, n\}$ of n jobs must be processed on any one of f factories from a set $F = \{F_k | k = 1, 2, \ldots, f\}$. Each factory F_k contains the same permutation flow shop with m machines from a set $M = \{M_i | i = 1, 2, \ldots, m\}$. Each job J_j has a list of m operations $\left\{O_{1,j}, O_{2,j}, \ldots, O_{m,j}\right\}$, which have to be sequentially processed one after another on m machines. There exists a buffer with a limited capacity $B_i > 0$ between any two successive machines M_i and $M_{i+1} (i \neq m)$ in each factory. Each job should pass through each buffer between adjacent machines and obey the First In First Out (FIFO) rule in each buffer. Precisely, when a job has been completed on the machine M_i, it will be passed to the next machine M_{i+1} (if the next machine is available), buffer B_i (if the current buffer is available and the next machine is unavailable), or blocked on the current machine M_i. Each operation $O_{i,j}$ can be processed by one machine M_i at a speed from a set of v $= \{v_{i,j} | (i = 1, 2, \ldots, m, j = 1, 2, \ldots, n)\}$. When $O_{i,j}$ is processed on the machine M_i at a speed $v_{i,j}$, the actual processing time of $O_{i,j}$ is denoted as $p_{i,j} = t_{i,j} / v_{i,j}$, where $t_{i,j}$ is the standard processing time of one operation $O_{i,j}$. In this optimization problem, a schedule contains three decisions: assigning jobs to factories, sequencing partial jobs in each factory, and selecting machine speeds for operations. This problem aims to find a set of non-dominated solutions or schedules so that the maximum completion time (i.e., makespan) and total energy consumption (TEC) are simultaneously minimized. Besides, some assumptions are considered as follows:

- The setup and transport times are negligible.
- Neither preemption nor interrupt is allowed, once jobs are being processed on machines.
- All jobs are available at time zero and there is no precedence constraint.
- One job cannot be handled on more than one machine at any time, and one machine cannot handle more than one job at any time.
- Once a job is allocated to a certain factory, all operations of this job must be handled in this assigned factory and cannot be transferred to another factory.

9.2.2 Mathematical Modeling

In this problem, jobs in each factory are denoted as $\pi = \{\pi_k | k = 1, \cdots, f\}$, where $\pi_k = \left[J_{\pi_k(1)}, \cdots, J_{\pi_k(h)}, \cdots, J_{\pi_k(n_k)}\right]$ is the partial schedule of n_k jobs allocated to the factory $F_k \in \mathcal{F}$, $\pi_k(h) \in \{1, 2, \ldots, n\}$. $d_{i,\pi_k(h)}$ represents the departure time of operation $O_{i,\pi_k(h)}$ from a machine M_i and can be defined as follows:

$$d_{1,\pi_k(1)} = p_{1,\pi_k(1)} \tag{9.1}$$

$$d_{i,\pi_k(1)} = d_{i-1,\pi_k(1)} + p_{i,\pi_k(1)} \quad \forall i = 2, \cdots, m. \tag{9.2}$$

$$d_{1,\pi_k(h)} = d_{1,\pi_k(h-1)} + p_{1,\pi_k(h)} \quad \forall h = 2, \cdots, B_1 + 1 \tag{9.3}$$

$$\begin{aligned} d_{i,\pi_k(h)} &= \max\left(d_{i-1,\pi_k(h)}, d_{i,\pi_k(h-1)}\right) \\ &+ p_{i,\pi_k(h)} \quad \forall h = 2, \cdots, B_i + 1, \forall i = 2, \cdots, m-1 \end{aligned} \tag{9.4}$$

$$d_{i,\pi_k(h)} = \max\left(d_{1,\pi_k(h-1)} + p_{1,\pi_k(h)}, d_{2,\pi_k(h-B_1-1)}\right) \forall h > B_1 + 1 \tag{9.5}$$

$$d_{i,\pi_k(h)} = \max\left(\max\left(d_{i-1,\pi_k(h)}, d_{i,\pi_k(h-1)}\right) + p_{i,\pi_k(h)}, d_{i+1,,\pi_k(h-B_i-1)}\right)$$
$$\forall h > B_i + 1, i = 2, \cdots, m-1 \tag{9.6}$$

$$d_{m,\pi_k(h)} = \max\left(d_{m,\pi_k(h-1)}, d_{m-1,\pi_k(h)}\right) + p_{m,\pi_k(h)} \quad \forall h = 2, \cdots, n_k \tag{9.7}$$

In the above recursive formulations, for a partial job schedule π_k in the factory F_k the departure time of the first job on each machine is computed at first, then the second job, and so on until the last job. Equations (9.1) and (9.2) calculate the departure time of the first job in π_k from the first machine to the last machine in each factory F_k. Equations (9.3) and (9.4) define the departure time of job $J_{\pi_k(h)}$ on the first machine ($h = 2, 3, \cdots, B_1 + 1$) or on the i-th machine ($i = 2, 3, \ldots, m-1$; $h = 2, 3, \ldots, B_i + 1$) in the factory F_k. Equations (9.5) and (9.6) specify the departure time of the job $J_{\pi_k(h)}$ on the first machine $h > B_1 + 1$ or on the i-th machine ($h > B_i + 1, i = 2, 3, \ldots, m-1$) in the factory F_k. In this case, the limited buffer capacities should be considered. In Eq. (9.6), $\max\left(d_{i,\pi_k(h-1)}, d_{i-1,\pi_k(h)}\right) + p_{i,\pi_k(h)}$ denotes the completing time of operation $O_{i,\pi_k(h)}$, and $d_{i+1,\pi_k(h-B_i-1)}$ is related to the buffer capacities. Equation (9.7) defines the departure time of job $J_{\pi_k(h)}(h = 2, 3, \ldots, n_k)$ on the last machine in the factory F_k. It is noted that this problem will become a classical distributed permutation flow shop scheduling problem when $B_i \geq n_k - 1$.

The objectives of this problem are to minimize makespan and TEC. The first objective makespan $C_{\max}$ is given as follows:

$$\min f_1 = C_{\max} = \max_{k=1}^{f}\left\{d_{m,\pi_k(n_k)}\right\} \tag{9.8}$$

For the convenience of calculation, the second objective TEC consists of three parts: the run energy consumption EC_r, the storage energy consumption EC_s, and the idle energy consumption EC_d. Therefore, TEC is also written as follows:

$$\min f_2 = \mathrm{TEC} = \mathrm{EC}_r + \mathrm{EC}_s + \mathrm{EC}_d \tag{9.9}$$

The run energy consumption EC_r is defined as follows:

$$\mathrm{EC}_r = \sum_{k=1}^{f}\sum_{i=1}^{m}\sum_{h=1}^{n_k} P_i^{\mathrm{run}} \cdot p_{i,\pi k(h)} \tag{9.10}$$

where $p_{i,\pi k(h)}$ denotes the actual processing time of operation $O_{i,\pi k(h)}$.

The storage energy consumption EC_s is defined as follows:

$$\mathrm{EC}_s = \sum_{k=1}^{f} \sum_{i=1}^{m-1} \sum_{h=1}^{n_k} P_s \cdot \mathrm{BT}_{i,\pi k(h)} \tag{9.11}$$

where $\mathrm{BT}_{i,\pi_k(h)}$ is the storage time of job $J_{\pi_k(h)}$ in the buffer B_i at the factory F_k.

The idle energy consumption EC_d is defined as follows:

$$\mathrm{EC}_d = \sum_{k=1}^{f} \sum_{i=1}^{m} \left(d_{i,\pi_k(n_k)} - \sum_{h=1}^{n_k} p_{i,\pi_k(h)} \right) \cdot P_i^{\mathrm{idle}} \tag{9.12}$$

where $d_{i,\pi_k(n_k)} - \sum_{h=1}^{n_k} p_{i,\pi_k(h)}$ represents the total idle time of the machine M_i in the factory F_k.

Example 9.1 To illustrate the calculation of makespan and TEC, define a problem with $n = 8, m = 3, f = 2$, and $B_1 = B_2 = 1$. Let a schedule be $\pi = \{\pi_1, \pi_2\}$ with $\pi_1 = [J_1, J_2, J_3, J_4]$ and $\pi_2 = [J_5, J_6, J_7, J_8]$. Each machine processing speed $V_{i,j}$ is set to 1 for all operations. In this case, the actual processing time of each operation is the standard processing time. Suppose that the actual processing time $p_{2,1} = 3$, $p_{2,2} = p_{3,2} = p_{3,3} = p_{2,5} = p_{2,6} = p_{3,6} = p_{3,7} = 2$, $p_{i,j} = 1$ for other operations, $P_i^{\mathrm{run}} = 3$, $P_i^{\mathrm{idle}} = 1$, and $P_s = 0.5$. Figure 9.1 depicts a Gantt chart of this solution. At the same time, it can be calculated that $C_{\max}(\pi, \gamma)$ is equal to 11 time units and TEC is 120.5 energy consumption units.

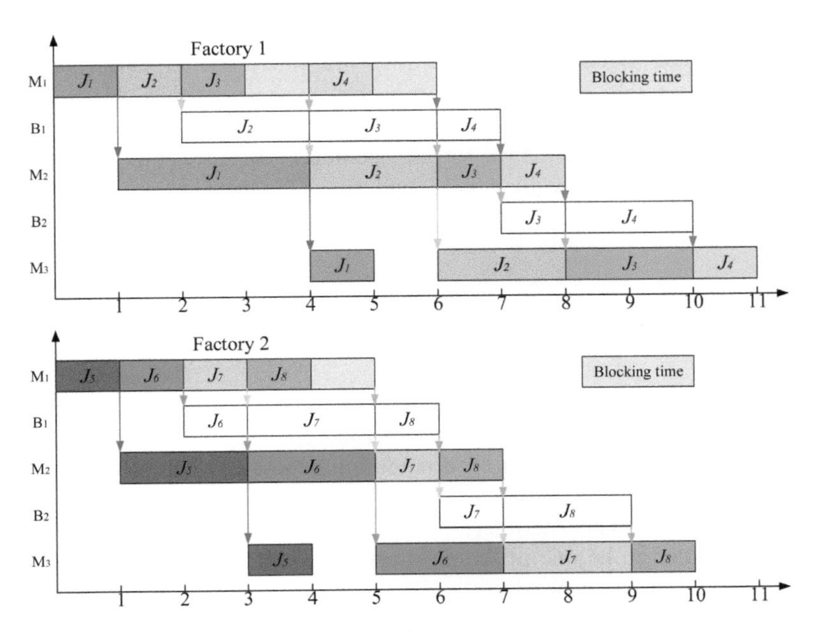

Fig. 9.1 Gantt chart of the solution in example 1

9.3 Proposed Algorithm

Because the flow shop scheduling problem with limited buffers is known to be NP-hard, the energy-efficient DPFSP_LB can be inferred to be NP-hard. That is, exact algorithms cannot find all Pareto optimal solutions within an acceptable computation time. To solve this DPFSP_LB, we proposed a Pareto-based collaborative multi-objective optimization algorithm (CMOA). The following factors motivated the development of CMOA for DPFSP_LB: (1) different search operators can contribute to search diversity; (2) high-performing operators can be designed using problem-specific properties; (3) the Pareto-based method is effective in solving MOPs. Figure 9.2 shows the proposed CMOA framework. In the following subsections, we will discuss some of the mathematical properties of this problem before delving into the main or innovative components of this algorithm design, namely the solution representation, initialization, collaborative search operator, and local search.

9.3.1 Solution Representation

This solution should include three decisions based on the characteristics of this problem: (1) assigning jobs to factories; (2) determining the job processing sequence in each factory; (3) choosing a machine processing speed for each operation. Consequently, one solution can be encoded as $\prod = (\pi; \gamma) = \left(\pi_1, \cdots, \pi_k, \cdots \pi_f; \left(v_{i,j}\right)_{m\times n}\right)$, where π indicates the assignment of jobs to factories and the sequence in which jobs are processed in each factory, π_k denotes the processing sequence of partial jobs in the factory F_k, namely $\pi_k = \left[J_{\pi_k(1)}, \cdots, J_{\pi_k(h)} \cdots, J_{\pi_k(n_k)}\right]$, $\sum_{k=1}^{f} n_k = n$, and $\gamma = \left(v_{i,j}\right)_{m\times n}$, resents the machine processing speed matrix for all operations. Besides, Fig. 9.3 illustrates a solution representation with $n = 8$ and $f = 2$, where $\pi = \{\pi_1, \pi_2\}$ with $\pi_1 = [J_1, J_2, J_3, J_4], \pi_2 = [J_5, J_6, J_7, J_8]$ and $v_{i,j}$ is set to 1 for all operations.

9.3.2 Initialization

The population size (denoted as PS) is constant. Three heuristic rules are used to construct some individuals in order to generate an initial population with high quality and diversity. The rest individuals are randomly generated. The following are the descriptions of these three heuristics: The first heuristic (NEH3_en) has the potential to improve the makespan criterion [8]. As previously stated, the NEH2_en rule has been demonstrated to be one of the most effective heuristics for the DPFSP with the makespan criterion. Inspired by the NEH2_en, we propose a new heuristic called NEH3_en to generate some good solutions for makespan based on its characteristics. The core idea of NEH3_en is as follows: Create a speed matrix at random and

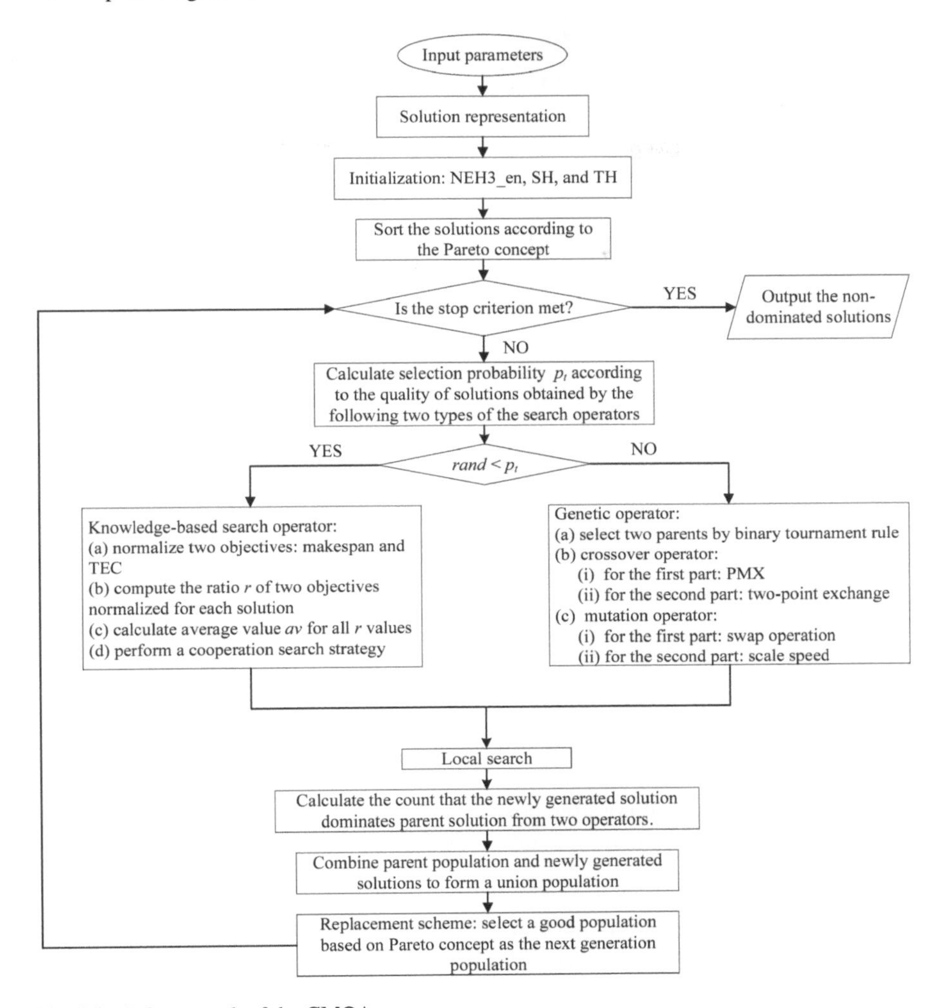

Fig. 9.2 A framework of the CMOA

calculate the actual processing time of each operation. Sort all jobs in reverse order of total actual processing time. After inserting one job into the best position among all factories, the precedent or successor of the job is removed and reinserted into the best position in the same factory. Algorithm 9.1 displays the NEH3_en pseudocode.

Algorithm 9.1 NEH3_en heuristic

1: **Input**: The present solution or schedule $\prod = (\pi; \gamma)$
2: **Output**: The revised solution or schedule
3: The machine processing speed matrix γ is randomly generated
4: Count the total really processing time $p_j = \sum_{i}^{m} p_{i,j}, \forall j$

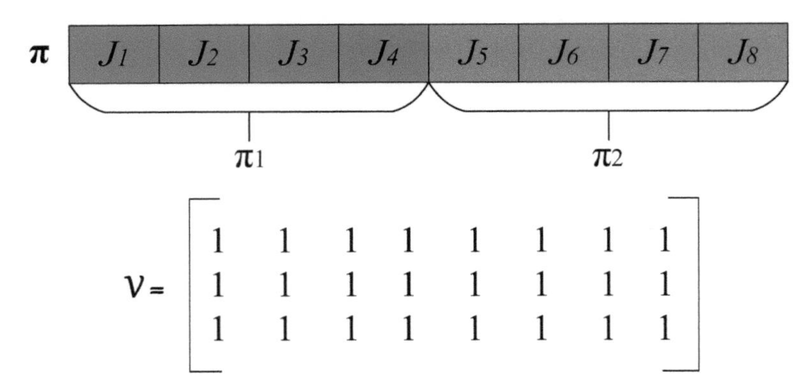

Fig. 9.3 Solution representation

5: All jobs in non-ascending order of the total actual processing times is sorted as π
6: **for** j = 1 to n **do**
7: **for** k = 1 to f **do**
8: check job J_j in all possible positions of π_k
9: Count the partial makespan $C_{\max}(\pi;\gamma)$
10: **end for**
11: $\pi_{k_{\min}} = \arg\left(\max_{k=1}^{f}(C_{\max}(\pi,\gamma))\right)$
12: Insert job J_j of $\pi_{k_{\min}}$ in the minimization $C_{\max}$ at the best position resulting
13: Extract a job J_h from the previous and next of job J_j in $\pi_{k_{\min}}$
14: Test this job J_h in all possible positions in $\pi_{k_{\min}}$
15: Insert the job J_h in the minimization $C_{\max}$ at the best position in $\pi_{k_{\min}}$
16: **end for**

The second heuristic (SH) generates solutions with the desired TEC criterion by selecting an appropriate speed for each operation. The main idea behind the SH rule is that TEC can be reduced by slowing down machine processing speed without affecting other operations. In this case, the solution's makespan remains constant while the TEC is reduced. In fact, slowing down is a type of energy-saving strategy because it reduces TEC to some extent. Algorithm 9.2 displays the detailed pseudocode of the SH rule.

The following is the third heuristic (TH) in favor of balancing factory workload: First, assign $[n/f]$ jobs to each factory, where $[n/f]$ represents the integer part of n/f (n and f are the numbers of jobs and factories, respectively), and then randomly assign the remaining jobs to factories one by one until all assignments are completed. This heuristic aims to distribute all jobs as evenly as possible across all factories. Algorithm 9.3 depicts the pseudocode for the TH rule.

Algorithm 9.2 SH heuristic

1: **Input**: The present solution or schedule $\prod = (\pi;\gamma)$

2: **Output**: The revised solution or schedule
3: **for** k = 1 to f **do**
4: **for** i = 1 to m − 1 **do**
5: **for** h = 1 to n_k − 1 **do**
6: **if** $S_{i,\pi_k(h)} + \frac{t_{i,\pi_k(h)}}{v_{i,\pi_k(h)}} < \min\{S_{i,\pi_k(h+1)}, S_{i+1,\pi_k(h)}, d_{i,\pi_k(h)}\}$ **then**
7: **while** $v_{i,\pi_k(h)} > v^{\min}_{i,\pi_k(h)} \& S_{i,\pi_k(h)} + \frac{t_{i,\pi_k(h)}}{v_{i,\pi_k(h)}-1} <$ $\min\{S_{i,\pi_k(h-1)}, S_{i-1,\pi_k(h)}, d_{i,\pi_k(h)}\}$ **do**
8: $v_{i,\pi_k(h)} = v_{i,\pi_k(h)}$-1
9: **end while**
10: $C_{i,\pi_k(h)} = S_{i,\pi_k(h)} + \frac{t_{i,\pi_k(h)}}{v_{i,\pi_k(h)}}$
11: **end if**
12: **end for**
13: **end for**
14: **end for**

Algorithm 9.3 TH heuristic

1: **Input**: The present solution or schedule $\prod = (\pi; \gamma)$
2: **Output**: The revised solution or schedule
3: Assign $[n/f]$ jobs to all factory and get a remainder number of n/f, denoted as rem
4: **while** rem > 0 **do**
5: Choose a factory randomly denoted as fr and define a set denoted as S_r
6: **if** S_r does not contain f_r **then**
7: put this f_r into a set S_r and rem - -
8: **end if**
9: **end while**

It is noted that we run three heuristic rules ten times to get some initial solutions with good convergence. Furthermore, one random initialization strategy is used to generate other initial solutions with a high diversity.

9.3.3 Collaborative Search Operator

Different search strategies can direct trial solutions toward various unknown regions of the solution space. As a result, we create a collaborative search operator by combining the advantages of various search operators. This collaborative operator in this paper contains two types of search operators: knowledge-based search operator and genetic operator.

(1) Knowledge-based search operator

The improvement of two objectives is the focus of the knowledge-based search operator. It is made up of seven different search operators that are based on the problem property. Let F_c and F_r stand for critical and random non-critical factories, respectively. J_c and J_r jobs are drawn at random from F_c and F_r, respectively. The following are the seven different types of search operators:

Outer-critical-factory insertion move (OCFI): Extract one job J_c at random from F_c and then place it in a random position in another factory F_r.

Outer-critical-factory swap move (OCFS): Take two jobs, J_c and J_r, at random from F_c. and F_r, and then swap their positions.

Inter-critical-factory insertion move (ICFI): Extract one job J_c at random from F_c and then insert it into one random position in the same factory F_c.

Inter-critical-factory swap move (ICFS): Extract two jobs at random from F_c and then swap their positions.

Decrease speed of non-critical-operation (DSNCO): Determine one non-critical operation that will have idle time after it, and then slow it down as much as possible without affecting the subsequent operations.

Decrease speed of the last operation blocked (DSLOB): Determine the last operation that was blocked on each machine, and then slow it down as much as possible without affecting subsequent operations.

Increase critical-operation speed (ICOS): Determine one critical operation from F_c and then accelerate this operation.

The seven search operators listed above each conduct a different search process in different directions. As a result of the work [9], the following cooperation strategy of multiple search operators is used:

Step 1: At each iteration, normalize two objectives for each solution, which is denoted as $\overline{C_{\max}} = \left(C_{\max} - C_{\max}^{\min}\right)/\left(C_{\max}^{\max} - C_{\max}^{\min}\right)$ and $\overline{\text{TEC}} = \left(\text{TEC} - \text{TEC}^{\min}\right)/\left(\text{TEC}^{\max} - \text{TEC}^{\min}\right)$, where C_{max}^{min} and C_{max}^{max} is the minimum makespan and maximum makespan at the current iteration, $\text{TEC}^{\min}$ and $\text{TEC}^{\max}$ represents the minimum TEC and maximum TEC at the current iteration, respectively.

Step 2: For each iteration, compute the value $r = \overline{\text{TEC}}/\overline{C_{\max}}$ of each solution, and then compute the average value av of all r values. When $r \leq$ av, it indicates that the makespan objective is larger than the TEC objective. As a result, the search operators associated with F_c (**OCFI, OCFS, ICFI, ICFS,** and **ICOS**) are executed. On the other hand, $r >$ av indicates that the TEC objective is larger than makespan. As a result, the search operators associated with F_c (i.e., **DSNCO** and **DSLOB**) are carried out. **Algorithm 9.4** contains the pseudocode for this knowledge-based search operator.

(2) Genetic operator

Crossover aims to explore the unknown regions of the solution search space. Partially matching crossover (PMX) can be used in the first part π based on the characteristics of the solution representation of DPFSP-LB. In the second Part γ, two-point crossover [10] is used. Mutation can help the algorithm in avoiding local optima. On the first part of one solution, we perform a swap operation.

The dynamic selection probability mechanism is used to select an appropriate search operator (either knowledge-based search operator or genetic operator) for updating the population based on the above two search operators. The core idea of this dynamic selection mechanism is that each operator's selection probability is proportional to the improved quality of solutions obtained by the corresponding operator. The first operator's specific selection probability value pt at each iteration t is defined as follows:

$$P_t = \begin{cases} \text{random}[0, 0.5), & \text{if } C_2 > C_1 \wedge t\%2 = 0; \\ \text{random}(0.5, 1], & \text{if } C_1 > C_2 \wedge t\%2 = 0; \\ 0.5, & \text{otherwise.} \end{cases} \tag{9.13}$$

where C_1 is the number of improvement solutions obtained by the knowledge-based search operator, C_2 is the number of improvement solutions obtained by the genetic operator, p_t represents the selection probability value of the first search operator, i.e., knowledge-based search operator, and $1 - p_t$ represents the selection probability value of the second search operator, i.e., genetic operator.

Algorithm 9.4 knowledge-based search operator

1: **Input**: The present solution or schedule $\prod = (\pi; \gamma)$
2: **Output**: The revised solution or schedule
3: Normalize the makespan and TEC of all solution and count $r = \overline{TEC}/\overline{C_{\max}}$ of all solution
4: Count the average value *av* of all *r* values
5: **while** r > PS **do**
6: **if** r > av **then**
7: **if** the number of jobs in F_c is greater than that in F_r with the second largest $C_{\max}^r$ **then**
8: conduct **OCFI** and **OCFS** in turn
9: **else if** the energy consumption of F_c in all factories is the smallest then
10: conduct **ICOS**
11: **else**
12: conduct **ICFI** and **ICFS** in turn
13: **end if**
14: **else**
15: conduct **DSNCO** and **DSLOB** in turn
16: **end if**
17: **end while**

9.3.4 *Local Search*

Metaheuristics, like other approximate algorithms, have good exploration performance but poor exploitation performance [11]. A knowledge-based local search heuristic is proposed and incorporated into our proposal to improve the local exploitation capability. This knowledge-based local search operator, inspired by the properties of this problem, can be applied to one non-dominated solution chosen at random from the incumbent population. Algorithm 9.5 contains the pseudocode for this proposed local search.

9.4 Experiments

To demonstrate the effectiveness of the CMOA, a considerable number of experiments were conducted in this section. All algorithms were implemented by Java. All experiments were performed on a PC with a 2.90 GHz Intel Core i7 and 16 GB RAM under a Windows 10 operating system.

9.4.1 *Instances and Performance Metrics*

Energy-efficient DPFSP_LB is a new problem, and there are no standard benchmarks for it. As a result, these instances are built by n jobs, m machines, and f factories, where $n = \{100, 150, 200\}$, $m = \{5, 10, 15\}$, and $f = \{3, 4, 5\}$. Each combination is made up of these various parameter configurations. Meanwhile, each combination has ten different instances with varying standard processing times, for a total of 270 instances. The discrete uniform distribution of the standard processing time $t_{i,j}$ is $[10, 100]$. The actual processing time is $p_{i,j} = t_{i,j}/v_{i,j}$, where $v_{i,j}$ is a random integer number from a discrete uniform distribution $[1, 4]$. The limited buffer is $B_i = 1$. The run energy power is $P_i^{\text{run}} = 1 \times v_{i,j}^2$. The idle energy power is $P_i^{\text{idle}} = 2$. The storage energy power is $P_s = 1$.

In this experiment, two performance metrics, generation distance (GD) [12] and inverted generation distance (IGD) [13], are used to evaluate the behavior of algorithms for solving these instances. GD is a convergence performance metric, and IGD is a comprehensive performance metric that takes both diversity and convergence into account. These metrics employ a normalization method.

9.4.2 Parameter Calibration

Parameter configurations would have a direct impact on the proposed CMOA's performance. These parameters primarily include population size (PS), crossover ratio (p_c), mutation ratio (p_m), and the number of local searches (LS). To determine the best parameter configuration on all instances, we used a Taguchi approach of design-of-experiments (DOE). There are four levels for each parameter. The specific level of each parameter of this CMOA is as follows: PS $=$ {50, 100, 150, 200}, $p_c = \{0.7, 0.8, 0.9, 1.0\}$, $p_m = \{0.1, 0.2, 0.3, 0.4\}$, and LS $=$ {4, 8, 12, 16}. In this experiment, one orthogonal array $L_{16}(4^4)$ is used. The proposed CMOA is run 30 times on each instance for each parameter configuration. For this calibration experiment, the maximum number of function evaluations (NFE) is set to 25,000. Figure 9.4 shows the main effect plot of four IGD metric parameters.

Algorithm 9.5 Local search operator

1: **Input**: A random non-dominated solution $\prod = (\pi; \gamma)$, set l $=$ 0 and w $=$ 0
2: **Output**: A new solution is generated below
3: **while** l $>$ LS **do**
4: **if** w $= =$ 0 **then**
5: conduct **ICFI** to obtain Π'
6: **else if** w $= =$ 1 **then**
7: conduct **ICFS** to obtain Π'
8: **else if** w $= =$ 2 **then**

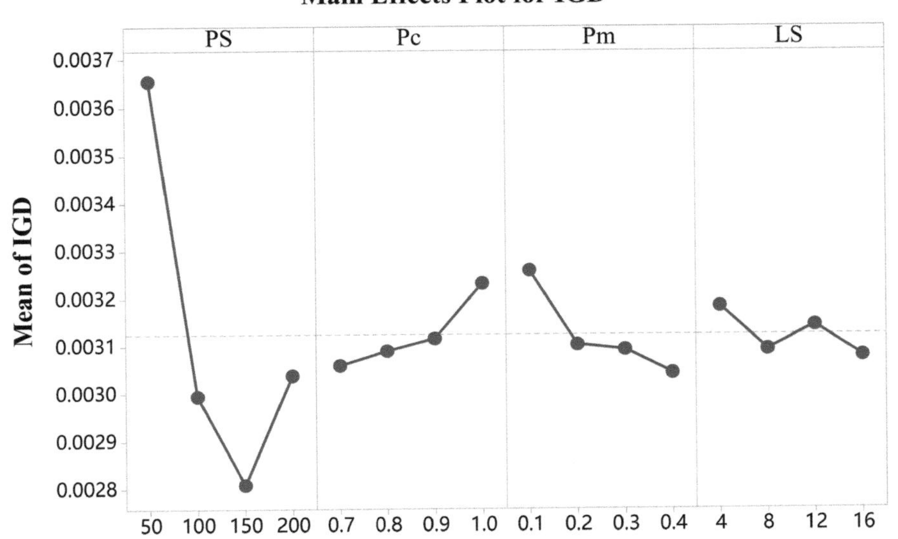

Fig. 9.4 Main effect plot of four parameters

9: conduct **DSNCO** to obtain Π'
10: **else**
11: conduct **DSLOB** to obtain Π'
12: **end if**
13: **if** $\Pi' \prec \Pi$ **then**
14: $\Pi \leftarrow \Pi'$
15: **end if**
16: w + +, and l + +
17: **if** w > 3 **then**
18: w = 0
19: **end if**
20: **end while**

The smaller the IGD value, the better for the multi-objective optimization algorithm. We can see from Fig. 9.4 that the IGD value decreases as PS increases. Similarly, $p_c = 0.9$ yields the best IGD value. The corresponding IGD is the smallest when $p_m = 0.1$. When LS = 12, the IGD is optimal. According to the findings, the best parameter configuration for CMOA is PS = 200, $p_c = 0.9$, $p_m = 0.1$, and LS = 12.

9.4.3 Effectiveness of Improvement Components of CMOA

This section aims to evaluate the contribution of each CMOA improvement component. The initialization strategy, collaborative search, and local search are among the enhancement components. As a result, four CMOA variants are produced: CMOA1, CMOA2, CMOA3, and CMOA4. The CMOA with a random initial population is represented by CMOA1. The CMOA with a pure genetic operator is represented by CMOA2. The CMOA with a pure knowledge-based search operator is denoted by CMOA3. The CMOA without a local search is denoted by CMOA4. All compared algorithms use the same stopping criterion, i.e., the maximum NFE is set to 50,000, to ensure a fair comparison. For 270 instances, we execute each algorithm 30 times independently. Tables 9.1 and 9.2 contain the mean values of GD and IGD obtained by various algorithms. Table 9.1 summarizes these results by number of factories. Table 9.2 summarizes these results by combining n and m. The best results are highlighted in **bold**.

First, the results from Tables 9.1 and 9.2 are analyzed. We can clearly see that the CMOA performs the best among all algorithms because it achieves lower mean values of two metrics (GD and IGD) on all instances than CMOA1, CMOA2, CMOA3, and CMOA4. With the exception of CMOA1, there is no absolute dominance advantage among CMOA variants. This means that each component of improvement contributes to the CMOA for solving the DPFSP_LB. The statistically significant difference between these results is then examined. A post-hoc Dunnett approach test is performed between the control method CMOA and its variants. Figure 9.5 depicts

Table 9.1 Average values of GD and IGD of variants from CMOA grouped by *f*

f	CMOA		CMOA1	CMOA2	CMOA3	CMOA4
	GD	IGD	GD IGD	GD IGD	GD IGD	GD IGD
3	**2.21e-3**	**5.51e-3**	6.00e-3 2.63e-2	2.88e-3 7.27e-3	3.38e-3 7.33e-3	2.79e-3 6.83e-3
4	**2.05e-3**	**7.10e-3**	8.14e-3 3.04e-2	2.48e-3 8.20e-3	3.53e-3 8.49e-3	2.59e-3 8.29e-3
5	**1.95e-3**	**1.01e-2**	9.92e-3 3.50e-2	2.70e-3 2.02e-2	3.99e-3 2.03e-2	2.63e-3 2.23e-2

Table 9.2 Average values of GD and IGD of variants from CMOA grouped by n and m

(n, m)	CMOA		CMOA1	CMOA2	CMOA3	CMOA4	
	GD	IGD	GD IGD	GD IGD	GD IGD	GD IGD	
(100,5)	**1.92e-3**	**8.33e-3**	6.92e-3 2.76e-2	2.48e-3 9.98e-3	2.95e-3 1.23e-2	2.35e-3	1.05e-2
(100,10)	**2.16e-3**	**8.30e-3**	8.58e-3 3.00e-2	2.66e-3 1.15e-2	3.60e-3 1.92e-2	2.43e-3	1.13e-2
(100,15)	**1.98e-3**	**9.33e-3**	9.34e-3 3.36e-2	2.52e-3 1.28e-2	2.98e-3 1.30e-2	2.49e-3	1.20e-2
(150,5)	**2.04e-3**	**6.06e-3**	6.30e-3 2.64e-2	3.01e-3 7.87e-3	3.23e-3 8.16e-3	2.55e-3	7.88e-3
(150,10)	**2.26e-3**	**7.56e-3**	8.46e-3 3.21e-2	3.31e-3 9.27e-3	3.27e-3 9.48e-3	2.67e-3	9.44e-3
(150,15)	**2.05e-3**	**7.53e-3**	9.35e-3 3.28e-2	2.83e-3 9.37e-3	3.57e-3 9.51e-3	2.85e-3	9.86e-3
(200,5)	**2.20e-3**	**7.03e-3**	5.57e-3 2.84e-2	3.15e-3 8.83e-3	3.28e-3 9.01e-3	2.53e-3	8.86e-3
(200,10)	**1.96e-3**	**7.60e-3**	8.37e-3 3.16e-2	3.02e-3 9.38e-3	3.52e-3 9.42e-3	2.39e-3	8.41e-3
(200,15)	**2.06e-3**	**7.32e-3**	9.27e-3 3.27e-2	3.80e-3 9.13e-3	4.02e-3 9.62e-3	2.62e-3	9.39e-3

a multiple comparison interval plot of two metrics for all algorithms compared. If an interval lacks a zero, the corresponding mean differs significantly from the control mean. As shown in Fig. 9.5, the confidence interval for the difference between CMOA and its variants contains only positive values for IGD and GD metrics. It also demonstrates that the CMOA outperforms its variants significantly. Furthermore, when compared to other improvement components, the initialization improvement strategy is more sensitive to the performance of the proposed CMOA. We can see that the other three improvement components contribute equally to the CMOA. According to the above analysis, it is possible to conclude that the integration of each component has a positive impact on the CMOA's performance.

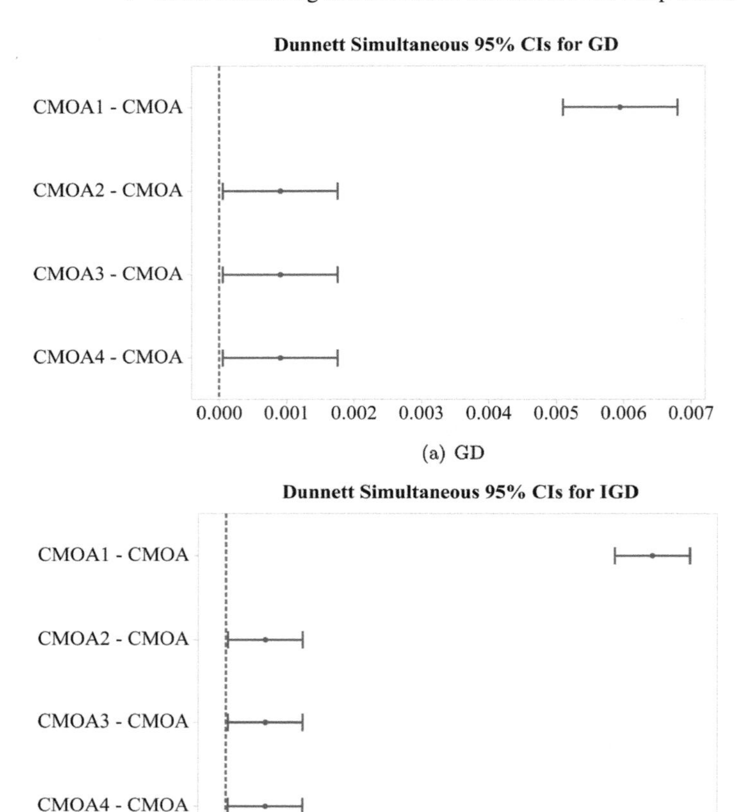

Fig. 9.5 Multiple comparison interval plot for two metrics: (a) GD, (b) IGD

9.4.4 *Effectiveness of Energy-Saving Strategy*

This section aims to demonstrate the efficacy of the energy-saving strategy. It should be noted that slowing down can reduce TEC. Slowing down the processing speed of a machine (speed scaling) is a type of energy-saving strategy. We conduct an experiment comparing CMOA to CMOA without the energy-saving strategy. CMOA_NE represents the CMOA without the energy-saving strategy in this section. We run each algorithm 30 times independently for each instance. Tables 9.3 and 9.4 summarize all mean GD and IGD values for both algorithms. Table 9.3 shows the mean results broken down by the number of factories. These mean results are shown in Table 9.4 and are organized by the combination of n and m.

The results of these tables show that CMOA outperforms CMOA_NE in all instances of GD and IGD. In addition, Fig. 9.6 shows a boxplot comparison of two

Table 9.3 Average values of GD and IGD of CMOA and CMOA_NE grouped by *f*

f	CMOA		CMOA_NE
	GD	IGD	GD IGD
3	**1.43e-3 4.28e-3**		3.53e-3 1.62e-2
4	**1.40e-3 4.88e-3**		4.88e-3 1.66e-2
5	**1.28e-3 5.77e-3**		6.23e-3 1.86e-2

Table 9.4 Average values of GD and IGD of CMOA and CMOA_NE grouped by n and m

(n, m)	CMOA		CMOA_NE	
	GD	IGD	GD	IGD
(100,5)	**1.55e-3**	**4.61e-3**	5.97e-3	1.55e-2
(100,10)	**1.42e-3**	**4.60e-3**	6.16e-3	1.64e-2
(100,15)	**1.21e-3**	**5.84e-3**	5.49e-3	1.80e-2
(150,5)	**1.36e-3**	**4.62e-3**	3.84e-3	1.60e-2
(150,10)	**1.16e-3**	**5.39e-3**	4.70e-3	1.77e-2
(150,15)	**1.34e-3**	**4.99e-3**	5.50e-3	1.81e-2
(200,5)	**1.84e-3**	**4.61e-3**	4.01e-3	1.72e-2
(200,10)	**1.18e-3**	**5.06e-3**	4.07e-3	1.77e-2
(200,15)	**1.27e-3**	**5.05e-3**	4.21e-3	1.77e-2

metrics between two algorithms. There is no overlapping interval between them, as shown in Fig. 9.6. Meanwhile, in terms of GD and IGD, the mean value of CMOA is significantly lower than that of CMOA_NE. It implies that CMOA is far superior to CMOA_NE. Furthermore, it demonstrates that the speed scaling strategy is an effective way to reduce energy consumption. The reason for its success is that slowing down machine processing speed can significantly reduce run energy consumption; at the same time, the idle time interval becomes narrower as the processing time of this non-critical operation increases. It will lead to a decrease in idle energy consumption. Finally, TEC is decreased. As a result, CMOA can benefit from the speed scaling strategy. CMOA_NE, on the other hand, does not include this strategy.

9.4.5 *Comparison of Algorithms*

This section aims to verify the efficacy of CMOA in these situations by conducting a comparison experiment. The CMOA is compared to the NSGA-II [7], MOEA/D [14], KCA [9], and MMOIG [15]. These compared algorithms are either classical ones or have only recently been published for solving shop scheduling problems. As a result, these algorithms are chosen for comparison. To ensure a fair comparison, the maximum NFE for all algorithms considered in this section is set to 25,000.

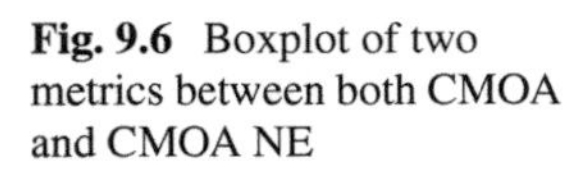

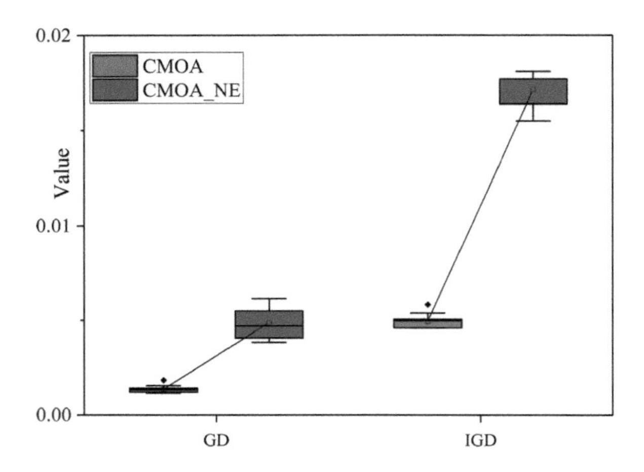

Fig. 9.6 Boxplot of two metrics between both CMOA and CMOA NE

Furthermore, we modified the compared algorithms to fit the problem's characteristics. In subsection 9.4.2, the parameters of these algorithms are calibrated as the same. Each algorithm was run 20 times on each instance independently. Table 9.5 displays all mean values for two metrics, sorted by factory number. Table 9.6 lists all mean values that are grouped by n and m.

First, the outcomes of various algorithms are examined. Tables 9.5 and 9.6 show that CMOA achieves the best total mean values on all instances of IGD, confirming its superior performance among all algorithms considered. Although CMOA's mean values for GD are slightly lower than MOEA/D's, CMOA is still competitive with MOEA/D and superior to other algorithms. Furthermore, using a post-hoc Dunnett test, Fig. 9.7 plots a multiple comparison interval plot of two metrics for all compared algorithms. The multiple comparison interval plot contains a zero between CMOA and MOEA/D, as shown in Fig. 9.7a. In terms of the GD metric, this means that there is no significant difference between CMOA and MOEA/D. In terms of the IGD metric, we can see from Fig. 9.7b that none of these interval plots contain a zero. It implies that CMOA outperforms its competitors statistically in terms of the IGD metric. As a result, the efficacy of CMOA is confirmed. The balance between collaborative-based global search and problem-knowledge local search is credited with CMOA's excellent performance.

Table 9.5 Average values of GD and IGD of comparison algorithms grouped by f

f	CMOA	NSGA-II	MOEA/D	KCA	MMOIG
	GD IGD	GD IGD	GD IGD	GD IGD	GD IGD
3	1.61e-3 **2.51e-3**	4.13e-3 1.42e-2	**1.09e-3** 1.50e-2	2.29e-3 1.05e-2	2.13e-3 1.41e-2
4	1 25e-3 **370e-3**	4 76e-3 157e-2	**1 04e-3** 166e-2	2 53e-3 114e-2	2 51e-3 155e-2
5	1.25e-3 **4.19e-3**	5.81e-3 1.69e-2	**1.11e-3** 1.79e-2	3.03e-3 1.23e-2	2.97e-3 1.66e-2

Table 9.6 Average values of GD and IGD of comparison algorithms grouped by n and m

(n, m)	CMOA		NSGA-II		MOEA/D		KCA		MMOIG	
	GD	IGD	GD	IGD	GD	IGD	GD	IGD	GD	IGD
(100,5)	1.54e-3	**3.42e-3**	5.58e-3	1.43e-2	**1.25e-3**	1.53e-2	2.52e-3	1.01e-2	2.73e-3	1.41e-2
(100,10)	1.36e-3	**3.62e-3**	8.55e-3	1.56e-2	**1.29e-3**	1.66e-2	3.07e-3	1.12e-2	4.03e-3	1.53e-2
(100,15)	1.02e-3	**4.06e-3**	6.20e-3	1.57e-2	**6.62e-4**	1.68e-2	2.87e-3	1.14e-2	3.39e-3	1.56e-2
(150,5)	1.54e-3	**3.38e-3**	4.43e-3	1.46e-2	**1.08e-3**	1.55e-2	2.45e-3	1.07e-2	2.03e-3	1.44e-2
(150,10)	1.32e-3	**3.81e-3**	4.56e-3	1.61e-2	**9.85e-4**	1.69e-2	2.56e-3	1.19e-2	2.34e-3	1.60e-2
(150,15)	1.31e-3	**3.44e-3**	4.77e-3	1.69e-2	**1.14e-3**	1.79e-2	2.92e-3	1.23e-2	2.85e-3	1.67e-2
(200,5)	1.42e-3	**2.72e-3**	4.48e-3	1.39e-2	**9.05e-4**	1.47e-2	2.63e-3	1.02e-2	1.63e-3	1.38e-2
(200,10)	1.23e-3	**3.35e-3**	4.39e-3	1.71e-2	**9.59e-4**	1.79e-2	2.71e-3	1.25e-2	2.34e-3	1.69e-2
(200,15)	1.28e-3	**3.40e-3**	4.41e-3	1.61e-2	**9.18e-4**	1.68e-2	2.58e-3	1.20e-2	2.49e-3	1.59e-2

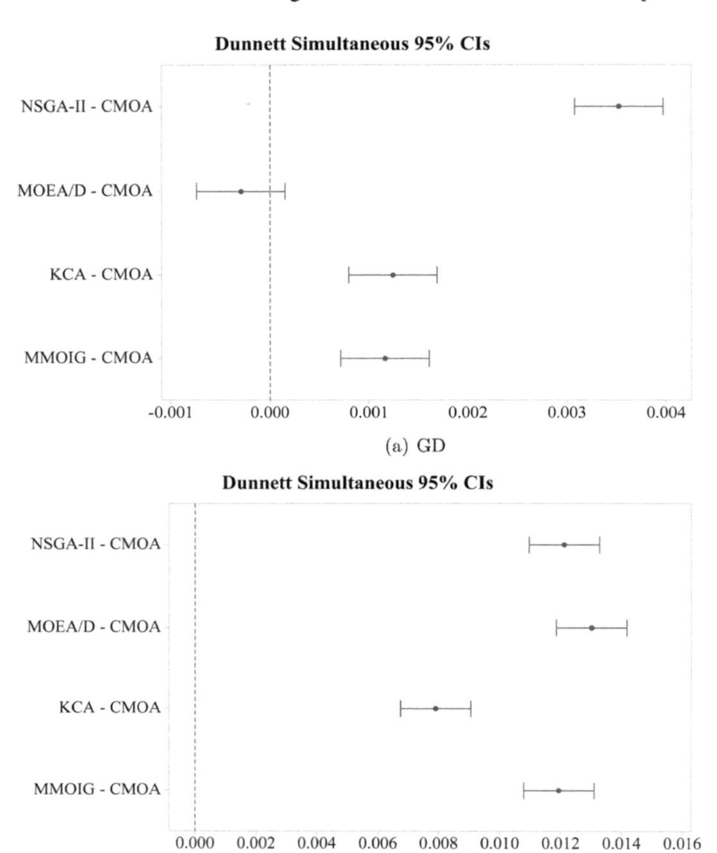

Fig. 9.7 Multiple comparison interval plot for two metrics: **a** GD and **b** IGD

Following that, the relationship between problem variables (i.e., n, f, and m) and algorithmic performance is investigated further. Figures 9.8 and 9.9 depict the interaction of problem variables and two metrics (i.e., GD and IGD). These illustrations, as seen in Fig. 9.8 and 9.9, confirm the previous conclusion about the CMOA's superiority. For the GD metric, Fig. 9.8a shows a significant deterioration with increasing f for NSGA-II, KCA, and MMOIG. MOEA/D remains unchanged as f increases. However, the performance of CMOA has a little improvement with the growth of f. Figure 9.8b shows that n has little effect on KCA and CMOA. Meanwhile, as n increases, the other algorithmic performance improves slightly. NSGA-II and MMOIG show a slight deterioration with increasing m in Fig. 9.8c. This has no effect on the KCA or MOEA/D, and the CMOA improves slightly as m grows. The interaction between problem variables and IGD is highlighted in Fig. 9.9. As shown in Fig. 9.9a, c, the performance of all algorithms degrades noticeably as f and m increase. Figure 9.9b shows that the performance of CMOA improves steadily as n

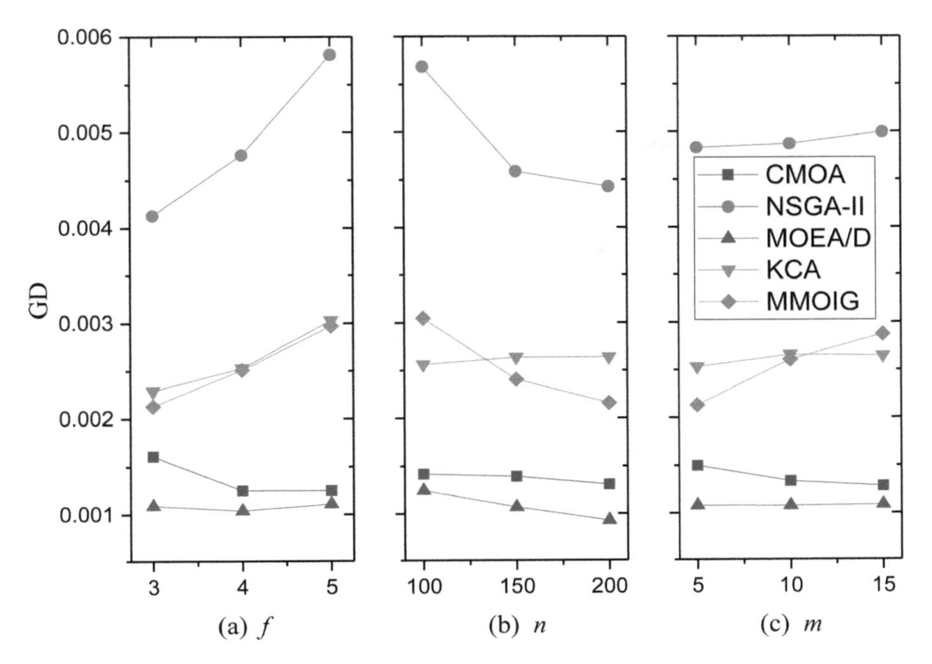

Fig. 9.8 Effect of problem variables on different algorithms concerning GD

increases; however, the performance of competitors exhibits the same behavior, with the IGD increasing slightly with the second level of n and remaining constant for the third level. Furthermore, Fig. 9.10 shows a Pareto front with the best IGD values from various algorithms on a single instance with $n = 150$, $m = 15$, and $f = 5$. We can see from Fig. 9.10 that two extreme objectives are significantly better than its competitors, indicating that CMOA has good coverage performance. In addition, a local area is enlarged to provide a clear view. This zoomed-in view demonstrates that the CMOA can produce acceptable non-dominated solutions in terms of convergence and coverage.

Based on the experimental results and analysis presented above, it is possible to conclude that the CMOA can effectively address the DPFSP_LB.

9.5 Chapter Conclusion

We investigate a new distributed permutation flow shop scheduling problem with limited buffers (DPFSP_LB) for minimizing makespan and total energy consumption in this paper (TEC). A CMOA is designed based on the characteristics of this problem to solve this energy-efficient DPFSP_LB. A cooperative initialization heuristic is proposed in this CMOA to generate high-quality initial solutions. An

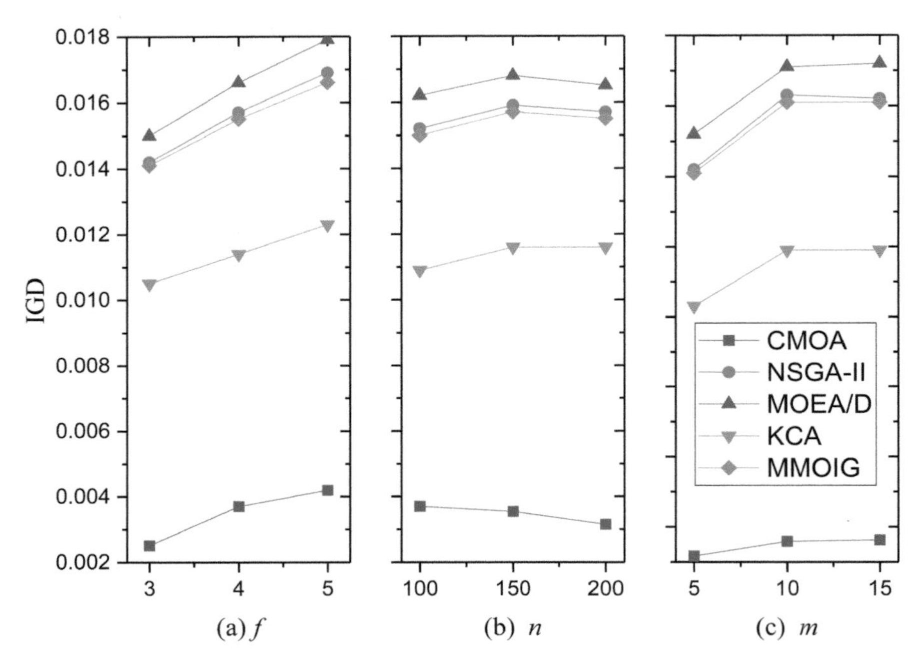

Fig. 9.9 Effect of problem variables on different algorithms concerning IGD

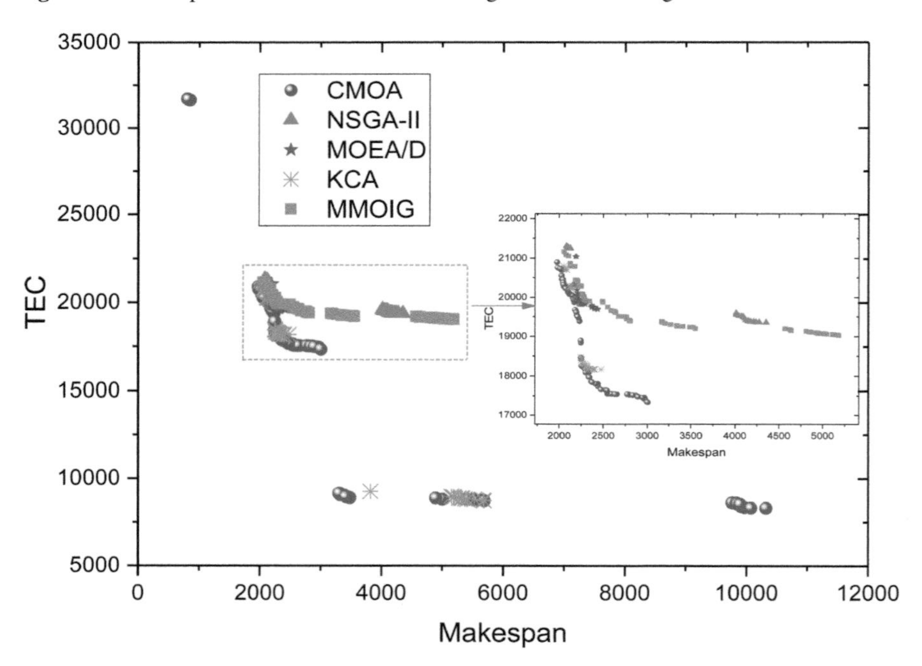

Fig. 9.10 Pareto front obtained by different algorithms on one instance with $n = 150$, $m = 15$, and $s = 5$

effective energy-saving strategy is applied to this problem in order to reduce TEC. A knowledge-based local search is presented and performed on one non-dominated solution to improve exploitation ability. Furthermore, the CMOA parameters are calibrated using a Taguchi design-of-experiment approach. Finally, the experiment is carried out by comparing the results of various well-known multi-objective optimization algorithms on 270 instances. On these instances, the empirical results show that CMOA outperforms the other compared multi-objective optimization algorithms.

Future research should look into distributed non-identical factories with hybrid flow shop, job shop, and dynamic shop scheduling problems. These issues are much more relevant to the real-world manufacturing system. Another future research direction is the development of high-performance optimization algorithms.

References

1. Naderi, B., Ruiz, R.: The distributed permutation flowshop scheduling problem. Comput. Oper. Res. **37**(4), 754–768 (2010)
2. Lu, C., et al.: Energy-efficient scheduling of distributed flow shop with heterogeneous factories: a real-world case from automobile industry in China. IEEE Trans. Industr. Inf. **17**(10), 6687–6696 (2021)
3. Bargaoui, H., Belkahla Driss, O., Ghédira, K.: A novel chemical reaction optimization for the distributed permutation flowshop scheduling problem with makespan criterion. Comput. Indus. Eng., **111**, 239–250 (2017)
4. Li, X., et al.: An effective multiobjective algorithm for energy-efficient scheduling in a real-life welding shop. IEEE Trans. Industr. Inf. **14**(12), 5400–5409 (2018)
5. Lu, C., et al.: A multi-objective cellular grey wolf optimizer for hybrid flowshop scheduling problem considering noise pollution. Appl. Soft Comput. **75**, 728–749 (2019)
6. Zhang, B., et al.: A three-stage multiobjective approach based on decomposition for an energy-efficient hybrid flow shop scheduling problem. IEEE Trans. Syst. Man Cybernetics: Syst. **50**(12), 4984–4999 (2020)
7. Deb, K., et al.: A fast and elitist multiobjective genetic algorithm: NSGA-II. IEEE Trans. Evol. Comput. **6**(2), 182–197 (2002)
8. Ruiz, R., Pan, Q., Naderi, B.: Iterated Greedy methods for the distributed permutation flowshop scheduling problem. Omega **83**, 213–222 (2019)
9. Wang, J., Wang, L.: A knowledge-based cooperative algorithm for energy-efficient scheduling of distributed flow-shop. IEEE Trans. Syst. Man Cybernetics: Syst. **50**(5), 1805–1819 (2020)
10. Lu, C., et al.: A multi-objective approach to welding shop scheduling for makespan, noise pollution and energy consumption. J. Clean. Prod. **196**, 773–787 (2018)
11. Han, Y., et al.: Evolutionary multiobjective blocking lot-streaming flow shop scheduling with machine breakdowns. IEEE Trans. Cybernetics **49**(1), 184–197 (2019)
12. Zitzler, E., Deb, K., Thiele, L.: Comparison of multiobjective evolutionary algorithms: empirical results. Evol. Comput. **8**(2), 173–195 (2000)
13. Zitzler, E., Thiele, L.: Multiobjective evolutionary algorithms: a comparative case study and the strength Pareto approach. IEEE Trans. Evol. Comput. **3**(4), 257–271 (1999)
14. Hui, Li, H.L., Qingfu Zhang, Q.Z.: Multiobjective optimization problems with complicated pareto sets, MOEA/D and NSGA-II. IEEE Trans. Evol. Comput. **13**(2), 284–302 (2009)
15. Ding, J., Song, S., Wu, C.: Carbon-efficient scheduling of flow shops by multi-objective optimization. Eur. J. Oper. Res. **248**(3), 758–771 (2016)

Chapter 10
Green Scheduling in Distributed Hybrid Flow Shop Environment

10.1 Brief Introduction

Hybrid flow shop scheduling problem (HFSP) with productivity objective had wide applicability to actual production process. In addition, in the context of economic globalization, the production extends from a single factory to a distributed production network [1]. Moreover, in order to achieve sustainable development, the scheduling model for green manufacturing has also received widespread attention [2, 3]. Therefore, this chapter studies a distributed hybrid process shop scheduling problem (DHFSP), and the problem contains two objectives: minimization the makespan and total energy consumption (TEC). At the same time, we combine the advantages of genetic operator and iterated greedy heuristic to design a Pareto-based multi-objective hybrid iterated greedy algorithm (MOHIG) to solve this problem.

In this MOHIG, first, in the initialization phase, we propose the strategy of cooperative initialization to improve the quality of the initial solution. Secondly, a knowledge-based multi-objective local search method is proposed to improve the search ability of the algorithm. Thirdly, an energy-saving technique is developed to decrease the idle energy consumption of machine tools. Compared with SPEA2, MOEA/D, NSGA-II, and other multi-objective optimization algorithms, the results show that MOHIG is superior to other algorithms in solving the above problems.

C. Lu et al., *Intelligence Optimization for Green Scheduling in Manufacturing Systems*, Engineering Applications of Computational Methods 18,
https://doi.org/10.1007/978-981-99-6987-6_10

10.2 Problem Statement and Modeling

10.2.1 Problem Statement

Here we describe DHFSP. It can be expressed as follows: There are F identical factories, with $M_{f,i}(M_{f,i} \geq 1)$ parallel machines at each stage $i(i = 1, 2, \cdots, S)$ of processing, each of which may have a different speed. Each factory f $(f = 1, 2, \cdots, F)$ has at least one processing stage with the parallel machine $M_{f,i} \geq 2$. N jobs are processed in these plants, and each job j $((j = 1, 2, \cdots, N)$ goes through all the processing stages. The job in the stage has a standard processing time denoted as $P_{i,j}$. The speed of the machine m $(m = 1, 2, \cdots, M_{f,i})$ in the ith stage is defined $V_{i,m}$. Therefore, we calculate the actual processing time of job j on machine f as $P_{i,j}/V_{i,m}$. . DHFSP is a multi-objective optimization problems (MOPs) [4]. On the basis of the above content, we can estimate the makespan and *TEC* possible conflict between relations, this is because if the job on the machine high speed processing, the processing time is shorter, and processing energy consumption will increase. Simply speaking, the maximum completion time (i.e., makespan) may be reduced, but the total energy consumption (*TEC*) will increase. Additionally, other assumptions are made as follows: After a job has been assigned to a specified plant, the job may not be transferred to another plant during the processing phase of the job. Each machine can only process one job at a time, and each job can only be handled by a machine at a time. The setup time and transportation time of job are negligible. The goal of the problem is simultaneously minimize the makespan and *TEC*.

Subsequently, an example is helpful to explain this problem, which has two factories, seven jobs, and eight machines. Table 10.1 shows the processing time of the job at each stage. Table 10.2 displays the speed and power of four machines in two stages.

Table 10.1 The processing time of jobs in different stage

Job	1	2	3	4	5	6	7	8
Stage1	4	4	3	2	4	4	2	1
Stage2	2	2	2	1	1	3	3	3

Table 10.2 The speed and power of different machines

	Stage 1		Stage 2	
	Machine 1	Machine 2	Machine 1	Machine 2
Speed	1	2	2	1
Idle power	2	8	8	2
Processing power	4	16	16	4

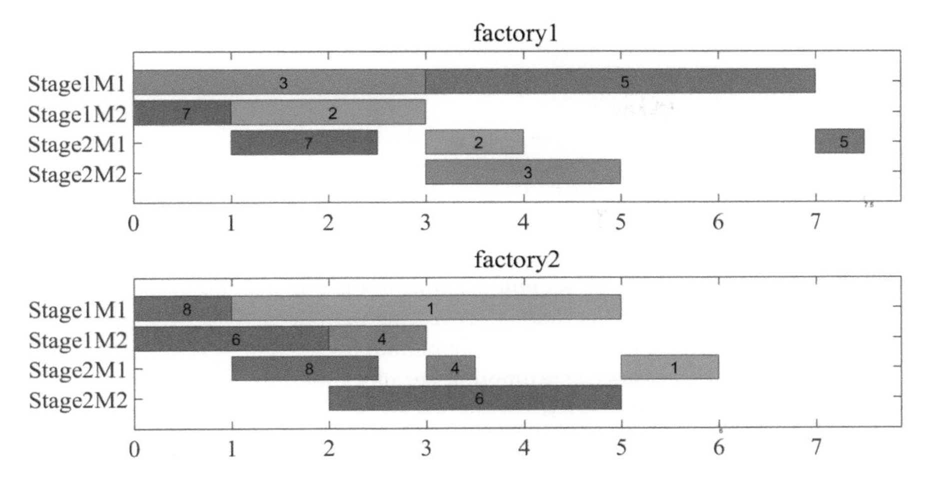

Fig. 10.1 Gantt chart of one solution

Let one solution be $\boldsymbol{\pi} = \{\boldsymbol{\pi}^1, \boldsymbol{\pi}^2\}$, where $\boldsymbol{\pi}^1 = \{3, 7, 2, 5\}$ and $\boldsymbol{\pi}^2 = \{8, 6, 1, 4\}$. The sequence of job numbers in the solution is the order in which these jobs are processed in the factory. Figure 10.1 shows the Gantt chart of this solution.

As can be seen from Fig. 10.1, the first objective makespan $C_{\max}$ is 7.5, the second objective *TEC* is the sum of machine idle energy consumption and the processing energy consumption. For the factory 1, the idle energy consumption is $4.5 \times 8 + 3 \times 2 = 42$ and the processing energy consumption is $7 \times 4 + 3 \times 16 + 3 \times 16 + 2 \times 4 = 132$. Similarly, the idle energy consumption of factory 2 is $3 \times 8 + 2 \times 2 = 28$, and the processing energy consumption of factory 2 is $5 \times 4 + 3 \times 16 + 3 \times 16 + 3 \times 4 = 128$. Based on the above calculation results, the total energy consumption can be calculated as 330.

10.2.2 Mathematical Modeling

There are some notations of the DHFSP are given as follows:

i	: index of stages, $i = 1, 2, \cdots, S$.
j	: index of jobs, $j = 1, 2, \cdots, N$.
f	: index of factories, $f = 1, 2, \cdots, F$.
m	: index of machines, $i = 1, 2, \cdots, M_{f,i}$.
q	: index of positions, $q = 1, 2, \cdots, N$.
$\boldsymbol{I}$	: set of stages, $\boldsymbol{I} = \{1, 2, \cdots, S\}$.
$\boldsymbol{J}$	: set of jobs, $\boldsymbol{J} = \{1, 2, \cdots, N\}$.
$\boldsymbol{K}$	: set of factories, $\boldsymbol{J} = \{1, 2, \cdots, F\}$.
$\boldsymbol{H}_{f,i}$	: set of machines, $\boldsymbol{H}_{f,i} = \{1, 2, \cdots, M_{f,i}\}$.
$\boldsymbol{P}$	: set of job positions, $\boldsymbol{P} = \{1, 2, \cdots, N\}$.

N : the number of jobs.
S : the number of stages.
F : the number of factories.
$M_{f,i}$: the number of machines in stage i of factory f.
M_f total number of machines in each factory.
$P_{i,j}$: the standard processing time of job j in stage i.
L : an infinitely positive number.
$\mathrm{MS}_{f,m,q}$: start time of the mth machine in the qth position in factory f.
$Mc_{f,m,q}$: completion time of the mth machine in the qth position in factory f.
EC_w : energy consumption during processing period.
EC_c : energy consumption during idle period.
$\mathrm{PW}_{i,m}$: processing energy consumption per unit time(processing power) of job on machine m in stage i.
$\mathrm{PI}_{i,m}$: idle energy consumption per unit time(idle power) of job on machine m in stage i.
$E_{\text{on\_off}}$: energy consumption for performing the turning on/off strategy.
T : maximum allowable number of switching off each machine.
$V_{i,m}$: processing speed on the mth machine in stage i.
$S_{i,j}$: start time of job j in stage i.
$C_{i,j}$: completion time of job j in stage i.
$C_{\max}$: the maximal completion time of all the factories.
TEC : total energy consumption.

$$X_{j,f}=\begin{cases} 1, & \text{if job } j \text{ is assigned to factory } f \\ 0, & \text{otherwise.} \end{cases}$$

$$Y_{j,q,m,i,f} = \begin{cases} 1, & \text{if job } j \text{ is on the } m\text{th machine in the } q\text{th position in stage } i \text{ of factory } f \\ 0, & \text{otherwise.} \end{cases}$$

$$Z_{j,m,i,f} = \begin{cases} 1, & \text{if job } j \text{ is processed on the } m\text{th machine in the } q\text{th position in stage } i \text{ of factory } f \\ 0, & \text{otherwise.} \end{cases}$$

$$\min \ \mathrm{OF} = [C_{\max}, \mathrm{TEC}] \tag{10.1}$$

$$\sum_{f=1}^{F} X_{j,f} = 1, \ j \in \boldsymbol{J}, \tag{10.2}$$

$$X_{j,f} = \sum_{m=1}^{M_{f,i}} \sum_{q=1}^{N} Y_{j,m,i,f}, \ i \in \boldsymbol{I}, \ j \in \boldsymbol{J}, \ f \in \boldsymbol{K}, \tag{10.3}$$

$$Z_{j,m,i,f} = \sum_{q=1}^{N} Y_{j,m,i,f}, i \in \boldsymbol{I}, j \in \boldsymbol{J}, f \in \boldsymbol{K}, m \in \boldsymbol{H}_{f,i} \tag{10.4}$$

$$\sum_{j=1}^{N} \mathrm{Y}_{j,q,m,i,f} \leq 1, i \in \boldsymbol{I}, q \in \boldsymbol{P}, f \in \boldsymbol{K}, m \in \boldsymbol{H}_{f,i} \tag{10.5}$$

$$\sum_{j=1}^{N} \mathrm{Y}_{j,q,m,i,f} \geq \sum_{j=1}^{N} \mathrm{Y}_{j,q+1,m,i,f},$$
$$i \in \boldsymbol{I}, q \in \{1, 2, \cdots, N-1\}, f \in \boldsymbol{K}, m \in \boldsymbol{H}_{f,i} \tag{10.6}$$

$$\mathrm{MS}_{f,m,q+1} \geq \mathrm{MC}_{f,m,q},$$
$$i \in \boldsymbol{I}, q \in \{1, 2, \cdots, N-1\}, f \in \boldsymbol{K}, m \in \boldsymbol{H}_{f,i} \tag{10.7}$$

$$\mathrm{MC}_{f,m,q} = \mathrm{MS}_{f,m,q} + \sum_{j=1}^{N} (\mathrm{Y}_{j,q,m,i,f} * \frac{P_{i,j}}{V_{i,m}})$$
$$i \in \boldsymbol{I}, q \in \{1, 2, \cdots, N-1\}, f \in \boldsymbol{K}, m \in \boldsymbol{H}_{f,i} \tag{10.8}$$

$$\mathrm{MS}_{f,m,q} \leq S_{i,j} + L\big(1 - Y_{j,q,m,i,f}\big),$$
$$i \in \boldsymbol{I}, j \in \boldsymbol{J}, q \in \boldsymbol{P}, f \in \boldsymbol{K}, m \in \boldsymbol{H}_{f,i} \tag{10.9}$$

$$\mathrm{MS}_{f,m,q} \geq S_{i,j} - L\big(1 - Y_{j,q,m,i,f}\big),$$
$$i \in \boldsymbol{I}, j \in \boldsymbol{J}, q \in \boldsymbol{P}, f \in \boldsymbol{K}, m \in \boldsymbol{H}_{f,i} \tag{10.10}$$

$$C_{i,j} = S_{i,j} + \sum_{j=1}^{M_{f,i}} (Z_{j,m,i,f} * \frac{P_{i,j}}{V_{i,m}}), i \in \boldsymbol{I}, j \in \boldsymbol{J}, f \in \boldsymbol{K}, \tag{10.11}$$

$$S_{i+1,j} \geq C_{i,j}, i \in \{1, 2, \cdots, S-1\}, j \in \boldsymbol{J} \tag{10.12}$$

$$S_{i,j} \geq 0, i \in \boldsymbol{I}, j \in \boldsymbol{J} \tag{10.13}$$

$$\mathrm{MS}_{f,m,q} \geq 0, \quad i \in \boldsymbol{I}, q \in \boldsymbol{P}, f \in \boldsymbol{K}, m \in \boldsymbol{H}_{f,i} \tag{10.14}$$

$$C_{\max} \geq C_{i,j}, \quad i \in \boldsymbol{I}, j \in \boldsymbol{J} \tag{10.15}$$

$$\mathrm{EC}_w = \sum_{i=1}^{S}\sum_{m=1}^{M_{f,i}}\sum_{j=1}^{N}\left(\left(C_{i,j} - S_{i,j}\right) * \mathrm{PW}_{i,m}\right) \tag{10.16}$$

$$\mathrm{EC}_s = \sum_{i=1}^{S}\sum_{f=1}^{F}\sum_{m=1}^{M_f}\sum_{q=1}^{N-1}\left(\left(\mathrm{MS}_{f,m,q+1} - \mathrm{MC}_{f,m,q}\right) * PI_{i,m}\right) + \sum_{i=1}^{S}\sum_{f=1}^{F}\sum_{m=1}^{M_f}\left(MS_{f,m,1} * PI_{i,m}\right) \tag{10.17}$$

$$\mathrm{TEC} = \mathrm{EC}_w + \mathrm{EC}_s \tag{10.18}$$

$$X_{j,f} \in \{0, 1\},\ j \in \boldsymbol{J},\ f \in \boldsymbol{K} \tag{10.19}$$

$$Y_{j,q,m,i,f} \in \{0, 1\},\ j \in \boldsymbol{J}, q \in \boldsymbol{P},\ f \in \boldsymbol{K}, m \in \boldsymbol{H}_{f,i} \tag{10.20}$$

$$Z_{j,m,i,f} i \in \boldsymbol{I},\ j \in \boldsymbol{J},\ f \in \boldsymbol{K}, m \in \boldsymbol{H}_{f,i} \tag{10.21}$$

Equation (10.1) is the objective function, indicating that makespan and TEC should be minimized simultaneously. Equation (10.2) defines that each job cannot be processed across the factories. Equation (10.3) requires that each job can only be processed on one machine at each stage in a factory. Equation (10.4) ensures that each job allocated in a certain factory is assigned to certain machines in each stage. Equation (10.5) means that each machine can only process one job at a time. Equation (10.6) requires that the previous position on one machine is occupied immediately before one job is placed in one position on this machine. Equation (10.7) indicates that the starting time of a later job on a machine is less than the finish time of the last job. Equation (10.8) is used to calculate the completion time of the job. Equation (10.9) and (10.10) impose the correspondence relation between machine position and job sequence. Equation (10.11) ensures that the starting time and actual processing time of an operation are added to determine its completion time. Equation (10.12) makes sure that the processing of the current operation can only begin after the previous operation has finished. The starting time of each job must not be less than zero according to Eq. (10.13). Equation (10.14) ensures that each machine's starting time will not be less than 0. The longest completion time is provided by Eq. (10.15). Equation (10.16) defines the processing energy consumption. Formula (10.17) defines the idle energy consumption. Formula (10.18) defines the TEC objective. Equation (10.19) to (10.21) show the value range of decision variables.

10.3 Proposed Algorithm

In order to solve this energy-efficient DHFSP, we design a Pareto-based multi-objective hybrid iterated greedy (MOHIG) algorithm. MOHIG combines genetic operators and iterative greedy heuristics [5, 6], and Algorithm 10.1 gives a pseudocode of the proposed MOHIG. The following is a specific description of the algorithm.

Algorithm 10.1: The MOHIG

1: **Input**: Population size(PS), maximum number of iterations (MNI), probability of crossover and mutation (pc and pm), and destruction scale (d).
2: **Output**: A set of non-dominated solutions
3: $P_0 \leftarrow$ Population initialization(PS)
4: $t \leftarrow 0$
5: **While** $t \leq MNI$ **do**
6: $S_t \leftarrow$ Selection (P_t)
7: $S_t' \leftarrow$ Crossover (S_t)
8: $S_t'' \leftarrow$ Mutation (S_t')
9: $S_t''' \leftarrow$ Destruction and construction (S_t'', d)
10: $Q_t \leftarrow$ Multi-Local Search (S_t''')
11: $C_t \leftarrow$ Combination (P_t, Q_t)
12: $\{F_1, F_2, \cdots\} \leftarrow$ Fast Non-Dominated Sort (C_t)
13: $P_{t+1} \leftarrow \emptyset$
14: $i \leftarrow 1$
15: **While** $|P_{t+1}| + |F_i| \leq PS$ **do**
16: Crowding Distance Assignment (F_i)
17: $P_{t+1} \leftarrow$ Combination (P_{t+1}, F_i)
18: $i++$
19: **End while**
20: $P_{t+1} \leftarrow$ Elitist Strategy (P_{t+1})
21: $t++$
22: **End While**

10.3.1 Encoding and Decoding

DHFSP consists of three sub-problems: (1) assign all jobs to different factories; (2) allocate partial jobs in each factory to appropriate machines in each stage; (3) determine a reasonable processing sequence of jobs on each machine. We adopt a permutation-based representation as the solution encoding method [7]. Specifically, using a list $\boldsymbol{\pi} = \{\pi^1, \cdots, \pi^f, \cdots \pi^F\}$ to represent a junction, π^f contains the jobs assigned to factory f, where the order of jobs is the processing order of the jobs in that factory. Jobs will be rearranged in the following stage in accordance with the stage's completion time. Additionally, if several jobs have the same start time in the current stage as they did in the previous stage, they will be sorted in that order.

10.3.2 Initialization

According to the characteristics of DHFSP multi-objective, we propose a cooperative initialization strategy. If we adopt the strategy of general initialization, we are difficult to obtain two optimal objectives (i.e., makespan and *TEC*) [8]. Therefore, in order to distribute partial initial solutions to the two extreme objectives as possible, this paper initializes three particular solutions using the cooperative initialization strategy. The cooperative initialization strategy is defined as follows: The first solution to reduce makespan is yielded by the NEH_2 rule [9]. For the second solution, each job is inserted into all possible positions among all factories, and then select the position with the smallest TEC. Similar to the previous strategy, the criteria for the third solution is making the position of minimum TEC and minimum large span. To ensure the diversity of solutions, other solutions are randomly generated. Algorithm 10.2 shows the pseudocode of this cooperative initialization strategy.

Algorithm 10.2: Collaborative Initialization

1: **Input**: A initial solution set $\boldsymbol{P} = \{\pi_1, \pi_2, \ldots, \pi_{PS}\}$, π is a solution and *PS* is the population size.
2: **Output**: An improved initial solution set $\boldsymbol{P}'$.
3: Randomly generate an initial solution set $\boldsymbol{P}$.
4: Sort the total processing time of all jobs in each stage in ascending order.
5: Get the first solution based on makespan:
6: **for** $j = 1$ to N **do.**
7: Insert the job j into all possible locations among all factories and find the location with the smallest makespan
8: Insert the job j into the position with the smallest makespan and update the factory f that records the insertion
9: **end for**
10: Get the second solution based on *TEC*:
11: **for** $j = 1$ to N **do**
12: Insert the job j into all possible locations among all factories and find the location with the smallest *TEC*
13: Insert the job j into the location with the smallest *TEC* and update the factory f that records the insertion
14: **end for**
15: Get the third solution based on makespan and *TEC*:
16: **for** $j = 1$ to N **do**
17: Insert the job j into all possible positions in all factories and find one or more non-dominated solutions based on makespan and *TEC*

18: Randomly select a position to insert the job into the non-dominated solution position and update the factory f that records the insertion
19: **end for**

10.3.3 Selection

Select two parents from one population by using the binary tournament selection. The steps of this selection are as follows:

Step 1: Two individuals $\boldsymbol{a}$ and $\boldsymbol{b}$ are randomly selected from the population.

Step 2: Determine whether the individual is better than $\boldsymbol{b}$:

(1) if $\boldsymbol{a}$ dominates $\boldsymbol{b}$ (i.e., $\boldsymbol{a} \prec \boldsymbol{b}$), then select $\boldsymbol{a}$.
(2) if $\boldsymbol{a}$ and $\boldsymbol{b}$ are non-dominated with each other, randomly selected an individual from $\boldsymbol{a}$ and $\boldsymbol{b}$.
(3) if b dominates $\boldsymbol{a}$, (i.e., $\boldsymbol{b} \prec \boldsymbol{a}$), then select $\boldsymbol{b}$.
(4) repeat the above step PS (i.e., population size) times to fill the next generation.

10.3.4 Crossover and Mutation

We propose a crossover operator based on multiple factories to search unknown regions. This crossover process is below:

(1) Select a position p among the two parent jobs in each factory obtained in the selection to divide the parent job, and put the jobs to the left of p into their respective set $\boldsymbol{R}$.
(2) Set up two children in the same order as their parents. Then delete jobs which belongs to the other parents set $\boldsymbol{R}$ for each child.
(3) Keep the order of the rest unchanged jobs for each child. The rest missing jobs in one child are supplemented by jobs from the other parents according to their original order. To better explain this step, Figs. 10.2, 10.3 and 10.4 shows an example. $\mathbf{P_1}$ and $\mathbf{P_2}$ are the two parents, and $\mathbf{C_1}$ and $\mathbf{C_2}$ are the corresponding children. The black triangle marks the crossover point.

After the above operation, we use a simple swap mutation operation, which helps to escape the local optimal. The method adopted is to randomly select two mutation points from a solution, and then exchange elements at the two mutation points to form a new solution.

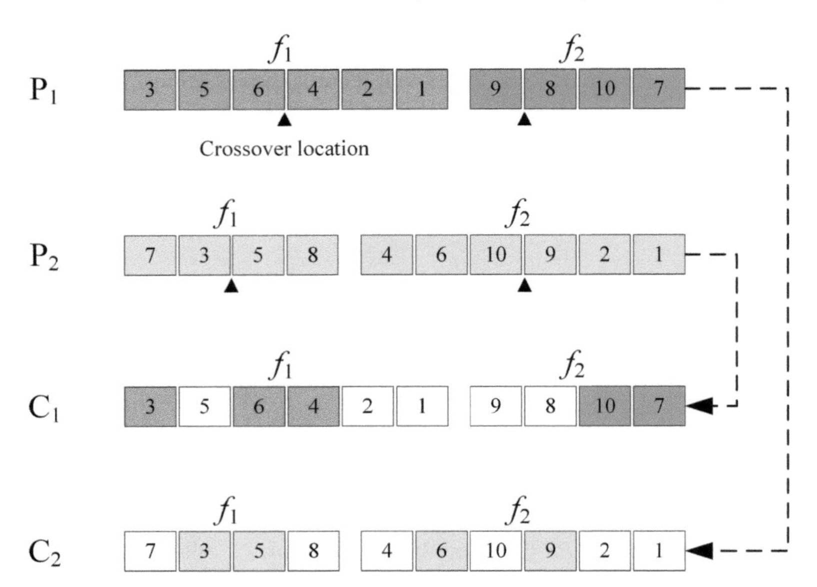

Fig. 10.2 Identify jobs to be deleted

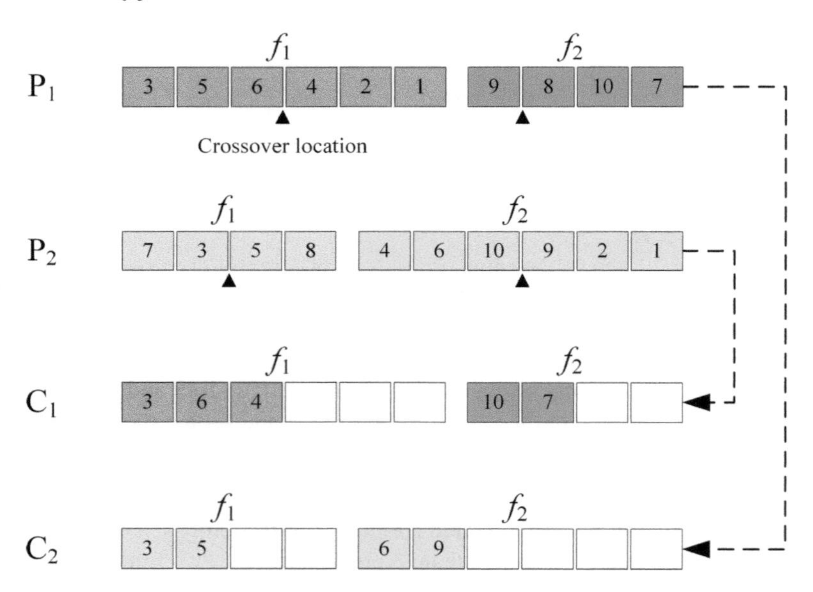

Fig. 10.3 Delete jobs and rearrange the remaining jobs in children

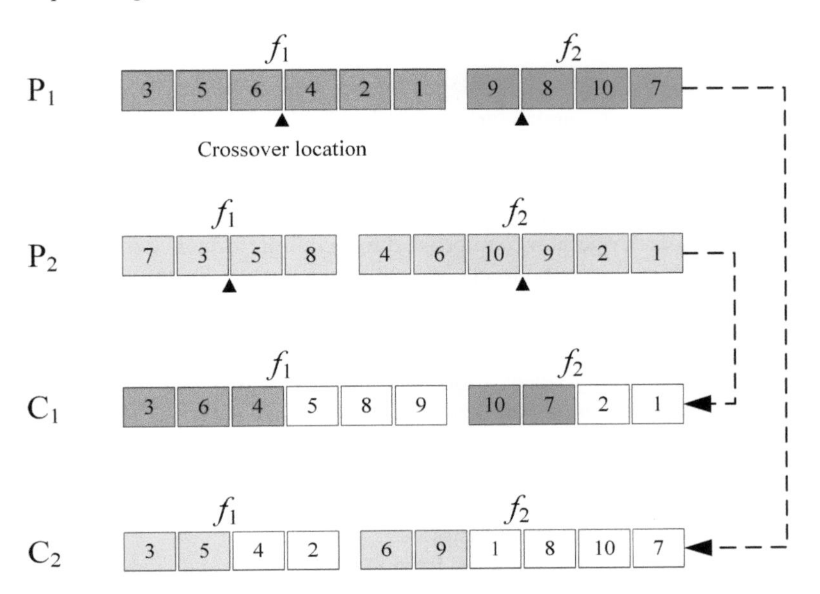

Fig. 10.4 Guarantee the feasibility of children

10.3.5 Destruction and Reconstruction

Destruction and reorganization is an important part of IG algorithm, which keeps the solution diverse. Destruction and reconstruction operator is given in Algorithm 10.3. The specific steps of this operator are as follows:

(1) Select d jobs at random from solution π, delete them from π, store them in π', and label the rest as π_1.
(2) For job in π', insert it into all possible positions in the partial jobs π_1 in turn.
(3) Select the position with the smallest makespan, and insert this job and update the partial jobs in π_1 one by one.

Algorithm 10.3: Destruction and Reconstruction

1: **Input**: A initial solution set π, d is the number of deleted jobs
2: **Output**: A new solution π'
3: In solution π, d jobs are selected randomly in turn, and get a partial solution π_1
4: **for** $j = 1$ to d **do**
5: Insert the job j into all possible positions of partial solution π_1 and record the makespan
6: Select the position with the smallest makespan, insert the job j, and update the partial solution π_1
7: **end for**
8: $\pi \leftarrow \pi'$

10.3.6 Local Search

In general, using a local search heuristic to solve single-objective optimization issues is a good approach. However, with MOPs, optimizing goal causes the other goal to deteriorate. Consequently, the previous local search algorithm for DHFSP cannot be simply applied to a single-objective problem. Here we use a knowledge-based multi-objective local search operator to improve the overall performance of the algorithm. Because the problem has two targets, we only have two key factories here, and two types of search operators are designed in F_c or F_e to adjust the factory allocation and the job sequence of each factory. The relevant local search operator is as follows:

(1) Fc_Insert: Randomly select a job from the key factory and reinsert it into all possible locations among all factories.
(2) Fc_Swap: Randomly select a job from the key factory and exchange it with other jobs.
(3) Fe_Insert: Randomly select a job from the key factory and reinsert it into all possible locations among all factories.
(4) Fe_Swap: Randomly select a job from the key factory and exchange it with all other jobs.

Through the above four local search operators, we can get a new solution, which is represented by π' and compared with the original solution. If a newly generated solution π' based on these local search heuristics dominates the solution π, π' will replace the π and will be inserted into a temporary archive $\boldsymbol{P}$. The pseudocode of this local search is given in Algorithm 10.4.

Algoorithm 10.4: Multi-Objective Local Search

1: **Input**: A solution π
2: **Output**: A new solution set $\boldsymbol{P}$
3: $Set_1 \leftarrow$ Execute Fc-Insert(π)
4: $Set_2 \leftarrow$ Execute Fc-Swap(π)
5: $Set_3 \leftarrow$ Execute Fe-Insert(π)
6: $Set_4 \leftarrow$ Execute Fe-Swap(π)
7: **for** $\pi' \in \{Set_1, Set_2, Set_3, Set_4\}$**do**
8: **if** $f(\pi') \prec f(\pi)$
9: $\pi \leftarrow \pi'$
10: Add π' to $\boldsymbol{P}$
11: **end if**
12: **end for**

10.3.7 Energy-Saving Strategy

A crucial environmental indicator is TEC. However, the ecology suffers as a result of the massive energy use. Consequently, it is essential to lower TEC without impacting finishing time. The strategy here is to shut down the machine when it is idle for a long time, which reduces the energy consumption of the machine. However, it is easy for the machine to be damaged if the machine is switched on and off too frequently. In order to solve this problem, we design the following energy-saving rules:

(1) Calculating the idle time of each machine at each stage of the factory.
(2) Sort these idle time values in a descending order, then select the largest T idle times, and the machine allowed to be shut down.
(3) If the idle time is one of the first T idle intervals and the energy consumption for turning on/off is less than the idle energy consumption of the machine, turn off this machine during the idle period.

Here we illustrate with an example: Suppose $T_{j,m_f,f}$ be a binary value. If the turning on/off strategy of job j is performed on the m_f th machine in factory f, it is 1; otherwise, it is 0. Let the energy consumption consumed of turning on/off be $E_{\text{on\_off}}$. Thus, when the turning on/off is introduced, the idle energy consumption of the machine can be calculated as follows:

$$\mathrm{EC}_s = \sum_{f=1}^{F}\sum_{m=1}^{M_f}\sum_{q=1}^{N-1}\left(\left(1 - T_{j,m_f,f}\right)\cdot\left(\mathrm{MS}_{f,m,q+1} - \mathrm{MC}_{f,m,q}\right)\cdot \mathrm{PI}_{i,m} + T_{j,m_f,f}\cdot E_{\text{on\_off}}\right)$$
$$+\sum_{f=1}^{F}\sum_{m=1}^{M_f}\left(\left(1 - T_{j,m_f,f}\right)\cdot \mathrm{MS}_{f,m,q+1}\cdot \mathrm{PI}_{i,m} + T_{j,m_f,f}\cdot E_{\text{on\_off}}\right) \tag{10.22}$$

10.3.8 Elitism Strategy

For elitist strategies, one low-rank solution is better than one high-rank solution. At the same rank, one solution with a larger crowding distance is better than one with a smaller crowding distance. A set of solutions with the lowest rank and a greater crowding distance is therefore considered to be a high-quality Pareto set. The parent population P_t and the progeny Q_t are combined, let's call it $Qt \cup P_t$ in iteration t. Then, one fast non-dominated sorting technique is executed on the merged population. Finally, by merging elitist strategies, the next generation population can be drawn from the merged population.

10.4 Experiments

10.4.1 Instances and Performance Metrics

To verify the performance of MOHIG, we generated e00 test instances in total. We define the number of jobs as N, the number of stages as S, and the number of factories as F, where $N = \{50, 100, 150, 200\}$,$S = \{2, 4, 6, 8, 16\}$, and $F = \{2, 3, 4, 5, 6\}$. The processing speed for each machine is $V_{i,m} = \{0.5,\ 1.0,\ 1.5\}$. The processing power $\mathrm{PW}_{i,m}$ is set to $4 \times V_{i,m} \times V_{i,m}$; the idle power $\mathrm{PI}_{i,m}$ is set to $0.25 \times PW_{i,m}$, $E_{\text{on\_off}}$ is set to 6, and is set to 4.

In the experiment, we adopted three different performance indexes, which are GD [10], IGD [11], and Spread [12].

10.4.2 Parameter Calibration

We adopted a Taguchi approach of design-of-experiments (DOE) experimental design the best parameter configuration of the algorithm. Among them, several main parameters include: population size (PS), crossover probability (pc), mutation probability (pm), and destruction scale (d). The level of each parameter is as follows:PS $= \{20, 30, 40, 50\}$, $d = \{20, 30, 40, 50\}$, $pc = \{0.6, 0.7, 0.8, 0.9\}$, pm $= \{0.2, 0.4, 0.6, 0.8\}$. An orthogonal array $16(4^4)$ is used, which includes 16 different combinations consisting of PS, d, pc, and pm. The levels of the four key parameters are described in Table 10.3. It shows that the response variable (RV) is IGD. The MOHIG is running e times for each combination. Among them, Table 10.4 gives the RV value of an orthogonal array. Table 10.5 provides the RV values and the significance rank, where Delta represents the maximum difference between their average RV values at different levels. Furthermore, Fig. 10.5 also shows the main effects diagram of four parameters of MOHIG.

As shown in Fig. 10.5, *PS* curve slope is steepest in all parameters, namely the *PS* for the biggest influence of the algorithm. The value of PS = 80 is better than that of other levels, because the diversity of solutions may be increased by a big population. If the value of *PS* remains high, the performance of MOHIG will decrease accordingly.

Table 10.3 Parameter values at each factor level

Parameter	Factor level			
	1	2	3	4
PS	20	30	40	50
d	2	3	4	5
pc	0.6	0.7	0.8	0.9
pm	0.2	0.4	0.6	0.8

Table 10.4 Orthogonal arrays and RV values

Number	*PS*	*pc*	*pm*	*d*	*RV*
1	1	1	1	1	2.84e-02
2	1	2	2	2	3.23e-02
3	1	3	3	3	3.57e-02
4	1	4	4	4	3.89e-02
5	2	1	2	3	3.55e-02
6	2	2	1	4	3.82e-02
7	2	3	4	1	3.43e-02
8	2	4	3	2	3.70e-02
9	3	1	3	4	3.95e-02
10	3	2	4	3	3.86e-02
11	3	3	1	2	3.67e-02
12	3	4	2	1	3.67e-02
13	4	1	4	2	3.72e-02
14	4	2	3	1	3.51e-02
15	4	3	2	4	1.64e-02
16	4	4	1	3	1.96e-02

Table 10.5 Response value of each parameter

Level	*PS*	*Pc*	*pm*	*d*
1	3.38e-02	3.52e-02	3.07e-02	3.36e-02
2	3.63e-02	3.61e-02	3.02e-02	3.58e-02
3	3.79e-02	3.08e-02	3.68e-02	3.24e-02
4	2.71e-02	3.31e-02	3.73e-02	3.33e-02
Delta	1.08e-02	5.27e-03	7.04e-03	3.43e-03
Rank	1	3	2	4

Similarly, we can determine the optimal configuration of the other parameters: PS = 80, $d = 4$, pc = 0.8, and pm = 0.4.

10.4.3 Validity of Initialization

This section compares the proposed cooperative initialization approach against the random initialization strategy in order to evaluate the validity of the proposed cooperative initialization technique. Tables 10.6, 10.7, 10.8 show the statistical results of e independent runs for two initialization strategies on each instance.

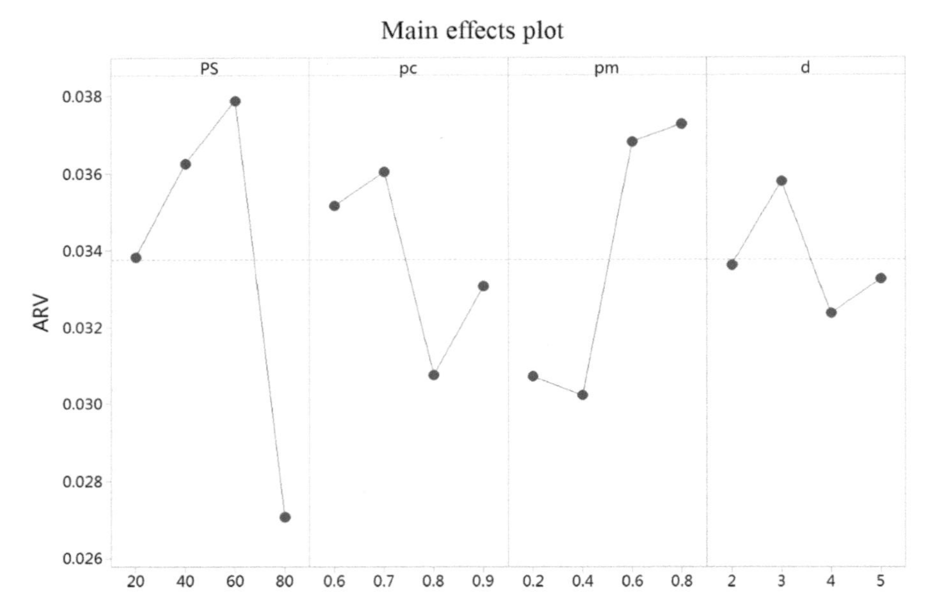

Fig. 10.5 Main effects plot of MOHIG

In Tables 10.6 and 10.7, most results by Coo_Init are one level lower than those by Ran_Init. For the Spread results, we found some anomalies at $n = 50$. These outliers, though, are also incredibly uncommon and only appear in jobs with small staff sizes. The effects on the Spread of Coo_Init are better in approximately 95% of the testing outcomes. These imply that the suggested initialization method is far superior to the random method. In addition, the Wilcoxon signed-rank test was displayed for the two strategies at the significance level of 0.05. Table 10.9 shows the relevant p values on instances grouped respectively by factories, stages, and jobs.

At the level of 0.05, there are remarkable differences between the Coo_Init and Ran_Init for the number of factories, stages, and jobs. According to above findings, this cooperative starting technique can produce high-quality solutions. To improve the performance of MOHIG, it is worthwhile to develop a cooperative initialization technique.

10.4.4 Effectiveness of Local Search

To demonstrate the validity of the proposed local search, we compared the knowledge-based multi-objective local search strategy (designated as MO_LS) with no local search strategy (designated as NO_LS). Tables 10.10, 10.11, and 10.12 show the mean values of IGD, GD, and Spread, respectively, where the optimal results are highlighted in bold.

Table 10.6 Comparison between IGD mean values by Ran_Init and Coo_Init

(n, s)	Ran_Init	Coo_Init	Ran_Init	Coo_Init	Ran_Init	Coo_Init	Ran_Init	Coo_Init	Ran_Init	Coo_Init
	F = 2		F = 3		F = 4		F = 5		F = 6	
(50, 2)	2.57e-02	**7.49e-03**	6.86e-02	**1.12e-02**	5.13e-02	**6.88e-03**	3.03e-02	**1.15e-02**	4.25e-02	**7.43e-03**
(50, 4)	3.52e-02	**1.08e-02**	4.11e-02	**7.86e-03**	4.52e-02	**5.22e-03**	5.16e-02	**7.55e-03**	4.24e-02	**7.04e-03**
(50, 6)	3.17e-02	**6.03e-03**	4.79e-02	**5.84e-03**	5.25e-02	**8.16e-03**	3.52e-02	**2.36e-03**	7.45e-02	**1.32e-02**
(50, 8)	4.24e-02	**9.03e-03**	4.98e-02	**7.88e-03**	5.24e-02	**3.35e-03**	3.94e-02	**1.92e-02**	6.25e-02	**6.46e-03**
(50, 10)	4.04e-02	**6.08e-03**	6.30e-02	**1.22e-02**	2.79e-02	**6.03e-03**	5.00e-02	**1.04e-02**	4.22e-02	**3.44e-03**
(100, 2)	9.39e-02	**1.53e-02**	5.32e-02	**9.11e-03**	4.15e-02	**9.92e-03**	4.43e-02	**8.67e-03**	6.08e-02	**8.18e-03**
(100, 4)	6.62e-02	**5.71e-03**	3.37e-02	**5.53e-03**	7.41e-02	**5.06e-03**	6.65e-02	**9.75e-03**	5.53e-02	**9.16e-03**
(100, 6)	4.54e-02	**2.69e-03**	7.78e-02	**7.10e-03**	6.35e-02	**4.65e-03**	6.65e-02	**8.27e-03**	7.25e-02	**1.04e-02**
(100, 8)	9.83e-02	**1.83e-02**	3.17e-02	**8.07e-03**	3.73e-02	**4.03e-03**	4.87e-02	**6.53e-03**	4.03e-02	**4.25e-03**
(100, 10)	8.60e-02	**7.85e-03**	3.71e-02	**3.46e-03**	1.13e-01	**1.20e-02**	4.12e-02	**5.52e-03**	8.43e-02	**4.04e-03**
(150, 2)	5.70e-01	**1.73e-01**	3.05e-01	**1.10e-02**	7.22e-02	**7.80e-03**	8.13e-02	**3.48e-03**	6.22e-02	**1.22e-02**
(150, 4)	4.81e-02	**5.79e-03**	2.29e + 00	**3.02e-03**	5.37e-02	**7.35e-03**	6.05e-02	**8.32e-03**	7.29e-02	**8.07e-03**
(150, 6)	6.46e-02	**4.58e-02**	3.52e-02	**5.15e-03**	2.68e-02	**3.07e-03**	4.97e-02	**6.61e-03**	5.00e-02	**8.78e-03**
(150, 8)	1.17e-01	**6.33e-03**	5.44e-02	**7.85e-03**	5.16e-02	**5.10e-03**	5.41e-02	**4.96e-03**	6.07e-02	**4.57e-03**
(150, 10)	4.17e-02	**9.81e-03**	8.26e-02	**1.33e-02**	4.47e-02	**5.13e-03**	5.09e-02	**7.68e-03**	2.20e-01	**6.44e-03**
(200, 2)	1.70e-01	**1.12e-02**	5.04e-02	**4.31e-03**	6.14e-02	**1.00e-02**	5.61e-02	**2.91e-03**	4.81e-02	**6.36e-03**
(200, 4)	8.71e-02	**6.01e-02**	4.62e-02	**1.01e-02**	4.45e-02	**7.86e-03**	7.40e-02	**1.49e-02**	4.68e-02	**4.96e-03**
(200, 6)	6.21e-02	**7.21e-03**	4.42e-02	**8.03e-03**	6.02e-02	**9.55e-03**	4.32e-02	**6.29e-03**	5.16e-02	**8.08e-03**
(200, 8)	4.75e-02	**1.15e-02**	4.20e-02	**6.86e-03**	5.24e-02	**8.41e-03**	9.98e-02	**1.04e-02**	7.63e-02	**9.97e-04**
(200, 10)	5.26e-02	**5.02e-03**	5.58e-02	**9.08e-03**	4.46e-02	**8.09e-03**	4.89e-02	**6.77e-03**	4.48e-02	**4.03e-03**

Table 10.7 Comparison between GD mean values by Ran_Init and Coo_Init

(n, s)	Ran_Init	Coo_Init	Ran_Init	Coo_Init	Ran_Init	Coo_Init	Ran_Init	Coo_Init	Ran_Init	Coo_Init
	F = 2		F = 3		F = 4		F = 5		F = 6	
(50, 2)	3.45e-02	**2.20e-02**	1.06e-01	**4.83e-02**	8.23e-02	**3.26e-02**	3.74e-02	**2.52e-02**	3.66e-02	**1.87e-02**
(50, 4)	6.94e-02	**5.54e-02**	8.51e-02	**3.00e-02**	7.78e-02	**2.00e-02**	4.53e-02	**2.19e-02**	5.63e-02	**2.23e-02**
(50, 6)	9.64e-02	**1.66e-02**	9.28e-02	**3.01e-02**	1.12e-01	**3.08e-02**	6.48e-02	**4.21e-03**	5.39e-02	**2.86e-02**
(50, 8)	6.25e-02	**2.56e-02**	1.08e-01	**3.22e-02**	1.30e-01	**1.91e-02**	1.04e-01	**2.68e-02**	2.08e-01	**2.87e-02**
(50, 10)	5.91e-02	**2.02e-02**	6.75e-02	**3.24e-02**	4.69e-02	**1.64e-02**	7.07e-02	**1.41e-02**	7.14e-02	**1.93e-03**
(100, 2)	1.75e-01	**4.50e-02**	7.23e-02	**3.94e-02**	5.90e-02	**3.02e-02**	4.44e-02	**2.15e-02**	7.71e-02	**2.98e-02**
(100, 4)	2.02e-01	**2.70e-02**	5.46e-02	**1.70e-02**	1.62e-01	**9.39e-03**	1.19e-01	**2.48e-02**	5.81e-02	**1.83e-02**
(100, 6)	1.49e-01	**7.13e-03**	2.15e-01	**4.09e-02**	3.67e-01	**1.46e-02**	6.65e-02	**2.90e-02**	1.28e-01	**4.20e-02**
(100, 8)	2.16e-01	**1.75e-02**	4.23e-02	**2.29e-02**	6.30e-02	**1.15e-02**	9.71e-02	**1.85e-02**	1.10e-01	**4.27e-03**
(100, 10)	4.29e-01	**2.73e-02**	9.57e-02	**8.13e-03**	1.68e-01	**3.57e-02**	8.65e-02	**2.32e-03**	3.24e-01	**7.80e-03**
(150, 2)	7.37e-01	**7.87e-02**	3.60e + 00	**0.00e + 00**	2.14e-01	**2.25e-02**	3.37e-01	**1.09e-02**	4.59e-02	**2.33e-02**
(150, 4)	1.65e-01	**2.49e-02**	1.76e + 01	**3.48e-02**	9.66e-02	**2.69e-02**	1.17e-01	**2.73e-02**	1.13e-01	**4.00e-02**
(150, 6)	1.01e-01	**4.09e-02**	6.34e-02	**1.80e-02**	3.80e-02	**1.63e-02**	1.13e-01	**1.95e-02**	1.31e-01	**2.64e-02**
(150, 8)	2.78e-01	**2.07e-02**	7.38e-02	**3.15e-02**	5.62e-02	**2.81e-02**	1.58e-01	**1.52e-03**	2.86e-01	**1.46e-03**
(150, 10)	1.20e-01	**2.65e-02**	1.89e-01	**3.22e-02**	3.93e-02	**2.42e-02**	7.05e-02	**2.77e-02**	1.40e + 00	**6.63e-03**
(200, 2)	6.72e-01	**2.81e-02**	6.57e-02	**2.48e-02**	1.36e-01	**3.98e-02**	2.41e-01	**1.49e-03**	7.26e-02	**2.48e-02**
(200, 4)	1.50e-01	**1.08e-01**	5.85e-02	**2.84e-02**	4.99e-02	**2.56e-02**	6.43e-02	**3.63e-02**	5.25e-02	**1.48e-02**
(200, 6)	1.58e-01	**4.05e-02**	1.16e-01	**1.68e-02**	6.70e-02	**3.55e-02**	6.19e-02	**2.02e-02**	5.05e-02	**2.71e-02**
(200, 8)	1.ee-01	**3.45e-02**	6.11e-02	**2.11e-02**	5.12e-02	**2.53e-02**	4.08e-01	**1.38e-02**	2.96e-01	**2.28e-03**
(200, 10)	2.44e-01	**1.81e-02**	1.21e-01	**4.14e-02**	5.72e-02	**2.80e-02**	5.16e-02	**1.83e-02**	5.72e-02	**1.40e-02**

Table 10.8 Comparison between Spread mean values by Ran_Init and Coo_Init

(n, s)	Ran_Init	Coo_Init	Ran_Init	Coo_Init	Ran_Init	Coo_Init	Ran_Init	Coo_Init	Ran_Init	Coo_Init
	F = 2		F = 3		F = 4		F = 5		F = 6	
(50, 2)	**9.92e-01**	1.02e + 00	1.12e + 00	**9.41e-01**	1.20e + 00	**9.61e-01**	1.02e + 00	**9.30e-01**	**9.65e-01**	9.71e-01
(50, 4)	1.06e + 00	**1.03e + 00**	1.14e + 00	**9.19e-01**	1.09e + 00	**9.33e-01**	1.05e + 00	**9.05e-01**	**1.00e + 00**	1.05e + 00
(50, 6)	1.11e + 00	**1.02e + 00**	1.02e + 00	**1.05e + 00**	1.04e + 00	**9.32e-01**	1.39e + 00	**9.81e-01**	**1.00e + 00**	1.02e + 00
(50, 8)	1.20e + 00	**9.73e-01**	1.05e + 00	**1.01e + 00**	1.30e + 00	**9.40e-01**	1.09e + 00	**9.11e-01**	1.12e + 00	**9.50e-01**
(50, 10)	1.21e + 00	**1.01e + 00**	9.98e-01	**9.27e-01**	1.21e + 00	**8.99e-01**	1.14e + 00	**8.78e-01**	1.31e + 00	**9.56e-01**
(100, 2)	1.30e + 00	**8.47e-01**	1.07e + 00	**8.93e-01**	1.03e + 00	**8.59e-01**	1.01e + 00	**9.01e-01**	9.67e-01	**9.25e-01**
(100, 4)	1.27e + 00	**1.02e + 00**	1.08e + 00	**9.10e-01**	1.31e + 00	**9.35e-01**	1.14e + 00	**8.61e-01**	1.01e + 00	**9.09e-01**
(100, 6)	1.39e + 00	**1.05e + 00**	1.19e + 00	**9.86e-01**	1.46e + 00	**9.30e-01**	9.96e-01	**8.60e-01**	9.64e-01	**9.22e-01**
(100, 8)	1.26e + 00	**9.30e-01**	1.17e + 00	**9.48e-01**	1.11e + 00	**9.62e-01**	1.05e + 00	**9.43e-01**	1.47e + 00	**9.03e-01**
(100, 10)	1.41e + 00	**1.01e + 00**	1.50e + 00	**9.02e-01**	1.27e + 00	**9.81e-01**	1.13e + 00	**1.00e + 00**	1.37e + 00	**9.29e-01**
(150, 2)	1.15e + 00	**9.96e-01**	1.65e + 00	**1.03e + 00**	1.23e + 00	**8.53e-01**	1.54e + 00	**9.22e-01**	1.04e + 00	**9.27e-01**
(150, 4)	1.23e + 00	**9.11e-01**	1.45e + 00	**1.08e + 00**	1.11e + 00	**9.77e-01**	1.09e + 00	**8.45e-01**	1.05e + 00	**9.11e-01**
(150, 6)	1.16e + 00	**1.06e + 00**	1.28e + 00	**9.61e-01**	1.24e + 00	**9.76e-01**	1.21e + 00	**9.38e-01**	1.40e + 00	**9.81e-01**
(150, 8)	1.40e + 00	**9.65e-01**	1.02e + 00	**1.01e + 00**	1.13e + 00	**9.57e-01**	1.47e + 00	**9.37e-01**	1.55e + 00	**8.51e-01**
(150, 10)	1.15e + 00	**1.02e + 00**	1.22e + 00	**1.04e + 00**	1.11e + 00	**9.42e-01**	1.13e + 00	**8.87e-01**	1.38e + 00	**1.01e + 00**
(200, 2)	1.30e + 00	**9.89e-01**	1.07e + 00	**9.42e-01**	1.17e + 00	**9.83e-01**	1.48e + 00	**8.51e-01**	9.87e-01	**8.83e-01**
(200, 4)	1.21e + 00	**9.71e-01**	1.02e + 00	**9.28e-01**	1.11e + 00	**9.80e-01**	9.84e-01	**9.11e-01**	1.17e + 00	**9.40e-01**
(200, 6)	1.17e + 00	**1.03e + 00**	1.25e + 00	**9.37e-01**	1.09e + 00	**9.69e-01**	1.01e + 00	**8.91e-01**	1.04e + 00	**1.03e + 00**
(200, 8)	9.88e-01	**9.51e-01**	1.09e + 00	**9.36e-01**	1.08e + 00	**9.89e-01**	1.30e + 00	**9.62e-01**	1.49e + 00	**9.12e-01**
(200, 10)	1.35e + 00	**9.64e-01**	1.22e + 00	**9.05e-01**	1.14e + 00	**8.81e-01**	1.05e + 00	**9.03e-01**	1.27e + 00	**9.12e-01**

Table 10.9 *p* values of all indicators for factories, stages, and jobs between Ran_Init and Coo_Init

	number	GD	IGD	Spread
Factory	2	8.86e-05	8.86e-05	1.03e-04
	3	8.86e-05	8.86e-05	1.20e-04
	4	8.86e-05	8.86e-05	8.86e-05
	5	8.86e-05	8.86e-05	8.86e-05
	6	8.86e-05	8.86e-05	4.49e-04
Stage	2	8.86e-05	8.86e-05	1.16e-03
	4	8.86e-05	8.86e-05	8.86e-05
	6	8.86e-05	8.86e-05	8.86e-05
	8	8.86e-05	8.86e-05	8.86e-05
	10	8.86e-05	8.86e-05	8.86e-05
Job	50	1.23e-05	1.23e-05	1.77e-05
	100	1.23e-05	1.23e-05	2.26e-05
	150	1.23e-05	1.23e-05	1.77e-05
	200	1.23e-05	1.23e-05	1.23e-05

The values of MO_LS are significantly lower than those of NO LS in all situations involving the IGD indication, as seen in Tables 10.10, 10.11 and 10.12. Similarly, MO_LS is less than NO_LS in terms of GD and Spread indicators on more than 95% of occasions. The *p* values for the two strategies on instances grouped by the number of factories, stages, and jobs are shown in Table 10.13. We can infer from this, in these circumstances, MO LS is statistically superior to NO LS. This further demonstrates the efficacy of knowledge-based multi-objective local search, which helps MOHIG function better. Consequently, it can be concluded that a knowledge-based multi-objective local search approach can effectively address this problem.

10.4.5 *Effectiveness of Energy-Saving Strategy*

This section's objective is to show how effective the suggested energy-saving plan is. As a result, an experiment is conducted to compare MOHIG with and without the proposed energy-saving technique (designated as Strategy_1 and Strategy_2, respectively). Table 10.14 shows the mean results of *TEC* obtained by Strategy_1 and Strategy_2 on all instances. It demonstrates how the suggested energy-saving method can successfully lower energy usage and support the sustainability of production system, the reason is that the energy-saving strategy can greatly reduce *TEC* without equipment depreciation.

Table 10.10 Comparison between IGD mean values by MO_LS and NO_LS

(n, s)	MO_LS	NO_LS	MO_LS	NO_LS	MO_LS	NO_LS	MO_LS	NO_LS	MO_LS	NO_LS
	F = 2		F = 3		F = 4		F = 5		F = 6	
(50, 2)	**2.89e-02**	5.77e-02	**8.55e-03**	1.17e-02	**3.94e-03**	7.21e-03	**5.45e-03**	9.63e-03	**1.91e-02**	2.98e-02
(50, 4)	**8.79e-03**	1.72e-02	**4.27e-03**	8.71e-03	**6.13e-03**	1.33e-02	**3.92e-03**	1.06e-02	**2.76e-02**	3.58e-02
(50, 6)	**3.30e-03**	6.70e-03	**5.44e-03**	1.13e-02	**3.25e-03**	9.09e-03	**4.43e-03**	8.56e-03	**3.51e-03**	8.12e-03
(50, 8)	**6.87e-03**	1.57e-02	**4.14e-03**	8.26e-03	**1.69e-03**	4.50e-03	**7.84e-02**	1.28e-01	**4.15e-03**	9.52e-03
(50, 10)	**3.16e-02**	6.24e-02	**3.92e-03**	8.35e-03	**2.52e-02**	3.63e-02	**3.68e-03**	5.52e-03	**2.95e-02**	4.50e-02
(100, 2)	**1.35e-01**	1.72e-01	**2.40e-03**	5.32e-03	**2.39e-02**	5.25e-02	**3.49e-03**	6.18e-03	**3.90e-03**	7.83e-03
(100, 4)	**6.93e-03**	1.18e-02	**3.78e-02**	1.03e-01	**2.71e-03**	5.79e-03	**3.43e-03**	7.74e-03	**3.59e-03**	7.28e-03
(100, 6)	**2.15e-03**	5.48e-03	**8.13e-03**	2.22e-02	**1.02e-01**	2.74e-01	**3.96e-03**	9.50e-03	**6.51e-03**	1.54e-02
(100, 8)	**8.39e-03**	1.07e-02	**6.73e-03**	1.03e-02	**2.52e-02**	4.08e-02	**7.99e-03**	1.23e-02	**4.16e-03**	6.65e-03
(100, 10)	**4.55e-02**	6.26e-02	**7.46e-03**	1.13e-02	**3.73e-03**	7.78e-03	**1.98e-03**	5.06e-03	**9.79e-03**	1.42e-02
(150, 2)	**6.49e-02**	9.68e-02	**3.67e-03**	3.97e-03	**3.73e-03**	6.01e-03	**6.94e-03**	8.12e-03	**4.34e-03**	8.77e-03
(150, 4)	**1.54e-02**	2.57e-02	**1.35e-02**	1.39e-02	**3.20e-03**	7.35e-03	**4.31e-03**	8.80e-03	**7.16e-03**	1.41e-02
(150, 6)	**4.28e-02**	7.38e-02	**7.27e-03**	1.60e-02	**2.29e-02**	3.92e-02	**5.86e-03**	1.15e-02	**8.36e-03**	1.16e-02
(150, 8)	**9.25e-03**	1.53e-02	**2.93e-03**	6.53e-03	**3.62e-03**	7.85e-03	**2.42e-03**	3.00e-03	**3.67e-03**	5.74e-03
(150, 10)	**1.27e-02**	3.56e-02	**3.77e-03**	7.81e-03	**2.71e-03**	5.40e-03	**3.75e-03**	8.13e-03	**4.11e-03**	4.86e-03
(200, 2)	**9.47e-03**	1.60e-02	**2.25e-03**	4.92e-03	**3.36e-03**	8.10e-03	**9.35e-03**	1.60e-02	**2.57e-03**	6.08e-03
(200, 4)	**5.06e-02**	9.78e-02	**3.90e-03**	7.47e-03	**6.77e-03**	1.26e-02	**2.01e-02**	4.53e-02	**3.39e-03**	4.48e-03
(200, 6)	**3.22e-03**	6.42e-03	**2.99e-03**	6.36e-03	**6.30e-03**	1.63e-02	**3.44e-03**	7.34e-03	**3.41e-03**	6.74e-03
(200, 8)	**4.75e-03**	1.01e-02	**3.48e-02**	5.82e-02	**3.26e-03**	7.62e-03	**3.58e-03**	7.03e-03	**7.83e-03**	1.14e-02
(200, 10)	**1.21e-02**	2.39e-02	**9.77e-03**	2.14e-02	**8.67e-03**	1.74e-02	**3.38e-03**	6.56e-03	**2.62e-03**	6.60e-03

Table 10.11 Comparison between GD mean values by MO_LS and NO_LS

(n, s)	MO_LS	NO_LS	MO_LS	NO_LS	MO_LS	NO_LS	MO_LS	NO_LS	MO_LS	NO_LS
	F = 2		F = 3		F = 4		F = 5		F = 6	
(50, 2)	**5.76e-02**	9.60e-02	**2.96e-02**	4.45e-02	**2.04e-02**	4.73e-02	**1.89e-02**	3.56e-02	**3.10e-02**	4.68e-02
(50, 4)	**4.11e-02**	6.20e-02	**2.58e-02**	6.24e-02	**1.68e-02**	5.40e-02	**1.46e-02**	3.90e-02	**1.23e-02**	2.74e-02
(50, 6)	**1.85e-02**	5.27e-02	**1.61e-02**	3.92e-02	**1.41e-02**	5.09e-02	**8.61e-03**	9.79e-03	**1.55e-02**	4.24e-02
(50, 8)	**2.81e-02**	5.72e-02	**1.95e-02**	4.82e-02	**1.24e-02**	3.32e-02	**9.19e-02**	1.82e-01	**2.49e-02**	7.29e-02
(50, 10)	**4.93e-02**	8.31e-02	**1.98e-02**	4.80e-02	**2.72e-02**	3.60e-02	**1.81e-02**	2.22e-02	1.33e-02	**9.21e-03**
(100, 2)	3.89e-01	**1.86e-01**	**2.20e-02**	5.20e-02	**2.99e-02**	6.76e-02	**2.06e-02**	4.42e-02	**1.61e-02**	3.73e-02
(100, 4)	**5.15e-02**	1.03e-01	**1.19e-01**	1.81e-01	**2.78e-02**	3.77e-02	**2.02e-02**	5.34e-02	**1.73e-02**	5.28e-02
(100, 6)	1.43e-02	**1.31e-02**	**4.15e-02**	1.44e-01	**2.94e-01**	3.83e-01	**1.80e-02**	5.53e-02	**3.15e-02**	9.61e-02
(100, 8)	**1.00e-01**	1.01e-01	**2.14e-02**	3.71e-02	**1.69e-02**	4.06e-02	**2.54e-02**	4.69e-02	1.46e-02	**1.05e-02**
(100, 10)	**4.33e-02**	4.59e-02	**2.79e-01**	4.91e-01	**4.27e-02**	7.58e-02	**1.21e-02**	5.99e-03	2.96e-02	**2.64e-02**
(150, 2)	**8.52e-02**	1.07e-01	**2.23e-02**	0.00e + 00	**4.00e-02**	4.73e-02	8.29e-03	**5.41e-03**	**2.66e-02**	5.55e-02
(150, 4)	**2.77e-01**	4.88e-01	**2.74e-01**	2.75e-01	**1.60e-02**	4.56e-02	**2.07e-02**	4.86e-02	**2.22e-02**	5.56e-02
(150, 6)	**4.24e-02**	8.00e-02	**2.06e-02**	4.92e-02	**2.80e-02**	4.58e-02	**2.33e-02**	4.47e-02	**3.55e-02**	3.95e-02
(150, 8)	2.79e-02	**2.75e-02**	**1.51e-02**	4.73e-02	**1.41e-02**	3.96e-02	6.04e-03	**3.65e-03**	**5.57e-02**	8.13e-02
(150, 10)	**1.34e-01**	2.11e-01	**4.00e-02**	7.78e-02	**1.18e-02**	3.66e-02	**3.09e-02**	7.37e-02	5.14e-03	**1.29e-03**
(200, 2)	**8.41e-02**	1.76e-01	**1.82e-02**	3.66e-02	**1.76e-02**	5.51e-02	**2.97e-02**	4.52e-02	**1.99e-02**	4.83e-02
(200, 4)	**9.43e-02**	2.20e-01	**1.94e-02**	3.68e-02	**2.44e-02**	5.13e-02	**2.34e-02**	5.91e-02	**1.24e-02**	1.83e-02
(200, 6)	**3.50e-02**	5.58e-02	**2.05e-02**	4.92e-02	**3.95e-02**	7.50e-02	**5.03e-02**	6.80e-02	**1.37e-02**	4.55e-02
(200, 8)	**2.67e-02**	7.18e-02	**4.56e-02**	6.83e-02	**2.09e-02**	5.51e-02	**3.01e-02**	3.27e-02	2.12e-02	**1.67e-02**
(200, 10)	**1.78e-01**	2.08e-01	**3.40e-02**	9.45e-02	**3.10e-02**	7.42e-02	**1.79e-02**	4.06e-02	**1.14e-02**	3.17e-02

Table 10.12 Comparison between Spread mean values by MO_LS and NO_LS

(n, s)	MO_LS	NO_LS	MO_LS	NO_LS	MO_LS	NO_LS	MO_LS	NO_LS	MO_LS	NO_LS
	F = 2		F = 3		F = 4		F = 5		F = 6	
(50, 2)	**9.62e-01**	1.10e + 00	**9.01e-01**	1.22e + 00	9.22e-01	**9.15e-01**	**8.71e-01**	1.00e + 00	**9.61e-01**	1.06e + 00
(50, 4)	**9.87e-01**	1.ee + 00	**9.82e-01**	1.ee + 00	9.17e-01	**8.85e-01**	**8.71e-01**	1.13e + 00	**9.86e-01**	1.09e + 00
(50, 6)	**9.40e-01**	1.05e + 00	**9.46e-01**	1.01e + 00	**9.30e-01**	1.16e + 00	**1.18e + 00**	1.33e + 00	**9.41e-01**	1.10e + 00
(50, 8)	**1.08e + 00**	1.33e + 00	**9.38e-01**	1.04e + 00	**1.14e + 00**	1.32e + 00	**9.43e-01**	1.08e + 00	**9.79e-01**	1.02e + 00
(50, 10)	**9.95e-01**	1.15e + 00	**9.48e-01**	1.06e + 00	**9.79e-01**	1.14e + 00	**1.03e + 00**	1.28e + 00	**1.07e + 00**	1.32e + 00
(100, 2)	**1.02e + 00**	1.15e + 00	**1.04e + 00**	1.17e + 00	**9.12e-01**	9.53e-01	**1.01e + 00**	1.11e + 00	**1.02e + 00**	1.11e + 00
(100, 4)	**9.43e-01**	1.26e + 00	**1.01e + 00**	1.07e + 00	**1.13e + 00**	1.38e + 00	**9.94e-01**	1.27e + 00	**8.45e-01**	1.07e + 00
(100, 6)	**1.13e + 00**	1.44e + 00	**9.56e-01**	1.16e + 00	**9.83e-01**	1.15e + 00	**1.02e + 00**	1.07e + 00	**1.02e + 00**	**9.51e-01**
(100, 8)	**1.18e + 00**	1.35e + 00	1.03e + 00	**1.02e + 00**	**1.04e + 00**	1.14e + 00	**9.35e-01**	9.94e-01	**1.22e + 00**	1.47e + 00
(100, 10)	**8.82e-01**	1.32e + 00	1.19e + 00	**1.15e + 00**	**9.98e-01**	1.17e + 00	**1.21e + 00**	1.46e + 00	**1.24e + 00**	1.51e + 00
(150, 2)	**1.03e + 00**	1.24e + 00	**1.16e + 00**	1.38e + 00	**1.04e + 00**	1.52e + 00	**1.31e + 00**	1.47e + 00	**9.40e-01**	1.14e + 00
(150, 4)	**1.01e + 00**	1.15e + 00	1.26e + 00	**1.23e + 00**	**9.99e-01**	1.11e + 00	**8.97e-01**	1.20e + 00	**8.95e-01**	1.10e + 00
(150, 6)	**8.76e-01**	1.07e + 00	**1.01e + 00**	1.02e + 00	**9.75e-01**	1.03e + 00	**1.06e + 00**	1.17e + 00	**9.18e-01**	1.29e + 00
(150, 8)	**1.14e + 00**	1.47e + 00	**1.00e + 00**	1.11e + 00	**9.15e-01**	1.07e + 00	**1.22e + 00**	1.45e + 00	**1.20e + 00**	1.47e + 00
(150, 10)	**9.95e-01**	1.14e + 00	**1.02e + 00**	1.28e + 00	**1.02e + 00**	1.09e + 00	**9.49e-01**	9.90e-01	**1.23e + 00**	1.25e + 00
(200, 2)	**1.01e + 00**	1.15e + 00	1.07e + 00	**1.05e + 00**	**9.44e-01**	1.21e + 00	**1.15e + 00**	1.35e + 00	**9.12e-01**	1.00e + 00
(200, 4)	**1.01e + 00**	1.16e + 00	**9.23e-01**	1.04e + 00	**9.88e-01**	1.04e + 00	**9.67e-01**	9.83e-01	**1.09e + 00**	1.17e + 00
(200, 6)	**9.68e-01**	1.19e + 00	**1.03e + 00**	1.35e + 00	**1.09e + 00**	1.22e + 00	**9.71e-01**	1.08e + 00	**9.88e-01**	1.09e + 00
(200, 8)	1.13e + 00	**9.50e-01**	**9.82e-01**	1.03e + 00	1.05e + 00	**1.02e + 00**	**1.15e + 00**	1.45e + 00	**1.21e + 00**	1.33e + 00
(200, 10)	**1.04e + 00**	1.26e + 00	**1.04e + 00**	1.24e + 00	**9.75e-01**	1.07e + 00	**1.05e + 00**	1.19e + 00	**1.08e + 00**	1.22e + 00

Table 10.13 *p* values of all indicators for factories, stage, and job by MO_LS and NO_LS

	number	GD	IGD	Spread
Factory	2	2.50e-03	8.86e-05	4.49e-04
	3	2.19e-04	8.86e-05	5.93e-04
	4	8.86e-05	8.86e-05	2.19e-04
	5	5.17e-04	8.86e-05	8.86e-05
	6	1.16e-03	8.86e-05	1.40e-04
Stage	2	8.86e-05	8.86e-05	1.40e-04
	4	2.20e-03	8.86e-05	1.63e-04
	6	8.03e-03	8.86e-05	2.19e-04
	8	2.19e-04	8.86e-05	1.40e-04
	10	1.20e-04	8.86e-05	1.02e-03
Job	50	3.22e-05	1.23e-05	1.77e-05
	100	6.65e-04	1.23e-05	1.77e-05
	150	1.40e-04	1.23e-05	8.09e-05
	200	1.57e-05	1.23e-05	3.22e-05

10.4.6 Comparison of MOHIG and Other Algorithms

In this section, we compare MOHIG with several common algorithms to measure their performance. The algorithms we used for comparison are SPEA2 [13], NSGA-II [14], and MOEA/D [15]. They apply the same termination criteria across all algorithms to provide fair comparison. We also use the coding and decoding systems mentioned above, as well as crossover, mutation (if applicable), and energy-saving techniques. A Taguchi approach is strictly used to calibrate these multi-objective optimization algorithm parameters. For NSGA-II, the values of pc and pm are equal to 0.9 and 0.2, respectively. For SPEA2, the values of pc and pm are equal to 0.9 and 0.3. The archive size is 40. For MOEA/D, the values of pc and pm are set to 0.8 and 0.2 respectively, and the neighborhood size is 20. The population of all algorithms is set to 80 in the experiment. Tables 10.15, 10.16 and 10.17 provides the mean values among different algorithms. Experimental results show that MOHIG is superior to other algorithms on every occasion. These data are put through a Wilcoxon signed-rank test with a significance level of 0.05 to see if variations between statistical results are noteworthy. Table 10.18 summarizes the relevant *p* values among the comparison algorithms on instances classified by factories, stages, and jobs. As shown in Table 10.18, the advantages of the proposal are overwhelming from the experimental results.

To more clearly compare the differences between algorithm performance, Figs. 10.6, 10.7 and 10.8 present the interval diagram of the comparison algorithms regarding three indicators. There are no regions where MOHIG and its compared algorithms overlap. Clearly, MOHIG outperforms its competitors by a wide margin.

Table 10.14 Comparison between *TEC* mean values by strategy_1 and strategy_2

(n, s)	Strategy_1	Strategy_2	Strategy_1	Strategy_2	Strategy_1	Strategy_2	Strategy_1	Strategy_2	Strategy_1	Strategy_2
	F = 2		F = 3		F = 4		F = 5		F = 6	
(50, 2)	**3.43e + 02**	3.54e + 02	**2.75e + 02**	3.20e + 02	**3.37e + 02**	3.88e + 02	**3.50e + 02**	3.93e + 02	**2.49e + 02**	2.94e + 02
(50, 4)	**7.01e + 02**	7.26e + 02	**7.19e + 02**	8.14e + 02	**7.06e + 02**	7.36e + 02	**6.34e + 02**	7.22e + 02	**7.74e + 02**	8.13e + 02
(50, 6)	**1.05e + 03**	1.07e + 03	**1.13e + 03**	1.15e + 03	**1.13e + 03**	1.24e + 03	**8.37e + 02**	8.45e + 02	**9.64e + 02**	1.08e + 03
(50, 8)	**1.64e + 03**	1.72e + 03	**1.38e + 03**	1.44e + 03	**1.47e + 03**	1.56e + 03	**8.77e + 02**	8.77e + 02	**1.44e + 03**	1.55e + 03
(50, 10)	**7.29e + 02**	9.00e + 02	**7.29e + 02**	8.81e + 02	**8.57e + 02**	1.12e + 03	**7.25e + 02**	1.09e + 03	**6.39e + 02**	1.33e + 03
(100, 2)	**2.17e + 03**	2.78e + 03	**1.83e + 03**	2.09e + 03	**1.62e + 03**	1.91e + 03	**1.80e + 03**	2.31e + 03	**1.71e + 03**	2.00e + 03
(100, 4)	**3.15e + 03**	3.93e + 03	**3.12e + 03**	3.36e + 03	**2.44e + 03**	2.98e + 03	**2.79e + 03**	3.36e + 03	**2.82e + 03**	3.55e + 03
(100, 6)	**3.52e + 03**	3.79e + 03	**3.90e + 03**	4.33e + 03	**4.23e + 03**	5.59e + 03	**4.01e + 03**	4.46e + 03	**4.07e + 03**	4.73e + 03
(100, 8)	**1.41e + 03**	2.01e + 03	**1.27e + 03**	1.84e + 03	**1.30e + 03**	1.78e + 03	**1.26e + 03**	2.07e + 03	**1.31e + 03**	2.26e + 03
(100, 10)	**2.83e + 03**	3.22e + 03	**2.79e + 03**	3.17e + 03	**2.81e + 03**	4.17e + 03	**2.99e + 03**	3.59e + 03	**2.59e + 03**	4.82e + 03
(150, 2)	**5.14e + 03**	6.00e + 03	**5.43e + 03**	6.71e + 03	**5.19e + 03**	7.08e + 03	**4.46e + 03**	5.70e + 03	**3.85e + 03**	5.16e + 03
(150, 4)	**6.54e + 03**	7.11e + 03	**5.62e + 03**	6.29e + 03	**5.75e + 03**	6.30e + 03	**5.28e + 03**	6.09e + 03	**5.89e + 03**	6.79e + 03
(150, 6)	**1.79e + 03**	2.53e + 03	**1.52e + 03**	2.82e + 03	**1.81e + 03**	2.40e + 03	**1.78e + 03**	2.96e + 03	**1.55e + 03**	2.73e + 03
(150, 8)	**4.90e + 03**	6.31e + 03	**3.96e + 03**	4.86e + 03	**3.86e + 03**	5.02e + 03	**3.43e + 03**	6.28e + 03	**3.61e + 03**	6.09e + 03
(150, 10)	**6.39e + 03**	6.93e + 03	**7.40e + 03**	9.62e + 03	**5.87e + 03**	6.64e + 03	**5.59e + 03**	6.67e + 03	**5.22e + 03**	7.82e + 03
(200, 2)	**8.72e + 03**	9.71e + 03	**8.02e + 03**	8.99e + 03	**7.95e + 03**	8.74e + 03	**9.31e + 03**	1.22e + 04	**7.79e + 03**	9.27e + 03
(200, 4)	**2.52e + 03**	4.32e + 03	**2.25e + 03**	3.26e + 03	**2.12e + 03**	3.61e + 03	**1.94e + 03**	3.46e + 03	**2.12e + 03**	4.07e + 03
(200, 6)	**5.10e + 03**	6.00e + 03	**4.68e + 03**	6.07e + 03	**4.57e + 03**	6.23e + 03	**4.60e + 03**	5.80e + 03	**4.75e + 03**	6.58e + 03
(200, 8)	**7.77e + 03**	8.40e + 03	**7.91e + 03**	9.80e + 03	**7.69e + 03**	9.28e + 03	**7.72e + 03**	1.22e + 04	**6.67e + 03**	1.09e + 04
(200, 10)	**1.16e + 04**	1.35e + 04	**1.00e + 04**	1.19e + 04	**1.03e + 04**	1.13e + 04	**1.00e + 04**	1.13e + 04	**8.93e + 03**	1.04e + 04

Table 10.15 Comparison between IGD mean values by different algorithms

(n, s)	MOHIG	MOEA/D	SPEA2	NSGA-II	MOHIG	MOEA/D	SPEA2	NSGA-II
	F = 2				F = 3			
(50, 2)	**1.13e-02**	2.33e-02	2.60e-02	2.15e-02	**8.74e-03**	3.35e-02	3.08e-02	2.69e-02
(50, 4)	**2.01e-02**	3.19e-02	2.99e-02	2.85e-02	**2.66e-02**	6.97e-02	6.80e-02	5.91e-02
(50, 6)	**1.08e-02**	2.50e-02	3.10e-02	2.61e-02	**1.21e-02**	4.31e-02	4.00e-02	3.88e-02
(50, 8)	**1.14e-03**	3.90e-02	3.31e-02	3.00e-02	**9.92e-03**	4.00e-02	4.43e-02	4.20e-02
(50, 10)	**1.23e-02**	2.68e-02	2.33e-02	2.20e-02	**2.10e-02**	5.15e-02	5.28e-02	4.75e-02
(100, 2)	**2.85e-02**	6.84e-02	8.00e-02	7.85e-02	**1.87e-02**	6.00e-02	6.51e-02	5.26e-02
(100, 4)	**1.93e-02**	1.42e-01	1.30e-01	1.35e-01	**1.11e-03**	2.92e-02	3.35e-02	2.61e-02
(100, 6)	**7.38e-04**	3.39e-02	2.80e-02	2.77e-02	**1.47e-03**	5.18e-02	5.26e-02	5.25e-02
(100, 8)	**5.48e-03**	2.62e-02	2.82e-02	2.73e-02	**1.33e-02**	3.04e-02	3.28e-02	3.01e-02
(100, 10)	**4.52e-03**	1.48e-02	1.90e-02	1.58e-02	**5.60e-03**	2.21e-02	2.76e-02	2.25e-02
(150, 2)	**8.76e-02**	4.25e-01	5.09e-01	4.56e-01	**1.01e-02**	4.57e-02	6.48e-02	5.29e-02
(150, 4)	**6.16e-02**	1.44e-01	1.67e-01	1.42e-01	**3.52e-03**	7.70e-02	9.10e-02	7.62e-02
(150, 6)	**2.16e-02**	7.23e-02	9.76e-02	6.92e-02	**1.56e-02**	4.07e-02	4.52e-02	3.61e-02
(150, 8)	**6.09e-02**	1.81e-01	2.16e-01	2.02e-01	**2.40e-02**	8.48e-02	6.98e-02	7.14e-02
(150, 10)	**2.38e-02**	5.68e-02	5.37e-02	4.76e-02	**2.27e-02**	7.81e-02	8.72e-02	7.81e-02
(200, 2)	**3.81e-02**	2.41e-01	2.75e-01	2.60e-01	**3.00e-03**	3.52e-02	3.07e-02	3.08e-02
(200, 4)	**5.61e-02**	1.32e-01	1.60e-01	1.43e-01	**1.24e-03**	2.33e-02	2.76e-02	2.35e-02
(200, 6)	**1.80e-02**	8.72e-02	8.71e-02	8.31e-02	**2.22e-02**	5.03e-02	6.83e-02	5.12e-02
(200, 8)	**1.09e-03**	2.59e-02	1.76e-02	1.90e-02	**3.50e-02**	1.89e-01	1.60e-01	1.41e-01
(200, 10)	**3.04e-02**	8.98e-02	1.07e-01	9.39e-02	**4.64e-02**	1.56e-01	1.91e-01	1.50e-01

Table 10.16 Compare between IGD mean values by different algorithms

(n, s)	MOHIG	MOEA/D	SPEA2	NSGA-II	MOHIG	MOEA/D	SPEA2	NSGA-II
	F = 4				F = 5			
(50, 2)	**1.06e-02**	2.75e-02	2.91e-02	2.71e-02	**1.11e-02**	3.13e-02	3.10e-02	2.64e-02
(50, 4)	**1.27e-02**	3.84e-02	3.55e-02	3.16e-02	**8.16e-03**	3.97e-02	3.74e-02	3.47e-02
(50, 6)	**6.25e-03**	5.67e-02	5.97e-02	5.33e-02	**9.36e-04**	4.46e-02	3.00e-02	3.04e-02
(50, 8)	**1.65e-03**	3.38e-02	3.58e-02	3.24e-02	**2.10e-02**	2.34e-02	2.44e-02	1.99e-02
(50, 10)	**1.19e-02**	2.85e-02	3.76e-02	3.29e-02	**9.92e-03**	3.92e-02	3.80e-02	3.57e-02
(100, 2)	**1.34e-02**	3.39e-02	4.26e-02	3.65e-02	**1.05e-02**	5.12e-02	5.09e-02	4.83e-02
(100, 4)	**1.19e-03**	3.64e-02	3.30e-02	3.13e-02	**1.75e-03**	4.47e-02	4.22e-02	4.02e-02
(100, 6)	**1.18e-03**	3.72e-02	3.99e-02	4.01e-02	**1.86e-02**	6.47e-02	8.38e-02	6.50e-02
(100, 8)	**1.06e-02**	3.ee-02	3.20e-02	2.80e-02	**1.24e-02**	5.38e-02	5.89e-02	5.34e-02
(100, 10)	1.26e-02	6.48e-02	6.86e-02	6.13e-02	3.32e-03	3.13e-02	3.28e-02	3.10e-02
(150, 2)	7.05e-03	4.29e-02	5.45e-02	4.32e-02	1.78e-03	4.24e-02	4.79e-02	4.59e-02
(150, 4)	1.78e-03	3.46e-02	3.30e-02	2.99e-02	1.99e-02	1.17e-01	1.00e-01	1.08e-01
(150, 6)	1.70e-02	3.95e-02	4.15e-02	3.44e-02	7.79e-03	3.77e-02	4.27e-02	3.48e-02
(150, 8)	1.46e-03	2.83e-02	2.69e-02	2.47e-02	1.96e-03	4.36e-02	4.19e-02	3.74e-02
(150, 10)	3.17e-03	3.06e-02	2.77e-02	3.08e-02	2.27e-02	7.81e-02	8.72e-02	7.81e-02
(200, 2)	1.76e-03	4.05e-02	4.10e-02	3.16e-02	3.00e-03	3.52e-02	3.07e-02	3.08e-02
(200, 4)	1.92e-03	2.99e-02	2.73e-02	2.50e-02	1.24e-03	2.33e-02	2.76e-02	2.35e-02
(200, 6)	2.25e-02	6.25e-02	6.37e-02	6.ee-02	2.22e-02	5.03e-02	6.83e-02	5.12e-02
(200, 8)	1.96e-03	3.16e-02	2.96e-02	3.11e-02	3.50e-02	1.89e-01	1.60e-01	1.41e-01
(200, 10)	1.55e-03	4.18e-02	4.69e-02	4.43e-02	4.64e-02	1.56e-01	1.91e-01	1.50e-01

Table 10.17 Compare between IGD mean values by different algorithms

(n, s)	MOHIG	MOEA/D	SPEA2	NSGA-II
	F = 6			
(50, 2)	**1.25e-02**	3.66e-02	3.82e-02	3.39e-02
(50, 4)	**7.66e-03**	2.99e-02	2.68e-02	2.53e-02
(50, 6)	**1.23e-02**	5.86e-02	5.93e-02	5.45e-02
(50, 8)	**1.05e-03**	2.72e-02	3.09e-02	3.05e-02
(50, 10)	**1.ee-02**	5.32e-02	4.68e-02	4.38e-02
(100, 2)	**1.07e-02**	5.18e-02	5.61e-02	5.05e-02
(100, 4)	**1.89e-02**	1.04e-01	1.12e-01	8.51e-02
(100, 6)	**1.49e-03**	3.42e-02	3.83e-02	3.57e-02
(100, 8)	**7.32e-03**	5.58e-02	5.09e-02	4.52e-02
(100, 10)	**1.53e-03**	6.59e-02	6.27e-02	6.16e-02
(150, 2)	**1.97e-02**	1.04e-01	1.15e-01	1.04e-01
(150, 4)	**2.08e-02**	1.18e-01	1.35e-01	1.20e-01
(150, 6)	**1.08e-02**	4.89e-02	5.07e-02	4.78e-02
(150, 8)	**1.86e-02**	8.65e-02	1.17e-01	9.94e-02
(150, 10)	**2.73e-02**	1.03e-01	1.11e-01	9.15e-02
(200, 2)	**1.59e-02**	6.39e-02	6.45e-02	5.95e-02
(200, 4)	**7.16e-04**	2.09e-02	1.99e-02	1.77e-02
(200, 6)	**1.19e-03**	3.27e-02	3.33e-02	3.04e-02
(200, 8)	**2.02e-02**	1.36e-01	1.53e-01	1.30e-01
(200, 10)	**1.43e-03**	3.43e-02	3.50e-02	3.15e-02

It further supports MOHIG's preferable search methodology. The reason is that the cooperative initialization strategy can make some initial solutions evenly distributed in different regions, and improve the convergence and diversity of solutions in the initial stage. In addition, a knowledge-based multi-objective local search heuristic can effectively generate high-quality solutions in the later stage of the search. Therefore, it can be said that the suggested MOHIG can effectively handle DHFSP based on the aforementioned statistical findings and analysis.

10.5 Chapter Conclusion

In this research, we investigated an energy-efficient distributed hybrid flow shop scheduling problem (DHFSP). First, a mathematical model of this issue was developed with goals of minimizing makespan and overall energy use. In order to resolve this energy-efficient DHFSP, a Pareto-based multi-objective hybrid iterated greedy algorithm (MOHIG) was then suggested. We also suggested a collaborative initiation

Table 10.18 *p* values of all indicators of the algorithm compared by different factories, stages, and jobs

Comparison algorithm	IGD			GD			Spread		
	MOHIG vs MOEA/D	MOHIG vs. SPEA2	MOHIG vs. NSGA-II	MOHIG vs. MOEA/D	MOHIG vs. SPEA2	MOHIG vs. NSGA-II	MOHIG vs. MOEA/D	MOHIG vs. SPEA2	MOHIG vs. NSGA-II
F = 2	8.86e-05	8.86e-05	8.86e-05	1.03e-04	8.86e-05	8.86e-05	1.03e-04	8.86e-05	8.86e-05
F = 3	8.86e-05	8.86e-05	8.86e-05	1.02e-03	8.86e-05	8.86e-05	1.02e-03	8.86e-05	2.19e-04
F = 4	8.86e-05	8.86e-05	8.84e-05	2.19e-04	8.86e-05	8.86e-05	2.19e-04	8.86e-05	2.54e-04
F = 5	8.86e-05	8.86e-05	1.03e-04	8.86e-05	8.86e-05	8.86e-05	8.86e-05	8.86e-05	1.03e-04
F = 6	8.86e-05	8.86e-05	8.86e-05	1.03e-04	8.86e-05	8.86e-05	1.03e-04	8.86e-05	8.86e-05
S = 2	8.86e-05	8.86e-05	1.03e-04	2.93e-04	8.86e-05	8.86e-05	2.93e-04	8.86e-05	8.86e-05
S = 4	8.86e-05	8.86e-05	8.86e-05	5.93e-04	8.86e-05	8.86e-05	5.93e-04	8.86e-05	1.89e-04
S = 6	8.86e-05	8.86e-05	8.86e-05	8.86e-05	8.86e-05	8.86e-05	8.86e-05	8.86e-05	1.63e-04
S = 8	8.86e-05	8.86e-05	8.86e-05	1.40e-04	8.86e-05	8.86e-05	1.40e-04	8.86e-05	2.93e-04
S = 10	8.86e-05	8.86e-05	8.86e-05	1.20e-04	8.86e-05	8.86e-05	1.20e-04	8.86e-05	8.86e-05
N = 50	1.23e-05	1.23e-05	1.23e-05	5.13e-05	1.23e-05	1.23e-05	5.13e-05	1.23e-05	1.23e-05
N = 100	1.23e-05	1.23e-05	1.23e-05	1.39e-05	1.23e-05	1.23e-05	1.39e-05	1.23e-05	1.77e-05
N = 150	1.23e-05	1.23e-05	1.23e-05	4.57e-05	1.23e-05	1.23e-05	4.57e-05	1.23e-05	1.74e-04
N = 200	1.23e-05	1.23e-05	1.39e-05	1.77e-05	1.23e-05	1.23e-05	1.77e-05	1.23e-05	1.23e-05

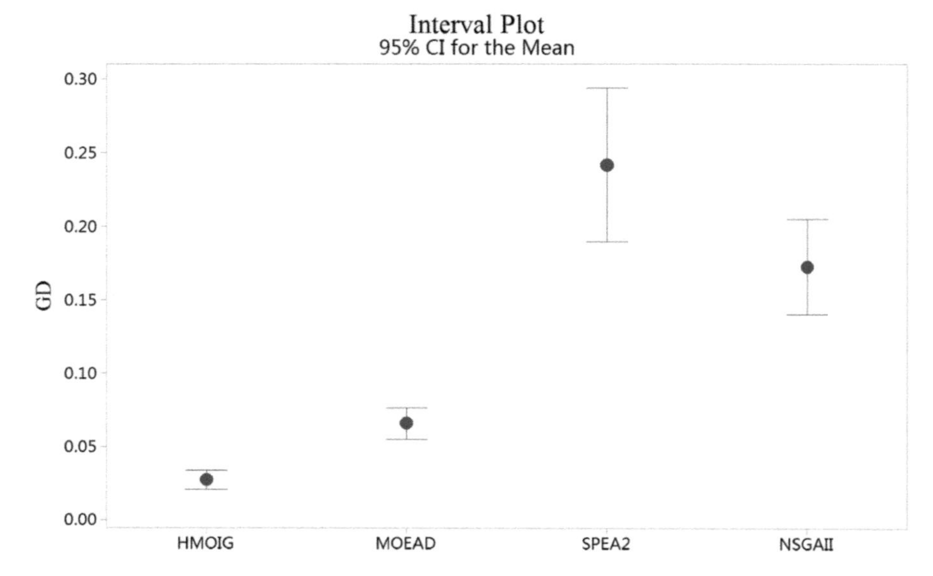

Fig. 10.6 Interval graph of GD indicator of test algorithms

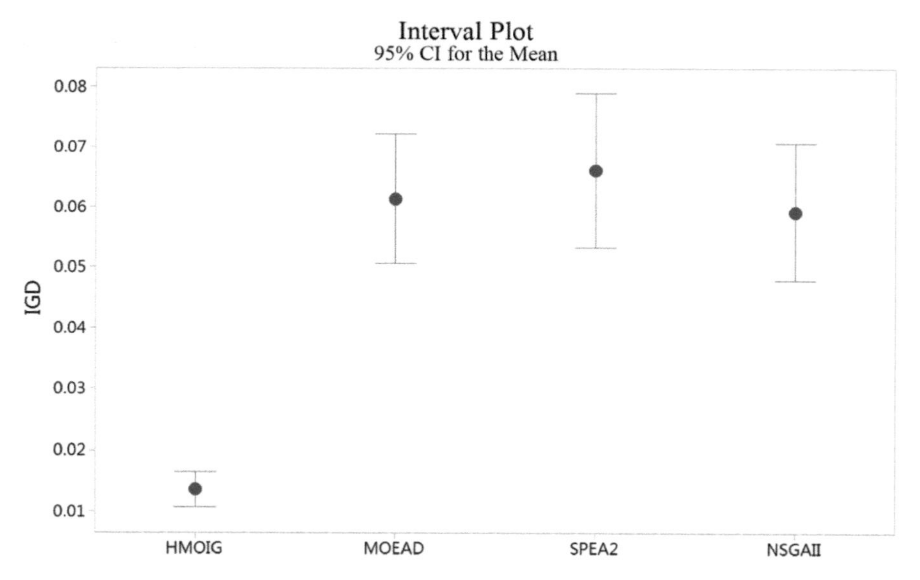

Fig. 10.7 Interval graph of IGD indicator of test algorithms

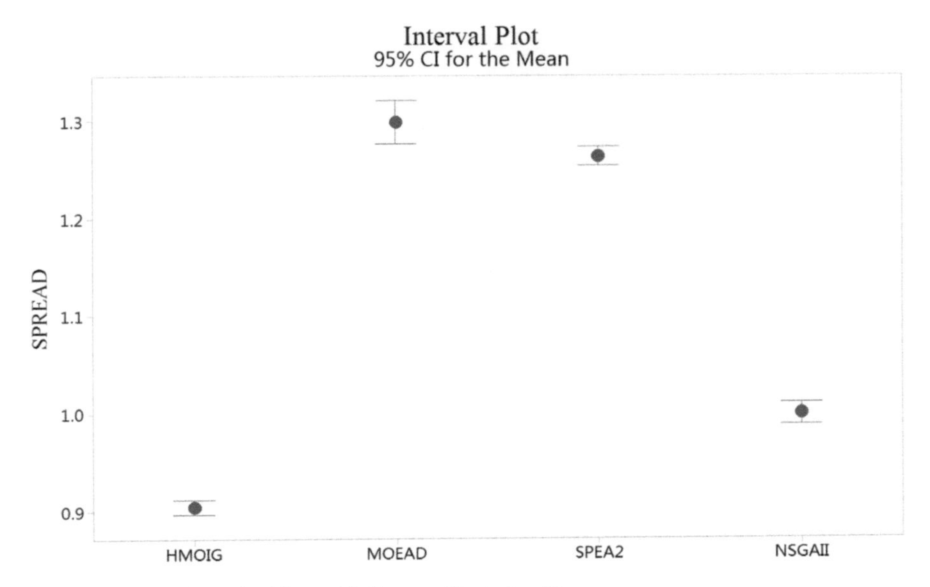

Fig. 10.8 Interval graph of Spread indicator of test algorithms

technique to enhance the initial solution based on the problem properties. In addition to creating an efficient energy-saving approach, we optimized our projected energy consumption objective. To enhance the exploitation capabilities, a knowledge-based multi-objective local search strategy was also suggested. Subsequent experiments prove that MOHIG is superior to other multi-objective optimization algorithms in solving energy-efficient DHFSP. Distributed production environment has important research significance. In addition to the above static research, we can extend it to distributed heterogeneous process shop scheduling problem with uncertainty. As far as the problem goal of production scheduling is concerned, environmental benefits can also be included in the evaluation index of enterprise production. In terms of algorithms, in order to create effective hybrid meta-heuristics, we can examine collaborative optimization strategies with global exploration and local exploitation. In the end, instead of letting flow shop scheduling research remain purely academic, we should implement it into real-world flow shop production systems. It also has very broad implications for how we may improve scheduling theory and practice.

References

1. Naderi, B., Ruiz, R.: The distributed permutation flowshop scheduling problem[J]. Comput. Oper. Res. **37**(4), 754–768 (2010)
2. Li, R., Gong, W., Lu, C.: Self-adaptive multi-objective evolutionary algorithm for flexible job shop scheduling with fuzzy processing time[J]. Comput. Ind. Eng. **168**, 108099 (2022)

3. Lu, C., Gao, L., Li, X., et al.: Energy-efficient permutation flow shop scheduling problem using a hybrid multi-objective backtracking search algorithm[J]. J. Clean. Prod. **144**(FEB.15), 228–238 (2017)
4. Liu, H., Gu, F., Zhang, Q.: Decomposition of a multiobjective optimization problem into a number of simple multiobjective subproblems[J]. IEEE Trans. Evol. Comput. **18**(3), 450–455 (2014)
5. Nagano, M.S., Almeida, F.S.D., Miyata, H.H.: An iterated greedy algorithm for the no-wait flowshop scheduling problem to minimize makespan subject to total completion time[J]. Eng. Optim. **4**, 1–19 (2020)
6. Zhao, Z.Y., Zhou, M.C., Liu, S.X.: Iterated greedy algorithms for flow-shop scheduling problems: a tutorial[J]. IEEE Trans. Autom. Sci. Eng. **PP**(99), 1–19 (2021)
7. Liu, H., Liang, G., Pan, Q.: A hybrid particle swarm optimization with estimation of distribution algorithm for solving permutation flowshop scheduling problem[J]. Expert Syst. Appl. **38**(4), 4348–4360 (2011)
8. Lu, X.: A multi-objective cellular grey wolf optimizer for hybrid flowshop scheduling problem considering noise pollution[J]. Appl. Soft Comput., **75** (2019)
9. Ruiz, R., Pan, Q.K., Naderi, B.: Iterated Greedy methods for the distributed permutation flowshop scheduling problem[J]. Omega. **83**(MAR.), 213–222 (2019)
10. Liu Y, Wei J, Li X, et al. Generational Distance Indicator-Based Evolutionary Algorithm With an Improved Niching Method for Many-Objective Optimization Problems[J]. IEEE Access. 2019, PP(99): 1.
11. Zhang, W., Chen, J., Wang, H., et al.: Analyses of inverted generational distance for many-objective optimisation algorithms[J]. Int. J. Bio-Inspired Comput. **14**(1), 62 (2019)
12. Wang, L., Zhiwen, et al.: Influence spread in geo-social networks: a multiobjective optimization perspective.[J]. IEEE Trans. Cybernetics (2019)
13. Zitzler, E.: Multiobjective evolutionary algorithms: a comparative case study and the strength pareto approach[J]. IEEE Trans. Evol. Comput. **3** (1999)
14. Deb, K., Pratap, A., Agarwal, S., et al.: A fast and elitist multiobjective genetic algorithm: NSGA-II[J]. IEEE Trans. Evol. Comput. **6**(2), 182–197 (2002)
15. Hui, L., Zhang, Q.: Multiobjective optimization problems with complicated Pareto sets, MOEA/D and NSGA-II[J]. IEEE Trans. Evol. Comput. **13**(2), 284–302 (2009)